Time-Resolved Spectroscopy

An Experimental Perspective

Textbook Series in Physical Sciences

Series Editor: Lou Chosen

Understanding Nanomaterials
Malkiat S. Johal

Concise Optics: Concepts, Examples, and Problems
Ajawad I. Haija, M. Z. Numan, W. Larry Freeman

A Mathematica Primer for Physicists
Jim Napolitano

Understanding Nanomaterials, Second Edition
Malkiat S. Johal, Lewis E. Johnson

Physics for Technology, Second Edition
Daniel H. Nichols

Time-Resolved Spectroscopy: An Experimental Perspective
Thomas Weinacht, Brett J. Pearson

For more information about this series, please visit:
www.crcpress.com/Textbook-Series-in-Physical-Sciences/book-series/TPHYSCI

Time-Resolved Spectroscopy

An Experimental Perspective

Thomas C. Weinacht
Stony Brook University, Stony Brook
New York

Brett J. Pearson
Dickinson College, Carlisle
Pennsylvania

CRC Press
Taylor & Francis Group
Boca Raton London New York

CRC Press is an imprint of the
Taylor & Francis Group, an **informa** business

CRC Press
Taylor & Francis Group
6000 Broken Sound Parkway NW, Suite 300
Boca Raton, FL 33487-2742

First issued in paperback 2020

ISBN-13: 978-1-4987-1673-4 (hbk)
ISBN-13: 978-0-367-78040-1 (pbk)

Library of Congress Cataloging-in-Publication Data

Names: Weinacht, Thomas, author. | Pearson, Brett J., author.
Title: Time-resolved spectroscopy : an experimental perspective / Thomas Weinacht, Brett J. Pearson.
Description: Boca Raton : CRC Press, Taylor & Francis Group, 2019.
Identifiers: LCCN 2018022261 | ISBN 9781498716734 (hardback : alk. paper)
Subjects: LCSH: Time-resolved spectroscopy. | Spectrum analysis.
Classification: LCC QP519.9.T59 W45 2018 | DDC 610.28–dc23
LC record available at https://lccn.loc.gov/2018022261

Visit the Taylor & Francis Web site at
http://www.taylorandfrancis.com

and the CRC Press Web site at
http://www.crcpress.com

To our children:
Réka, Dani, and Nora

Contents

Preface

This book grew out of a desire to connect experimental measurements in the lab to simple, theoretical descriptions that students in our research groups could understand and utilize themselves. Although the book is written from an experimental perspective, as people working in the field know, experiments are typically accompanied by significant theoretical efforts to both design and interpret the measurements. The main goal of the book is to provide a framework for how experimental measurements can be combined with analytic and/or numerical calculations to provide insight into the motion of electrons and nuclei.

We focus on relatively simple models, and this book certainly does not encompass all of the theory required. We aim to show that relatively straightforward models, backed up by high-level theory, can provide a significant degree of physical insight into experimental results. We hope that this book helps bridge the often-present gap that exists between experiment and theory and illustrate how, in our opinion, it is the two together that provide the greatest insight into quantum dynamics.

The format of the book is designed to help reinforce this perspective, with the text divided into five parts. Part I provides the fundamentals, including an introduction to time-resolved spectroscopy and the four, prototypical experimental techniques that we build upon throughout the text. Part I also presents the basic physics and chemistry underlying the experiments, such as atomic and molecular structure and light-matter interaction. The remaining four parts are grouped in two pairs. In each pair, the first part develops the models and equations necessary for understanding the experiments, while the second part presents a number of experimental results, along with their physical and mathematical interpretations based on the models developed in the previous part. Parts II and III consider atomic and molecular dynamics in only one dimension. While most systems of interest are, of course, multidimensional, starting in one dimension should allow readers to develop experience with the different techniques and interpretations while studying systems that appear relatively familiar and straightforward. Parts IV and V go on to examine multidimensional systems, building on the formalism developed in earlier chapters.

We have tried to make the book as self-contained as possible by starting with first principles whenever feasible, with a goal of making the book accessible to physics or chemistry students in the first year of graduate school (or advanced students in the final year of an undergraduate program). For typical students in this demographic, we would expect to devote approximately $2 - 4$ weeks to the fundamentals in Part I (readers with significant prior exposure to the concepts should be able to move more quickly through this material). The amount of time spent on the remaining parts depends on both the students' experiences and the instructor's interests. For example, one could make computation a real emphasis of the course by spending a significant amount of time on the numerical propagation routines of Chapters 5 and 6. Alternatively, some instructors may wish to move more quickly through this part, focusing instead on the wide variety of experimental approaches. Homework problems appear at the end of the chapters that develop the models and equations. Appendix C contains additional problems, including ones that build on results from the experimental chapters.

A number of discussions in the book came out of really great questions we have been asked by students in the lab. The conversations that followed these questions (often continued over days or weeks), led us to appreciate not only the helpfulness of a good, physical model, but also the inevitable subtleties that come along with it. We hope that readers will find these discussions useful and that the book has at least a flavor of addressing those "questions you always wanted to know the answer to, but were afraid to ask."

We gratefully acknowledge grants from the National Science Foundation during the course of writing the manuscript. Many people helped make this book possible. We wish to thank a number of students who contributed in different ways. Spencer Horton, Yusong Liu, Chuan Cheng, and Brian Kaufman read through sections of the book and provided feedback and figures. Kyle Liss and Sahil Nayaar developed simulations whose results appear in the book. This book contains some mathematical treatments that arose from the Ph.D thesis work of Chien-Hung Tseng and Carlos Trallero. We appreciate the students at Stony Brook University and Dickinson College who enthusiastically participated in classes that used portions of this book while still in development. We benefited from numerous discussions with colleagues, including Phil Allen, Tom Allison, Dominik Schneble, and Tzu-Chieh Wei (all at Stony Brook University), Philip Bucksbaum (Stanford University), Robert Jones (University of Virginia), C.D. Lin (Kansas State University), Spiridoula Matsika (Temple University), William McCurdy (University of California, Davis), and Carlos Trallero (University of Connecticut). We are also indebted to two anonymous reviewers whose comments on an initial draft of the manuscript greatly improved the final product. We thank Spencer Horton for his work on the cover design.

See the book's dedicated webpage at the publisher's site for online supplements, including full-color images available to download: www.crcpress.com/9781498716734.

Tom Weinacht
Brett Pearson

Authors

Thomas Weinacht is a Professor of Physics at Stony Brook University in New York. He received his B.S. in physics from the University of Toronto 1995 and a Ph.D. in physics from the University of Michigan in 2000. He started his position at Stony Brook University in 2002. His research focuses on controlling and following molecular dynamics with strong-field ultrafast laser pulses. He has published extensively in both physics and chemistry journals, with an emphasis on interpreting experimental measurements. His research group has developed a number of experimental techniques, and he has organized multiple international conferences and workshops in the field of time-resolved spectroscopy. He is a fellow of the American Physical Society.

Brett Pearson is an Associate Professor of Physics and Astronomy at Dickinson College in Carlisle, PA. He obtained a B.A. in physics in 1997 from Grinnell College and then a Ph.D. in physics from the University of Michigan in 2004. He was a postdoctoral fellow at Stony Brook University before moving to his current position. At Dickinson, Brett teaches across the curriculum and works with undergraduate students on research related to both ultrafast pulse shaping and single-photon quantum mechanics.

PART I

Introduction and Background

CHAPTER 1

Introduction

1.1 WHAT IS TIME-RESOLVED SPECTROSCOPY?

Our experience of life is time dependent—things move, changing their position or velocity with time. Yet quantum mechanics, which governs life at the atomic and molecular level and is one of the most successful theories of modern physics, is generally formulated in terms of time-independent quantities. Most early tests of quantum mechanics aimed to measure these time-independent quantities such as the energy levels of atoms or molecules. This approach is quite sensible when looking for analytic solutions to the differential equations that govern quantum mechanics, particularly for simple, isolated systems. The differential equations are often separable into temporal and spatial components, allowing one to find time-independent solutions relatively easily. The full time-dependent solutions are then built from linear combinations of the stationary results.

However, if we are interested in following dynamics and understanding how systems evolve, a time-dependent perspective is more useful. Time-resolved measurements have the potential to provide the most direct information about dynamics, particularly for many-body systems and those not isolated from their environment. The focus of this book is therefore to study physical systems from a time-dependent perspective: how do systems evolve in time, and how can we measure this evolution?

We are specifically interested in studying fundamental systems at the atomic and molecular level, where the laws of quantum mechanics dictate the dynamics. In quantum mechanics, the complex wave function $\Psi(\mathbf{r},t)$ provides a complete description of the system; at any given time, knowledge of $\Psi(\mathbf{r},t)$ yields information about an object's dynamical variables such as position, momentum, or energy. These quantities are called observables, as one can measure (observe) their values in a given system. *Time-resolved spectroscopy involves measuring observables for quantum systems whose wave functions are prepared in a well-defined initial state and subsequently evolve in time.*

While the dream of time-resolved measurements is to make real-time "molecular movies" of dynamics, it is rare that one can directly measure evolution of the molecular structure or wave function. Rather, insight is typically gained by comparing experimental measurements with theoretical calculations of observables. The experimentally tested calculations can then be used to generate the molecular movie. Thus, an important criterion when evaluating different measurement approaches is how easily they can be compared with calculations of the observable. Throughout the book, we highlight how experimental techniques are closely coupled to the theory and calculations used to interpret the measurements.

The perspective we take for most of the book is that the fundamental quantity associated with an object is its wave function, which evolves in time according to the time-dependent Schrödinger equation (TDSE).[1] The wave function determines everything that is possible to measure: position, momentum, energy, etc. For a one-dimensional system with wave function $\Psi(x,t)$ and potential energy $V(x,t)$, the TDSE reads

$$i\hbar\frac{\partial\Psi}{\partial t} = -\frac{\hbar^2}{2m}\frac{\partial^2\Psi}{\partial x^2} + V\Psi, \tag{1.1}$$

where $\hbar$ is the reduced Planck's constant (equal to one in atomic units) and m is the mass. In other words, $\Psi(x,t)$ is a *time-dependent* solution to the TDSE, and measurable quantities are represented by time-independent operators that act on the wave function. This is known as the Schrödinger picture, and it contrasts with the Heisenberg formulation of quantum mechanics, where the time dependence of measurable quantities is described by time-dependent operators.[2]

The wave function evolves according to the TDSE in a manner that is quite intuitive if one is familiar with the behavior of classical waves. Of course, the TDSE is first order in time and second order in space, whereas the wave equation for light, derived from Maxwell's equations, is second order in both space and time (the consequences of this are discussed in Chapter 5). Subsequent chapters highlight the similarities between quantum-mechanical wave functions and classical light waves. For example, wave functions that are initially localized tend to spread; waves can be split, interfere, and diffract; and there is an inherent relationship between the spread of the wave function in the "conjugate variables" (e.g. $\Delta x \cdot \Delta p \geq \hbar/2$). This last point is the essence of the Heisenberg uncertainty principle (see Section 5.3.1).

The conceptual leap in going from classical to quantum waves is that the quantum case describes the properties of particles, for which classical mechanics produces deterministic equations and predictions. In quantum mechanics, the wave function can only be used to calculate *probabilities* for different experimental measurements. For instance, if one knows the initial wave function for a particle at time $t = 0$, the TDSE determines the probability of finding the particle at different locations (or with different energies) for all subsequent times t. Although the trajectories for a classical particle can be known for all times given some initial conditions, quantum mechanics only provides prescriptions for calculating probabilities based on the time-dependent wave function.

---

[1] For systems with an extremely large number of degrees of freedom, it becomes impractical to write down a wave function for the full system. This is addressed in Chapter 11, where we consider a system coupled to a bath and introduce the density matrix. For building up both a mathematical description and our physical intuition of time-resolved spectroscopy, we find it most useful to follow the wave function approach until it becomes necessary to use other methods.

[2] While both formulations lead to the same results for observables, the Schrödinger picture is a natural choice for this book, in which we consider the fundamental quantity of interest to be the time-dependent wave function. It is worth noting here that not only are there other valid approaches to calculating observables, but also that the Schrödinger equation itself is only valid in the nonrelativistic limit. The more general Dirac equation can be used to describe relativistic spin-$1/2$ particles. However, the Dirac equation is typically much more difficult to solve, and since we are generally interested in nonrelativistic dynamics, unless otherwise mentioned we assume that the state of the system is given by the nonrelativistic Schrödinger wave function $\Psi(\mathbf{r},t)$.

While $\Psi(\mathbf{r},t)$ is ultimately what we would like to determine, it is not possible to measure $\Psi(\mathbf{r},t)$ directly, and we must satisfy ourselves with projections onto basis functions of different measurement operators (e.g. position, momentum, energy). From these results, we can infer the entire wave function using multiple measurements of identically prepared systems known as "ensembles." Much of experimental quantum mechanics relies on making these types of incomplete, or projected, measurements. By combining a series of measurements on collections of identically prepared quantum systems, one can extract information about the full state of the system. The details of these projections depend intimately on the experimental parameters. One of the primary goals of this book is to discuss how various approaches allow access to different pieces of information about the wave function and, depending on the goal of the experiment, why one may choose one technique over another.

1.2 WHY DO TIME-RESOLVED SPECTROSCOPY?

While it is intellectually satisfying to explicitly follow quantum dynamics in time using time-resolved spectroscopy, it is worth considering whether the time-domain approach can, in principle, yield different or new information beyond what is available from frequency-domain or time-independent spectroscopy. For example, by performing frequency-domain measurements such as absorption spectroscopy, it is possible to map out the eigenenergies of a quantum system by measuring energy differences between states. If the eigenfunctions can also be measured or calculated, the time evolution of the system for a given initial state can be determined without necessarily making any time-domain measurements.

For problems where a frequency-domain description of the dynamics involves a superposition of a large number of eigenstates, it can certainly be argued that a time-domain representation is simply much more convenient. In other words, the time-domain picture provides a better *basis* for the problem, similar to how describing a series of pulses in time is more convenient using Gaussian functions rather than superpositions of sine waves. However, the advantages of time-resolved spectroscopy over the time-independent approach really come to the fore when one considers a system with sufficient complexity, such that its spectrum cannot be resolved into a series of discrete lines. In a system with this many degrees of freedom, it makes sense to separate it into a portion that is described explicitly ("the system") and a portion that is treated in aggregate and interacts only weakly with the system ("the bath"). Typical examples include systems that are not well isolated from their environment.

As the number of degrees of freedom in the system increases, the level of precision with which it is possible to determine the eigenenergies and calculate the dynamics decreases. Eventually, it becomes impossible to rely on a combination of time-independent measurements and calculation to infer the time evolution of the system. For small molecules ($2-3$ atoms) in the gas phase, it is reasonable to calculate the dynamics in great detail, checking the results by comparing calculated eigenenergies to measured spectra. However, when one moves to molecules with ~ 10 atoms or more, there are many dynamics that cannot be reliably inferred from the combination of time-independent spectroscopy and calculation. Furthermore, when the system of interest is not isolated from its environment (e.g. condensed phase), time-resolved spectroscopy provides dynamical information that is not available in any other way.[3]

Consider the case of a time-dependent response to impulsive photoexcitation of a molecule in solution. In contrast to the absorption spectrum of an atom or small molecule in the gas phase, it is generally impossible to infer the dynamics simply based on a measurement of the absorption spectrum, as solution-phase absorption spectra are typically broad, continuous, and lacking in distinct features. Figure 1.1 shows the physical environment, absorption spectrum, and inferred dynamics for both an isolated molecule (top row) and a molecule in solution (bottom row). For the case of an isolated molecule in gas phase, the absorption spectrum consists of a series of distinct peaks at frequencies corresponding to energy differences between bound states of the system. From this, one can infer the dynamics for a particular coherent super-position of eigenstates (an oscillating wave packet). However, for a molecule in solu-tion, the absorption due to each transition between bound states is inhomogeneously broadened due to the variation in local environments experienced by each molecule in solution. This results in a broad, featureless absorption spectrum. This same spec-trum could also correspond to a continuum associated with an unbound, or dissocia-tive, final state (as shown in bottom-right panel). Thus, it is difficult to tell whether the broad absorption spectrum that one measures for condensed-phase systems arises from intermolecular interactions (inhomogeneous broadening) or continuum dynamics. As will be discussed, nonlinear spectroscopies such as two-dimensional electronic spec-troscopy can discriminate between homogeneous and inhomogeneous broadening and enable a much deeper understanding of dynamics in complex systems or molecules that interact strongly with their environment.

For example, the dynamics associated with human vision involve the isomerization of the 11-cis retinal chromophore embedded in the rhodopsin protein. The coupling between the chromophore and the protein is an essential part of vision, but it pre-cludes insight into the dynamics using time-independent spectroscopy. However, time-resolved measurements have determined the ultrafast cis-trans isomerization occurs in

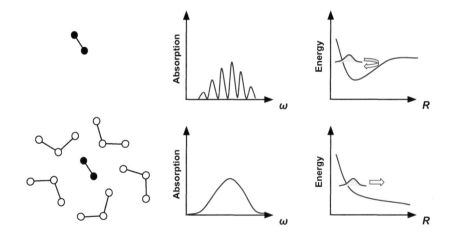

Figure 1.1 Physical environment (left), absorption spectrum (cen-ter), and possible dynamics inferred from the spectra (right) for isolated molecules (top row) and molecules in solution (bottom row). One can-not necessarily infer dynamics (such as the ones shown in the bottom right panel) from absorption spectra for molecules in solution.

[3]As noted earlier, in this limit, the molecular wave function does not accurately represent all of the molecules in the ensemble, and the density matrix is a better description of the full system. Neverthe-less, the wave function helps provide physical insight, and we stick with it whenever possible, extending the approach as interactions with the environment become important.

under 200 fs with energy from the absorbed light driving conformational changes in the G protein-coupled receptor, eventually leading to visual perception [1, 2]. The remarkable high efficiency of the process ($\sim$65%) is a result of the rapid dynamics following light absorption, and thus unravelling the dynamics using time-resolved spectroscopy provides a deep understanding of the physical mechanisms behind this fundamental biological process.

While this discussion highlights the role of ultrafast dynamics in a biological system, it is natural to ask whether the dynamics initiated with a very short laser pulse have direct relevance to what happens in nature. After all, in processes such as vision (or photosynthesis), the molecular dynamics are a result of the system absorbing sunlight; sunlight is clearly not a femtosecond laser pulse! However, while sunlight itself is "always on," any given molecule absorbs the light one photon at a time in an event that takes place very rapidly. Therefore, thinking about an "instantaneous" absorption of an ultrashort pulse is a nice way to elucidate subsequent dynamics. While light from a continuous-wave laser would also be absorbed one photon at a time, it is challenging to measure this effect, given that the light is always on.[4] Formally, the TDSE is linear with respect to the wave function, and thus if one understands the response to impulsive excitation, one also understands the response to longer excitation fields. Performing measurements with pulses shorter than any of the subsequent dynamics leads to the simplest interpretation of the experimental data.

1.3 HOW DO YOU DO TIME-RESOLVED SPECTROSCOPY?

1.3.1 The Pump–Probe Approach

In time-resolved spectroscopy, we are chiefly concerned with making a series of measurements of a quantum system at different times. We will find it useful to think about a generic time-domain spectroscopy experiment as a "pump–probe" interrogation of the quantum system. An applied pump pulse prepares the system, originally in equilibrium, in some new state. In general, this state will be away from dynamic equilibrium (nonstationary), and so the system will evolve in time. The dynamics one wishes to measure are often on a timescale shorter than state-of-the-art electronics can achieve (e.g. photodetector rise time), and so one must "gate" the measurement using some other means.[5] Optical time resolution, given by the light pulse duration, has long surpassed electronic time resolution by several orders of magnitude, and so time-resolved experiments typically gate the measurement using a probe pulse. The probe pulse interrogates the system, leading to the detection of a final product such as a charged particle or scattered photon. Since most experimental measurements are destructive

---

[4]If one is a little more careful in using the language of quantum mechanics, the light from a continuous-wave source can be absorbed over an extended period of time, promoting some small portion of the molecular wave function to an excited state at each instant. A measurement of whether a given molecule is in the excited state or not yields a positive result with a probability proportional to the square of the excited state wave function amplitude at any given moment. This wave function on the excited state will evolve, but is very hard to follow if there is continuous additional excitation. Thus, it is *clearest* to excite the molecule with a pulse that is shorter than the ensuing excited-state dynamics.

[5]This is analogous to the challenge involved in measuring very short optical pulses. Since ultrafast laser pulses are much shorter than the response of any photodetector, they cannot be measured directly like a radio or microwave, and instead, one must use another optical pulse (or itself) to gate the detection.

(they modify the quantum state) and less than 100% efficient, they are carried out on ensembles of identically prepared molecules; a new measurement at each time delay between the pump and probe pulses is carried out on a new ensemble of molecules. A detailed discussion of ensemble measurements is provided in Section B.1.

Figure 1.2 illustrates a simple pump–probe approach for the case of a diatomic molecule. As will be discussed in more detail in Chapter 2, the force binding the two atoms together can be thought of as the gradient of a spatially varying potential energy. The attraction between the electrons and protons in the two atoms leads to a potential along the internuclear coordinate with a minimum at the equilibrium bond length. This potential energy curve (or surface for polyatomic molecules) dictates how the nuclei move.[6] The molecule starts in the lower (ground) state with an initial nuclear probability density shown by the wave function centered at the energy minimum along the internuclear coordinate. This potential has a barrier to dissociation, implying that the two atoms are originally in a bound state (i.e. a diatomic molecule). The pump pulse takes a portion of the nuclear wave function to the upper state; in general, this upper state will not have a minimum at the same location, resulting in a net force on the atoms that pushes them apart. Since this particular potential is unbound, the evolution of the wave function in this new state corresponds to an ever-increasing separation between the two atoms, and the molecule dissociates (shown by the separating atoms in the middle panel).

Of course, if all we did was send in the pump pulse, we would have no way of knowing that the wave function was evolving in this way and that the atoms were, in fact, separating. It takes a subsequent interaction with the system to measure the evolution of the state in some way. As will be discussed throughout the book, there are a variety of ways to accomplish this measurement. The simplest approach would be to simply make a rapid measurement of the time-dependent state prepared by the pump pulse with a "fast detector." Consider the example of molecular dissociation discussed in Figure 1.2; if the molecule fluoresced from the excited state and one could resolve the fluorescence both temporally and spectrally, then the dissociation dynamics could be recovered from the time- and frequency-resolved fluorescence after the pump pulse. By following the evolution of the fluorescence spectrum in time, one could determine how the wave packet moves along the excited state potential toward dissociation.

However, since natural molecular timescales (picoseconds for rotations, femtoseconds for vibrations, and attoseconds for electronic dynamics) tend to be much shorter than the time resolution achievable with electronic detectors, one needs to gate the dynamics initiated by the pump with a sufficiently fast probe pulse. In addition, the dynamics initiated by the pump pulse do not always lead to a measurable signal such as fluorescence, and so a probe pulse may be required to produce an observable. For both these reasons, time-resolved measurements are almost always carried out in a pump–probe configuration. The top panel of Figure 1.2 shows a generic probe pulse measurement where a time-dependent signal provides a signature of the dissociation (e.g. absorption of the probe pulse as a function of time delay, or "transient absorption"). The fact the measured signal changes with pump–probe delay indicates dynamics on the excited state. (In this configuration, "time-zero" is defined to be the time at which the pump pulse interacts with the system.)

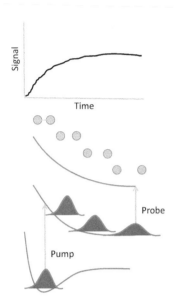

Figure 1.2   Illustration of a simple pump–probe measurement of molecular dissociation. The bottom panel shows the evolution of the nuclear wave function after the pump pulse transfers a portion of the ground-state wave function to a dissociative excited state. The middle panel shows how this corresponds to the atoms separating in time, and the top panel shows how a pump–probe measurement records this dissociation by measuring absorption, ionization, or some other observable from the excited state.

---

[6]Formally, potential energies as a function of a single coordinate are referred to as potential energy curves, whereas multidimensional potentials are called surfaces. For simplicity, we will always refer to them as potential energy surfaces, even for the one dimensional case.

The whole suite of time-resolved spectroscopic techniques can be thought of as different implementations of this basic approach. The pump and probe can be laser pulses in the infrared, visible, ultraviolet, or X-ray regions of the spectrum, or accelerated charged particles such as electrons that originate from either inside or outside the quantum system. The pump and probe can each involve single (linear) interactions between the applied field and quantum system, or can comprise multiple (nonlinear) interactions. Whether charged particles or photons, the interaction can always be described in terms of externally applied electric and magnetic fields. As almost all time-resolved spectroscopy measurements involve both pump and probe pulses, the final state of the system and any observables associated with it are nonlinearly dependent on the total applied field. Thus, time-resolved spectroscopies are generally referred to as nonlinear spectroscopies. We highlight several examples of nonlinear pump or probe interactions throughout the text.

1.3.2 Localized Wave Functions and Expectation Values

Many time-resolved measurements rely on changes in the expectation value of the positions of nuclei or electrons, and are therefore sensitive to the time dependence of $\langle R(t) \rangle$, where R might be the separation of two nuclei in a molecule, or the position of an electron relative to the center of a molecule. This matches our classical intuition of measuring changes in positions of particles in classical physics. However, in quantum mechanics, the wave function for a particle is usually delocalized, as the wave function naturally spreads in time for any anharmonic potential. Most time-resolved measurements aim to generate a localized initial wave function and then measure changes in the position of the relevant particles as a function of time.[7]

For a *completely* delocalized wave function, the measurable expectation values associated with position such as $\langle R(t) \rangle$ and $\langle R^2(t) \rangle$ do not vary with time, and it can be difficult to make a measurement that yields any sense of the motion. Indeed, one can ask the fundamental question whether any dynamics take place for such a completely delocalized wave function. Nevertheless, in bound systems, the wave function will tend to relocalize in what is known as a wave packet revival. When this occurs, the expectation value will again show a time dependence that can be followed in a pump–probe experiment.

Finally, we note that a *classical* description of light as an electromagnetic wave is sufficient to describe nearly all the interactions and dynamics of interest in this book. This is not surprising, since in the high intensities generated by pulsed lasers, the number of photons is large, and any effects due to the quantum nature of light can safely be ignored.[8] Nevertheless, we will regularly use the word "photon" throughout the book. In addition to its simplicity, it also helps indicate the discrete energy required for a quantum transition: the absorption of a photon of frequency ω from the light field transfers $\hbar\omega = E_2 - E_1$ amount of energy to the system, where E_2 and E_1 indicate the energies of the two quantum states involved. The term is also used frequently in the context of perturbative light–matter interactions, where it indicates the order of the process (e.g. a "one-photon transition" corresponds to a first-order process).

---

[7]We note that wave function localization and evolution could also occur in a different basis (e.g. momentum). For most of the examples we consider, position is the more natural basis, and we stick with the language of position here for clarity.

[8]For instance, in a visible laser pulse with 1 μJ of energy, there are over 10^{12} photons.

1.4 FORMAT OF THE BOOK

The primary aim of this book is to first identify the fundamental atomic and molecular dynamics we wish to understand and visualize, and then to describe how one makes measurements capturing these dynamics. We will develop both a qualitative and quantitative understanding of how the measurements connect to the dynamics of interest, progressively moving from simple systems with a single degree of freedom to more complex ones with many, coupled degrees of freedom.

The interaction of light and matter is at the core of time-resolved spectroscopy and is the focus of the background and introduction in Part I. Chapter 2 begins with a discussion of molecular structure, while Chapter 3 considers the basics of how molecules interact with light. In Chapter 4, we present an overview of measurement techniques in the form of four archetypal examples. We also briefly discuss general experimental considerations that are relevant across all forms of time-resolved spectroscopy.

Part II considers quantum dynamics in one dimension. Chapter 5 begins with a theoretical consideration of one-dimensional quantum systems evolving in time and how one represents, or pictures, them as wave functions. Chapter 6 goes on to develop the theoretical framework needed to describe the field-driven dynamics that typically occur in time-resolved spectroscopy.

In Part III, we transition to a description of how one experimentally measures a wave function in one dimension and what type of information is obtained. We discuss a variety of different implementations of the basic pump–probe framework, in each case emphasizing the experimental approach, the resulting data and interpretation, and the mathematical description of what is actually being measured. We consider three separate classes of experiments that build off the archetypal examples of Chapter 4: incoherent measurements (Chapter 7), coherent optical measurements (Chapter 8), and coherent diffractive measurements (Chapter 9).

Part IV moves on to discuss quantum dynamics in more than one dimension, explicitly considering systems where motion along one dimension is coupled to motion along others. This leads to rich dynamics such as mode coupling and internal conversion in larger systems. In Chapter 10, we consider dynamics in multiple dimensions using approaches that explicitly track the full evolution of the various degrees of freedom and how they couple to one another. As we shall see, it becomes convenient to focus on only a subset of the entire system, and Chapter 11 presents implicit approaches that do just this.

Finally, Part V extends the techniques developed in Part III to measure quantum dynamics in multiple dimensions. Chapters 12–14 follow the same progression as Chapters 7–9 did for the various experimental approaches in one dimension. Chapter 15 then examines a molecular system using multiple approaches to develop a complete picture of a femtosecond reaction.

The book also includes appendices designed to provide essential background for understanding the different time-resolved techniques and what they measure. The appendices are referenced throughout the chapters, and we encourage you to refer to them as needed. Depending on your background and level of familiarity with the topics, you will likely find that some sections require more of your attention than others.

CHAPTER 2

Molecular Structure

This chapter reviews and develops the underlying physics required for studying molecular dynamics using time-resolved spectroscopy. This material is treated in a time-independent manner, as it is the quantum properties of the molecules themselves that are relevant at this point, not the interaction with a time-dependent light field. While the details of the quantum structure do not always play a direct role in time-resolved experiments, an understanding of the material is essential for interpreting the dynamics. Here (and throughout), we assume a familiarity with the basic ideas of quantum mechanics.

In general, the total Hamiltonian for a molecule in a light field can be written as

$$\hat{H}_{\text{total}} = \hat{H}_{\text{light}} + \hat{H}_{\text{molecule}} + \hat{H}_{\text{interaction}}, \tag{2.1}$$

where $\hat{H}_{\text{light}}$ and $\hat{H}_{\text{molecule}}$ are the respective Hamiltonians for the light alone and molecule alone, while $\hat{H}_{\text{interaction}}$ contains all terms related to their interaction. We will generally ignore the first term on the right side ($\hat{H}_{\text{light}}$) and treat light fields as classical waves. This is an excellent approximation in the limit of large number of photons, as is typical for the time-resolved experiments we discuss. This chapter focuses on the molecular Hamiltonian independent of the applied light field ($\hat{H}_{\text{molecule}}$), while Chapter 3 concentrates on the interaction between the light and molecule ($\hat{H}_{\text{interaction}}$).

2.1 HYDROGEN ATOM

While the hydrogen atom is perhaps within the domain of standard, time-independent undergraduate quantum mechanics, it is often one of the last subjects covered. We begin with it as our first topic in an effort to connect the familiar with the new, as well as highlight various themes that will be revisited throughout the text.[1] While the Dirac equation would naturally account for relativity and spin–orbit coupling, it is generally much more difficult to solve, and we therefore restrict ourselves to the Schrödinger equation, adding in spin–orbit coupling at a later point.

---

[1] Starting with the hydrogen atom also permits consistent notation in atomic units, where $\hbar = e = m_e = k = 1$. Atomic units will be used throughout the book, and for those who have not encountered them before, it is probably helpful to start with familiar expressions.

We start with the time-independent Schrödinger equation (TISE):

$$\hat{H}\Psi = E\Psi, \tag{2.2}$$

where the Hamiltonian operator $\hat{H}$ for the hydrogen atom is given in atomic units as

$$\hat{H} = -\frac{\nabla^2}{2\mu} - \frac{Z}{r} \tag{2.3}$$

where r is the separation between the proton and electron and μ is the reduced mass: $\mu = \frac{m_{proton}m_{electron}}{m_{proton}+m_{electron}}$. Given the large difference between the proton and electron masses, we take $\mu \sim m_{electron} = 1$ in atomic units. While $Z = 1$ for hydrogen, we leave Z in the expression so that our results easily adapt to singly-ionized helium or other atomic ions. Thus, the TISE can be written as

$$\left(\frac{\nabla^2}{2} + \frac{Z}{r} + E \right)\Psi = 0. \tag{2.4}$$

The solution $\Psi = \Psi(\mathbf{r})$ is a three-dimensional function of space, and as with most differential equations involving multiple variables, it is desirable to seek *separable* solutions where Ψ is written as a product of lower-dimensional functions. Given the spherical symmetry of the potential, it is natural to look for separable solutions in spherical coordinates.[2] In spherical coordinates, the Laplacian is given by

$$\nabla^2 = \frac{1}{r^2}\frac{\partial}{\partial r}\left(r^2 \frac{\partial}{\partial r} \right) + \frac{1}{r^2 \sin\theta}\frac{\partial}{\partial \theta}\left(\sin\theta \frac{\partial}{\partial \theta} \right) + \frac{1}{r^2 \sin^2\theta}\frac{\partial^2}{\partial \phi^2}. \tag{2.5}$$

Based on the central potential, we look for solutions of the form:

$$\Psi(r,\theta,\phi) = R(r)Y(\theta,\phi). \tag{2.6}$$

Inserting this trial solution into the TISE and expanding out the radial derivatives yields

$$\left[\frac{1}{2}\left(\frac{\partial^2 R}{\partial r^2} + \frac{2}{r}\frac{\partial R}{\partial r} \right) + \frac{ZR}{r} + ER \right]Y$$

$$+ \frac{1}{2}\left[\frac{1}{r^2 \sin\theta}\frac{\partial}{\partial \theta}\left(\sin\theta \frac{\partial Y}{\partial \theta} \right) + \frac{1}{r^2 \sin^2\theta}\frac{\partial^2 Y}{\partial \phi^2} \right]R = 0, \tag{2.7}$$

which can be rewritten as

$$\left[\frac{\partial^2 R}{\partial r^2} + \frac{2}{r}\frac{\partial R}{\partial r} + \frac{2ZR}{r} + 2ER \right]\frac{r^2}{R}$$

$$= -\frac{1}{Y}\left[\frac{1}{\sin\theta}\frac{\partial}{\partial \theta}\left(\sin\theta \frac{\partial Y}{\partial \theta} \right) + \frac{1}{\sin^2\theta}\frac{\partial^2 Y}{\partial \phi^2} \right]. \tag{2.8}$$

---

[2]We note that separable solutions exist in both spherical and parabolic coordinate systems, the latter being particularly useful for an analytic treatment of the Stark effect.

Since the left and right sides of Equation 2.8 depend on different variables, they must each be equal to the same constant value for the equation to be satisfied (we will use the symbol λ to represent this separation constant). This leaves us with two, separate differential equations: one for $R(r)$ and one for $Y(\theta, \phi)$.

2.1.1 The Angular Equation

We tackle the angular expression first, using the tools of angular momentum to do so.[3] We begin by showing that the operator on the right side of Equation 2.8 corresponds to L^2, where $\vec{L}$ is the total angular momentum operator (dropping the "hats" for simplicity). We start by writing down the components of $\vec{L}$ with the assumption from classical mechanics that $\vec{L} = \vec{r} \times \vec{p}$, where $\vec{p} = -i\nabla$ is the momentum operator expressed in the position basis. In standard Cartesian coordinates, the components of L take the form

$$L_x = \frac{1}{i}\left[y\frac{\partial}{\partial z} - z\frac{\partial}{\partial y}\right], \tag{2.9a}$$

$$L_y = \frac{1}{i}\left[z\frac{\partial}{\partial x} - x\frac{\partial}{\partial z}\right], \tag{2.9b}$$

$$L_z = \frac{1}{i}\left[x\frac{\partial}{\partial y} - y\frac{\partial}{\partial x}\right]. \tag{2.9c}$$

We would like to express these in spherical coordinates so that we can compare L^2 in spherical coordinates to the angular component of Equation 2.8. The spherical coordinates relate to the Cartesian ones by

$$x = r\sin\theta\cos\phi, \quad y = r\sin\theta\sin\phi, \quad z = r\cos\theta. \tag{2.10}$$

Using this coordinate transformation, we relate the gradient vectors in Cartesian and spherical coordinates by the matrix transformation:

$$\begin{bmatrix} \frac{\partial}{\partial r} \\ \frac{\partial}{\partial \theta} \\ \frac{\partial}{\partial \phi} \end{bmatrix} = \begin{bmatrix} \sin\theta\cos\phi & \sin\theta\sin\phi & \cos\theta \\ r\cos\theta\cos\phi & r\cos\theta\sin\phi & -r\sin\theta \\ -r\sin\theta\sin\phi & r\sin\theta\cos\phi & 0 \end{bmatrix} \begin{bmatrix} \frac{\partial}{\partial x} \\ \frac{\partial}{\partial y} \\ \frac{\partial}{\partial z} \end{bmatrix} \tag{2.11}$$

The matrix in Equation 2.11 can be inverted to yield the inverse transformation:[4]

$$\begin{bmatrix} \frac{\partial}{\partial x} \\ \frac{\partial}{\partial y} \\ \frac{\partial}{\partial z} \end{bmatrix} = \begin{bmatrix} \sin\theta\cos\phi & \frac{\cos\theta\cos\phi}{r} & -\frac{\sin\phi}{r\sin\theta} \\ \sin\theta\sin\phi & \frac{\cos\theta\sin\phi}{r} & \frac{\cos\phi}{r\sin\theta} \\ \cos\theta & -\frac{\sin\theta}{r} & 0 \end{bmatrix} \begin{bmatrix} \frac{\partial}{\partial r} \\ \frac{\partial}{\partial \theta} \\ \frac{\partial}{\partial \phi} \end{bmatrix} \tag{2.12}$$

---

[3]Alternatively, one can further separate the angular solution $Y(\theta, \phi)$ into a product solution of functions that depend only on θ or ϕ. These two new differential equations can then be solved separately.

[4]One can verify that the matrix in Equation 2.12 is the inverse of the matrix in Equation 2.11 by multiplying them to yield the identity matrix.

Based on Equation 2.12 and the components of the angular momentum $\vec{L}$, we write L_x as

$$
\begin{aligned}
L_x &= \frac{1}{i}\left[y\frac{\partial}{\partial z} - z\frac{\partial}{\partial y}\right] \\
&= \frac{1}{i}\left[r\sin\theta\sin\phi\left(\cos\theta\frac{\partial}{\partial r} - \frac{1}{r}\sin\theta\frac{\partial}{\partial\theta}\right)\right. \\
&\quad \left. -r\cos\theta\left(\sin\theta\sin\phi\frac{\partial}{\partial r} + \frac{1}{r}\cos\theta\sin\phi\frac{\partial}{\partial\theta} + \frac{\cos\phi}{r\sin\theta}\frac{\partial}{\partial\phi}\right)\right] \\
&= \frac{1}{i}\left[-\sin\phi\frac{\partial}{\partial\theta} - \frac{\cos\phi}{\tan\theta}\frac{\partial}{\partial\phi}\right].
\end{aligned}
\tag{2.13}
$$

Similarly, we can find expressions for L_y and L_z:

$$
L_y = \frac{1}{i}\left[\cos\phi\frac{\partial}{\partial\theta} - \frac{\sin\phi}{\tan\theta}\frac{\partial}{\partial\phi}\right],
\tag{2.14a}
$$

$$
L_z = \frac{1}{i}\frac{\partial}{\partial\phi}.
\tag{2.14b}
$$

Based on these components, one can show that:

$$
L^2 = L_x^2 + L_y^2 + L_z^2 = -\left[\frac{1}{\sin\theta}\frac{\partial}{\partial\theta}\left(\sin\theta\frac{\partial}{\partial\theta}\right) + \frac{1}{\sin^2\theta}\frac{\partial^2}{\partial\phi^2}\right].
\tag{2.15}
$$

This is exactly the operator on the right side of Equation 2.8, and so $Y(\theta,\phi)$, the solution to the angular equation, is given by the eigenvectors of L^2 with eigenvalues $\lambda = l(l+1)$, where l is an integer.[5] The eigenvalue equation for Y_l (indexed by l) reads

$$
L^2 Y_l = l(l+1)Y_l.
\tag{2.16}
$$

The eigenfunctions of L^2 can also simultaneously be eigenfunctions of L_x, L_y or L_z, since each of these commute with L^2:

$$
\left[L_i, L^2\right] = 0 \ \ \text{for} \ \ i = x, y, z.
\tag{2.17}
$$

However, the individual components of the angular momentum operator do not commute with each other: $[L_i, L_j] \neq 0$ for $i \neq j$. Therefore, $Y(\theta,\phi)$ can be an eigenfunction of only one component of angular momentum. Choosing which coordinate defines the orientation of $Y(\theta,\phi)$, effectively making this axis different from the other two. Here, as in most other texts, we choose $Y(\theta,\phi)$ to be an eigenfunction of L_z as well as L^2.

The normalized eigenfunctions $Y(\theta,\phi)$ are the spherical harmonics $Y_{lm}(\theta,\phi)$, defined by the expression

$$
Y_{lm}(\theta,\phi) = \sqrt{\frac{(2l+1)(l-m)!}{4\pi(l+m)!}}P_l^m(\cos\theta)e^{im\phi},
\tag{2.18}
$$

where the $P_l^m(\cos\theta)$ are the associated Legendre polynomials. The eigenfunctions have well-defined angular momentum, with the numbers l and m being the angular

---

[5] Although the angular momentum commutation relations allow both integer and half-integer solutions, we restrict ourselves to integers since we will find the solutions to be the spherical harmonics.

momentum quantum numbers. The number l must be an integer starting from zero, while m takes on values $-l, -l+1, \ldots, l-1, l$ (for a total of $2l+1$ different m values for any l).

The spherical harmonics are also eigenfunctions of the L_z operator with eigenvalues m:

$$L_z Y_{lm} = m Y_{lm}. \tag{2.19}$$

The spherical harmonics have definite parity:

$$Y_{l,-m}(\theta, \phi) = (-1)^m Y_{lm}^*(\theta, \phi) = (-1)^m Y_{lm}(\theta, -\phi) \tag{2.20a}$$

$$Y_{lm}(\pi - \theta, \pi + \phi) = (-1)^l Y_{lm}(\theta, \phi). \tag{2.20b}$$

In addition, they satisfy the orthonormality condition:

$$\int \int d\Omega \, Y_{l'm'}^* Y_{lm} = \delta_{ll'} \delta_{mm'}. \tag{2.21}$$

2.1.2 The Radial Equation

We now turn our attention to the left side of Equation 2.8. It must also be equal to the constant λ, which we know is related to the angular momentum quantum number l:

$$\left[\frac{\partial^2 R}{\partial r^2} + \frac{2}{r} \frac{\partial R}{\partial r} + \frac{2ZR}{r} + 2ER \right] \frac{r^2}{R} = \lambda = l(l+1). \tag{2.22}$$

This can be rewritten as

$$\frac{1}{2} \frac{\partial^2 R}{\partial r^2} + \frac{1}{r} \frac{\partial R}{\partial r} + \left[E + \frac{Z}{r} - \frac{l(l+1)}{2r^2} \right] R = 0. \tag{2.23}$$

The first term in the square bracket represents the total energy of a given state, the second term is the Coulomb potential, and the third term is the "centrifugal barrier" resulting from angular momentum. At this point, it is useful to define a modified, or scaled, radial wave function given by

$$u(r) \equiv r R(r). \tag{2.24}$$

The differential equation for $u(r)$ is then given by

$$\frac{1}{2} \frac{d^2 u}{dr^2} + \left[E + \frac{Z}{r} - \frac{l(l+1)}{2r^2} \right] u(r) = 0. \tag{2.25}$$

In solving this differential equation, it is useful to consider the asymptotic behavior for large and small r ($r \to \infty$ and $r \to 0$). We first consider the large-r limit, in which the equation for u reduces to

$$\frac{1}{2} \frac{d^2 u}{dr^2} + E u \approx 0. \tag{2.26}$$

As this is a familiar differential equation, the solution is straightforward to simply write down (up to a constant):

$$u(r) = e^{\pm \sqrt{-2E} r}. \tag{2.27}$$

We note that for negative energies (bound states), the solution with the positive exponent blows up as $r \to \infty$ and is not physical. In the opposite limit ($r \to 0$), we have

$$\frac{1}{2}\frac{d^2u}{dr^2} - \frac{l(l+1)}{2r^2}u \approx 0. \tag{2.28}$$

This equation also lends itself to simple guesses for the solution:

$$u(r) = r^{l+1} \tag{2.29}$$

$$u(r) = r^{-l}. \tag{2.30}$$

The second of these is, of course, not physical as r tends to zero.

Based on the asymptotic behavior, we are motivated to try solutions for all r that combine features of the limiting cases. One guess for the form of such a solution uses polynomials as the connecting function:

$$u(r) = r^{l+1}e^{-\sqrt{-2E}r}\sum_{k=0}^{\infty}A_k r^k. \tag{2.31}$$

We can insert this solution into the differential equation for u and see if there exists an expression for A_k that satisfies the equation. This procedure yields the recursion relation:

$$A_k = -2A_{k-1}\left[\frac{Z-(l+k)\sqrt{-2E}}{(l+k)(l+k+1)-l(l+1)}\right]. \tag{2.32}$$

Our solution for u blows up at large r unless the series terminates. For our solution to be valid, there must be some integer k_{max} after which $A_{k_{max}+1} = 0$. This implies that

$$Z-(l+k_{max}+1)\sqrt{-2E}=0,$$

or

$$E = -\frac{Z^2}{2(l+k_{max}+1)^2}.$$

If we let $n = l+k_{max}+1$, our energy eigenvalues become[6]

$$E = -\frac{Z^2}{2n^2}, \tag{2.33}$$

where for $l = 0$, n can be $1,2,3\ldots$, while for $l = 1$, n can be $2,3,4\ldots$. In other words, for a given n value, l can go from 0 to $n-1$.

Therefore, we have a separable analytic solution, with the spherical harmonics constituting the angular part of the eigenfunction and a recursion relationship for the coefficients of a polynomial leading to the radial wave function. The recursion relationship given in Equation 2.32 relates to the associated Laguerre polynomials, and allows us to write the radial portion of the hydrogen eigenfunctions as

$$R_{nl}(r) = \sqrt{\frac{(n-l-1)!Z}{n^2[(n+1)!]^3}}\left(\frac{2Zr}{n}\right)^{l+1}\frac{e^{-Zr/n}}{r}L_{n+l}^{2l+1}\left(\frac{2Zr}{n}\right), \tag{2.34}$$

---

[6]Note that determining the eigenvalues is useful, even if one can't fully solve for the eigenvectors analytically. For a given eigenvalue, one could always integrate the TISE numerically to calculate the eigenvector associated with that particular eigenvalue.

where $L_\lambda^\mu(x)$ is an associated Laguerre function defined by

$$L_\lambda^\mu(x) = \frac{\partial^\mu}{\partial x^\mu} L_\lambda(x), \tag{2.35}$$

and $L_\lambda(x)$ being the Laguerre polynomial:

$$L_\lambda(x) = e^x \frac{\partial^\lambda}{\partial x^\lambda} \left(e^{-x} x^\lambda \right). \tag{2.36}$$

The R_{nl} are normalized

$$\int_0^\infty r^2 dr R_{nl}(r) = 1, \tag{2.37}$$

and the first few are given for hydrogen ($Z = 1$) by

$$R_{10}(r) = 2e^{-r} \tag{2.38a}$$

$$R_{20}(r) = \frac{1}{\sqrt{2}} e^{-r/2} \left(1 - \frac{r}{2} \right) \tag{2.38b}$$

$$R_{21}(r) = \frac{1}{2\sqrt{6}} e^{-r/2} r \tag{2.38c}$$

$$R_{30}(r) = \frac{2}{3\sqrt{3}} e^{-r/3} \left(1 - \frac{2r}{3} + \frac{2r^2}{27} \right). \tag{2.38d}$$

The following expectation values for powers of r are also useful:

$$\langle r \rangle = \frac{1}{2Z} \left[3n^2 - l(l+1) \right] \tag{2.39a}$$

$$\langle r^2 \rangle = \frac{n^2}{2Z^2} \left[5n^2 - 3l(l+1) + 1 \right] \tag{2.39b}$$

$$\left\langle \frac{1}{r} \right\rangle = \frac{Z}{n^2} \tag{2.39c}$$

$$\left\langle \frac{1}{r^2} \right\rangle = \frac{Z^2}{n^3(l+\frac{1}{2})} \tag{2.39d}$$

$$\left\langle \frac{1}{r^3} \right\rangle = \frac{Z^3}{n^3 l(l+1)(l+\frac{1}{2})}, \tag{2.39e}$$

where the expression for $\langle 1/r^3 \rangle$ is only valid for $l > 0$ (it diverges for $l = 0$).[7]

2.1.3 The Electron Spin

Thus far we have ignored the spin of the electron, which is a fermion with spin 1/2. Associated with this spin is a magnetic moment, $\vec{\mu}_e = -\frac{g}{2}\vec{S}$, where g is the electron g factor (≈ 2.002319). In the rest frame of the electron, the proton circles it, generating a magnetic field due to the moving charge. One can think of the electron's magnetic moment as interacting with the magnetic field of the moving proton (or more generally the positively charged nucleus). One method to estimate the magnetic field of the

---

[7]The expectation value calculated for solutions to the (relativistic) Dirac equation does not diverge because of small differences in the solutions to the Dirac and Schrödinger equations at small r.

nucleus as seen by the electron in its rest frame is to Lorentz transform the electric field of the nucleus in its own rest frame.[8] Electric and magnetic fields transform according to

$$\vec{E}'_{\text{parallel}} = \vec{E}_{\text{parallel}} \tag{2.40a}$$

$$\vec{B}'_{\text{parallel}} = \vec{B}_{\text{parallel}} \tag{2.40b}$$

$$\vec{E}'_{\text{perp}} = \gamma \left(\vec{E}_{\text{perp}} + \vec{v} \times \vec{B} \right) \tag{2.40c}$$

$$\vec{B}'_{\text{perp}} = \gamma \left(\vec{B}_{\text{perp}} - \frac{\vec{v}}{c^2} \times \vec{E} \right), \tag{2.40d}$$

where $\gamma = 1/\sqrt{1 - v^2/c^2}$. Thus, the electric field of the nucleus in its own rest frame, $E = Z\hat{r}/r^2$, transforms to[9]

$$\vec{B} = -\frac{\vec{v}}{c^2} \times \vec{E} = -Z \frac{\vec{v}}{c^2} \times \frac{\hat{r}}{r^2} = \frac{Z\vec{L}}{r^3 c^2}. \tag{2.41}$$

Thus, the interaction energy for a magnetic dipole moment, μ, in this magnetic field, B, is given by

$$U = -\vec{\mu} \cdot \vec{B} = \frac{gZ}{2r^3 c^2} \vec{L} \cdot \vec{S}. \tag{2.42}$$

This expression does not include the so called "Thomas correction" factor of 1/2, which takes into account the non-inertial reference frame. Including this extra factor of 1/2 yields a result consistent with the Dirac equation, from which spin orbit coupling naturally arises as a consequence of relativistic quantum mechanics. One can make use of first-order perturbation theory to find the so-called "spin–orbit" correction to the state energies by calculating $\langle U \rangle$. However, several points are clear immediately from Equation 2.42. The first is that the energy correction is small compared with the binding energy of an electron, since the value of c in atomic units is $1/\alpha \approx 137$. A second is that there is no spin–orbit coupling for s states, where $l = 0$. Another is that spin–orbit coupling is very sensitive to the radial distance, falling off as $1/r^3$. Finally, when considering the dependence on Z, there is both the explicit dependence on Z in Equation 2.42, as well as the implicit dependence on Z contained in the expectation value of $1/r^3$ (see Equation 2.39). This gives an additional factor of Z^3, leading to a total Z dependence of Z^4. This means that the spin–orbit coupling energy can become comparable to electronic energy splittings in heavy atoms. For example, in iodine, the spin–orbit shift for the ground state is on the order of 1 eV.

In general, it is not possible to fully separate the spin and spatial degrees of freedom due to this spin–orbit coupling, as the true eigenstates of the system are coupled functions of spin and position (the wave function is "entangled", and cannot be written as a product of spin and spatial functions). While the eigenstates cannot be expressed as a *single* product of spin and spatial functions, it is possible to write them as a sum of products, where the energies of these different product states will be slightly different than without the spin–orbit coupling.

Spin–orbit coupling generally leads to slow, intersystem crossing when exciting a molecule from an initial state with a well-defined total spin to a final state with the

---

[8] We note that this approach neglects the fact that we are transforming to a noninertial frame, and this leads to a factor of 2 error relative to the result from the Dirac equation.

[9] This can also be found using the Biot–Savart Law.

same total spin but which is not an eigenstate of the total Hamiltonian. For example, the final state can be described as a superposition of two eigenstates that each have a mixture of singlet and triplet character. Since the energy difference between these product states is small, the timescale over which a coherent superposition evolves is relatively long (this is the time over which the phase difference between states changes by an appreciable fraction of 2π). Thus, for the short timescales probed by most time-resolved experiments, spin–orbit coupling is usually neglected.

2.2 HELIUM ATOM

The next natural step in complexity is a system with one more electron: the helium atom. This a three-body problem (the nucleus plus two electrons), which even classically does not have a closed-form analytic solution. Thus, it should not be surprising to find that we cannot solve the TISE without making significant approximations. We will follow the general approach of starting with the simplest picture possible, later working to improve on the result using perturbation theory. Although very simple compared with most systems of interest in time-resolved spectroscopy, helium provides a nice introduction to the complexities that result with multiple electrons.

We start with the well-justified assumption that the center of mass is fixed at the position of the nucleus, as the mass of the nucleus is so much greater than the electron mass. In this case, the Hamiltonian for the system is given by

$$\hat{H} = -\frac{1}{2}\left(\nabla_1^2 + \nabla_2^2\right) - \frac{Z}{r_1} - \frac{Z}{r_2} + \frac{1}{r_{12}}, \tag{2.43}$$

where ∇_1^2 is the Laplacian with respect to the coordinates of the first electron (r_1), ∇_2^2 the Laplacian with respect to the coordinates of the second electron (r_2), and r_{12} represents the magnitude of the vector connecting the two electrons. We again keep $\hat{H}$ general for nuclear charge Z, although $Z = 2$ for helium of course.

It is worth noting that were it not for the interaction between the two electrons embodied in the $1/r_{12}$ term, Equation 2.43 would simply be the sum of two hydrogenic Hamiltonians with nuclear charge two, and the three-body problem could be reduced to two, separate two-body problems (which independently could be treated as one-body problems by working in the center-of-mass frame). In the case of no electron interaction, the eigenenergies of the system would simply be the sum of two hydrogenic energies (in atomic units):

$$E_n \approx -\frac{Z^2}{2n^2} - \frac{Z^2}{2n^2} = -\frac{Z^2}{n^2}, \tag{2.44}$$

which for $Z = 2$ and $n = 1$ yield a ground-state energy $E_1 = -4 = -109$ eV. The measured value is -79 eV, indicating that electron correlation plans an important role. Similarly, without the electron interaction, the eigenstates could be expressed as *products* of one-electron hydrogenic eigenstates for the independent electrons (still ignoring spin):

$$\Psi(\mathbf{r_1}, \mathbf{r_2}) \approx R_{n_1 l_1}(r_1) Y_{l_1 m_1}(\theta_1, \phi_1) R_{n_2 l_2}(r_2) Y_{l_2 m_2}(\theta_2, \phi_2). \tag{2.45}$$

Of course, we are ultimately interested in including the interaction term ($1/r_{12}$) as part of a three-body system containing two identical fermions. We therefore need to consider the symmetry properties of our total wave function. For example, it would be unphysical if the wave function did not yield the same probability under exchange of any two identical particles. Thus, the square magnitude of the wave function must be the same after exchanging the two electrons. This can be accomplished by having a total wave function that is either symmetric or antisymmetric with respect to exchange of any two identical particles (symmetric for the case of bosons and antisymmetric for fermions). For electrons, which are fermions, the form of the solution we seek must be antisymmetric with respect to exchange of the two particles, including their spin. This can be written as $\Psi(\mathbf{r_1}, \mathbf{r_2}, s_1, s_2) = -\Psi(\mathbf{r_2}, \mathbf{r_1}, s_2, s_1)$, and it will have a profound effect on how we construct solutions to the TISE, as we cannot immediately ignore the electron spin as with hydrogen.

2.2.1 Identical Particles

We begin by considering only the spin portion of the wave function for a system of two identical electrons. In particular, there are four possible spin states that have definite symmetry:

$$\chi(s_1, s_2) = \begin{cases} |\uparrow\uparrow\rangle \\ |\downarrow\downarrow\rangle \\ \frac{1}{\sqrt{2}}(|\uparrow\downarrow\rangle + |\downarrow\uparrow\rangle) \\ \frac{1}{\sqrt{2}}(|\uparrow\downarrow\rangle - |\downarrow\uparrow\rangle), \end{cases} \tag{2.46}$$

where the first arrow indicates the spin of electron one, while the second arrow indicates the spin of electron two. The spin wave functions in Equation 2.46 are the only combinations of spin up and down for the two electrons that satisfy the requirement of having definite parity with respect to exchange of the two electrons (i.e. symmetric or antisymmetric). "Asymmetric" wave functions such as $\chi(s_1, s_2) = |\uparrow\downarrow\rangle$ (electron 1 spin up, electron 2 spin down) are unphysical since the two electrons are identical and cannot be distinguished in this way.

To physically interpret the spin states of Equation 2.46, it is useful to calculate and compare the expectation values for the total spin, its projection on the z-axis, and the symmetry of the state. In order to do this, we recall that for a spin-$\frac{1}{2}$ particle in the s_z basis:

$$s_z = \frac{1}{2} \begin{bmatrix} 1 & 0 \\ 0 & -1 \end{bmatrix}, \tag{2.47}$$

and make use of the following:

$$\begin{aligned} s_z &= s_{1z} + s_{2z} \\ s^2 &= (\vec{s}_1 + \vec{s}_2)^2 = s_1^2 + s_2^2 + 2\vec{s}_1 \cdot \vec{s}_2 \\ &= s_1^2 + s_2^2 + 2s_{1x}s_{2x} + 2s_{1y}s_{2y} + 2s_{1z}s_{2z} \\ \langle s^2 \rangle &= s(s+1). \end{aligned} \tag{2.48}$$

The spin-state properties are tabulated in Table 2.1. Based on the total spin in the last column, the first three states are known as triplet states (having three possible projections onto the z-axis), whereas the final state is a singlet state (with only one possible projection onto the z-axis of zero).

Table 2.1 Spin-state properties for two spin-$\frac{1}{2}$ particles.

χ	**Symmetry**	$\langle s_z \rangle$	$\langle \mathbf{s}^2 \rangle$	**s**
$|\uparrow\uparrow\rangle$	+	1	2	1
$|\downarrow\downarrow\rangle$	+	−1	2	1
$\frac{1}{\sqrt{2}}\left(|\uparrow\downarrow\rangle + |\downarrow\uparrow\rangle\right)$	+	0	2	1
$\frac{1}{\sqrt{2}}\left(|\uparrow\downarrow\rangle - |\downarrow\uparrow\rangle\right)$	−	0	0	0

2.2.2 Multiparticle Wave Functions and the Slater Determinant

We now look for multiparticle solutions that involve *products* of spatial and spin wave functions. Given the earlier antisymmetric requirement, if we consider the total wave function to be a single product of $\psi(\mathbf{r}_1, \mathbf{r}_2)$ and $\chi(s_1, s_2)$, either $\psi(\mathbf{r}_1, \mathbf{r}_2)$ or $\chi(s_1, s_2)$ must be antisymmetric with respect to exchange of the electrons (while the other must be symmetric). If we allow for a sum of products, then a simple way to ensure that the total wave function is antisymmetric with respect to particle exchange is to write it as a difference between two products of single-particle spin and spatial wave functions (or "orbitals"), with the particle indices exchanged for the two terms. This guarantees that the total wave function is antisymmetric with respect to particle exchange:

$$\Psi(\mathbf{r}_1, \mathbf{r}_2, s_1, s_2) = \frac{1}{\sqrt{2}}\left[\psi_1(\mathbf{r}_1)\psi_2(\mathbf{r}_2)\,\chi_1(s_1)\chi_2(s_2) \right. \tag{2.49}$$
$$\left. - \psi_1(\mathbf{r}_2)\,\psi_2(\mathbf{r}_1)\chi_1(s_2)\chi_2(s_1)\right],$$

where $\psi_i(\mathbf{r}_j)$ is the ith spatial wave function for electron j (similarly for χ with spins). Note that the difference of products shown in Equation 2.49 can be expressed mathematically as the determinant of a matrix:

$$\Psi(\mathbf{r}_1, \mathbf{r}_2, s_1, s_2) = \frac{1}{\sqrt{2}}\begin{vmatrix} \psi_1(\mathbf{r}_1)\chi_1(s_1) & \psi_1(\mathbf{r}_2)\chi_1(s_2) \\ \psi_2(\mathbf{r}_1)\chi_2(s_1) & \psi_2(\mathbf{r}_2)\chi_2(s_2) \end{vmatrix} \equiv |\psi_1\chi_1\psi_2\chi_2|_{\text{SD}}, \tag{2.50}$$

where $|\psi_1\chi_1\psi_2\chi_2|_{\text{SD}}$ is shorthand for the full determinant. This is known as a Slater determinant ("SD") and can easily be generalized to the N-electron case (see Section 2.6). We note that if $\psi_1 = \psi_2$ *and* $\chi_1 = \chi_2$, then $\Psi(\mathbf{r}_1, \mathbf{r}_2, s_1, s_2) = -\Psi(\mathbf{r}_1, \mathbf{r}_2, s_1, s_2)$, which implies that the total wave function must be zero. This is the essence of the Pauli exclusion principle, stating that the two electrons cannot occupy the exact same state (spatial orbital and spin).

2.2.3 Corrections to the Energies

Having discussed the implications of a system with multiple identical particles, we are in a position to calculate corrections to the energies given in Equation 2.44 due to the $1/r_{12}$ term in the Hamiltonian. Specifically, in perturbation theory the first-order correction comes from the expectation value of $1/r_{12}$ using the properly symmetrized wave function without the interaction term (i.e. the first-order correction to the energies uses the zero-order wave functions). We therefore use the properly symmetrized, single Slater determinant state of Equations 2.49 and 2.50 in the correction to the energies:

$$E^{(1)} = \left\langle |\psi_1\chi_1\psi_2\chi_2|_{\text{SD}} \left| \frac{1}{r_{12}} \right| |\psi_1\chi_1\psi_2\chi_2|_{\text{SD}} \right\rangle. \tag{2.51}$$

Note that since $1/r_{12}$ does not affect the spin portion of the wave function, we really only need worry about the spatial portion when calculating $\langle 1/r_{12} \rangle$. Nevertheless, for completeness we write down the spin portion of the wave function as well before moving on to calculate the energy correction.

We start by looking for solutions where both electrons occupy the hydrogen-like $1s$ orbital. Substituting $\psi_1 = \psi_2 = \psi_{1s}$ in Equation 2.49 allows us to factor out the spatial portion:

$$
\begin{aligned}
\Psi_{GS}(\mathbf{r_1}, \mathbf{r_2}, s_1, s_2) &= \frac{1}{\sqrt{2}} \left[\psi_{1s}(\mathbf{r_1})\psi_{1s}(\mathbf{r_2})\chi_1(s_1)\chi_2(s_2) \right. \\
&\quad \left. - \psi_{1s}(\mathbf{r_2})\psi_{1s}(\mathbf{r_1})\chi_1(s_2)\chi_2(s_1) \right] \\
&= \frac{1}{\sqrt{2}} \psi_{1s}(\mathbf{r_1})\psi_{1s}(\mathbf{r_2}) \left[\chi_1(s_1)\chi_2(s_2) - \chi_1(s_2)\chi_2(s_1) \right],
\end{aligned} \tag{2.52}
$$

where Ψ_{GS} indicates we are considering the ground-state wave function. In this factored form, it is clear that the spatial portion of the wave function is symmetric. We therefore know the spin portion must be antisymmetric (the spin-singlet state). From the notation of Equation 2.52, it must be that $\chi_1 = \uparrow$ and $\chi_2 = \downarrow$. Ordering the factors in the spin portion such that the spin of particle one always comes first (like in Equation 2.46), the final form of the ground-state wave function is

$$
\Psi_{GS}(\mathbf{r_1}, \mathbf{r_2}, s_1, s_2) = \psi_{1s}(\mathbf{r_1})\psi_{1s}(\mathbf{r_2}) \frac{1}{\sqrt{2}} \left(|\uparrow\downarrow\rangle - |\downarrow\uparrow\rangle \right), \tag{2.53}
$$

where we have grouped the $1/\sqrt{2}$ factor with the spin portion so that the spin wave function remains normalized.

Proceeding with only the spatial portion of the wave function, the first-order correction to the energy is simply

$$
E^{(1)} = \left\langle \psi_{1s}(\mathbf{r_1})\psi_{1s}(\mathbf{r_2}) \left| \frac{1}{r_{12}} \right| \psi_{1s}(\mathbf{r_1})\psi_{1s}(\mathbf{r_2}) \right\rangle. \tag{2.54}
$$

The most straightforward way to deal with the integral of $1/r_{12}$ when working in individual electron coordinates ($\mathbf{r_1}$ and $\mathbf{r_2}$) is to use the fact that it can be expressed in terms of the Legendre polynomials P_l:

$$
\begin{aligned}
\frac{1}{r_{12}} &= \sum_l \frac{r_2^l}{r_1^{l+1}} P_l(\cos\theta_{12}), \quad \text{for } r_2 < r_1 \\
&= \sum_l \frac{r_1^l}{r_2^{l+1}} P_l(\cos\theta_{12}), \quad \text{for } r_1 < r_2,
\end{aligned} \tag{2.55}
$$

where θ_{12} is the angle between $\mathbf{r_1}$ and $\mathbf{r_2}$. Furthermore, the Legendre polynomials can be written in terms of the spherical harmonics as

$$
P_l(\cos\theta_{12}) = \frac{4\pi}{2l+1} \sum_m Y_{lm}^*(\theta_1, \phi_1) Y_{lm}(\theta_2, \phi_2). \tag{2.56}
$$

Using these to express $1/r_{12}$ in terms of r_1 and r_2, we can write Equation 2.54 as

$$
E^{(1)} = \sum_l \sum_m \frac{4\pi}{2l+1} \left[\int_0^\infty dr_1 \int_0^{r_1} dr_2 r_1^2 r_2^2 R_{10}^2(r_1) \frac{r_2^l}{r_1^{l+1}} R_{10}^2(r_2) \right.
$$

$$+ \int_0^\infty dr_1 \int_{r_1}^\infty dr_2 r_1^2 r_2^2 R_{10}^2(r_1) \frac{r_1^l}{r_2^{l+1}} R_{10}^2(r_2) \Bigg]$$

$$\times \int_1 \int_2 d\Omega_1 d\Omega_2 Y_{00}^*(\theta_1,\phi_1) Y_{lm}^*(\theta_1,\phi_1) Y_{00}(\theta_1,\phi_1) \qquad (2.57)$$

$$[3pt] \quad \times Y_{00}^*(\theta_2,\phi_2) Y_{lm}(\theta_2,\phi_2) Y_{00}(\theta_2,\phi_2)$$

$$= F(r_1,r_2) G(\theta_1,\phi_1,\theta_2,\phi_2). \qquad (2.58)$$

We compute the radial and angular integrals separately. With the leading factor and sums, the angular integrals are

$$G(\theta_1,\phi_1,\theta_2,\phi_2) = \sum_l \frac{4\pi}{2l+1} \sum_m \int_1 \int_2 Y_{00}(\theta_1,\phi_1) Y_{00}(\theta_2,\phi_2)$$

$$\times Y_{lm}^*(\theta_1,\phi_1) Y_{lm}(\theta_2,\phi_2) Y_{00}(\theta_1,\phi_1) Y_{00}(\theta_2,\phi_2) \, d\Omega_1 d\Omega_2. \qquad (2.59)$$

There are a total of four Y_{00}'s. The product of two Y_{00}'s simply brings out a factor of $1/4\pi$, canceling the leading factor. Grouping the remaining two Y_{00}'s into integrals based on electron coordinates yields

$$G(\theta_1,\phi_1,\theta_2,\phi_2) = \sum_l \frac{1}{2l+1} \sum_m \int_1 Y_{lm}^*(\theta_1,\phi_1) Y_{00}(\theta_1,\phi_1) \, d\Omega_1$$

$$\times \int_2 Y_{lm}(\theta_2,\phi_2) Y_{00}(\theta_2,\phi_2) \, d\Omega_2. \qquad (2.60)$$

By virtue of the orthonormality of the spherical harmonics given in Equation 2.21, both of these integrals yield $\delta_{l0}\delta_{m0}$. The δ_{m0} removes the sum over m, while the δ_{l0} reduces the leading fraction (and the entire angular portion) to one. The δ_{l0} also sets $l = 0$ in the radial portion.

We are left with the radial integral, which must be evaluated in two pieces due to the different functional forms of $1/r_{12}$ for $r_2 < r_1$ and $r_1 < r_2$:

$$E^{(1)} = \Bigg[\int_0^\infty dr_1 \int_0^{r_1} dr_2 r_1^2 r_2^2 R_{10}^2(r_1) \frac{1}{r_1} R_{10}^2(r_2)$$

$$+ \int_0^\infty dr_1 \int_{r_1}^\infty dr_2 r_1^2 r_2^2 R_{10}^2(r_1) \frac{1}{r_2} R_{10}^2(r_2) \Bigg]. \qquad (2.61)$$

Inserting the analytic form of $R_{10}(r) = 2Z^{3/2} e^{-Zr}$ yields the following:

$$E^{(1)} = 16Z^6 \Bigg[\int_0^\infty dr_1 \int_0^{r_1} dr_2 r_1 r_2^2 e^{-2Zr_1} e^{-2Zr_2}$$

$$+ \int_0^\infty dr_1 \int_{r_1}^\infty dr_2 r_1^2 r_2 e^{-2Zr_1} e^{-2Zr_2} \Bigg]. \qquad (2.62)$$

The integrals required to evaluate $E^{(1)}$ can be done by parts, since they involve only products of polynomials and exponentials. The result (in atomic units) is

$$E^{(1)} = Z\left(1! - \frac{2!}{16} - \frac{1!}{4}\right) = \frac{5Z}{8}. \qquad (2.63)$$

Combining the first-order correction with the zeroth-order energy for the ground state (Equation 2.44) yields for $Z \to 2$:

$$E = E^{(0)} + E^{(1)} = -Z^2 + \frac{5Z}{8} = -4 + \frac{5}{4} \equiv -74.8 \text{ eV}. \tag{2.64}$$

Comparison with the measured value of 78.8 eV shows a significant improvement over the zeroth-order energies (109 eV). Calculating higher-order corrections to the energy using perturbation theory allows one to get within a fraction of a percent of the experimental value, illustrating the power of perturbation theory even in a case when the correction is not strictly perturbative (the $1/r_{12}$ term is not an order of magnitude smaller than all other terms in the Hamiltonian).

Before moving on, we note that for excited states, the wave function does not factor as simply as Equation 2.53. The first-order corrections to the energies contain multiple terms, typically grouped together into what are called the "direct" (or Coulomb) and "exchange" terms.

2.3 MORE THAN TWO ELECTRONS: LITHIUM AND BEYOND

As we saw in helium, the presence of multiple, identical electrons necessitated a more careful approach when constructing the wave function. The situation only becomes more complicated as additional electrons are added. To keep the interpretation manageable, we continue to seek solutions that are factorizable into products of one-electron functions (or orbitals). Of course, one can always write the total wave function as a sum of products of one-electron wave functions. But for this solution to be useful, the sum should converge as quickly as possible (ideally after the first term!).

A natural starting point is to again use one-electron orbitals based on the hydrogen atom ($1s$, $2s$, $2p$, etc.). However, one clearly must adjust these orbitals for the case of lithium. For example, the $1s$ orbital used to describe the ground state should be scaled to account for the larger number of protons in the nucleus relative to hydrogen. In addition, the different angular momentum states of the outer orbitals penetrate the inner-electron orbitals (the "core" region) to a different extent. For instance, an s orbital has nonzero probability at $r = 0$, and therefore sees a deeper potential for small r than a p orbital (which has zero probability for being at $r = 0$ and sees a "screened" central core).

We begin by writing the electronic wave function as a sum of products of individual-electron orbitals, with *each product* antisymmetrized using a Slater determinant. Since the full expression can get quite cumbersome, for compactness we suppress the spin portion of the wave function and do not write out all the terms of each Slater determinant. The wave function is then

$$\Psi(\mathbf{r_1}, \mathbf{r_2}, \mathbf{r_3}) = \sum_i a_i \left| \tilde{\psi}_{\alpha_i}(\mathbf{r_1}) \tilde{\psi}_{\beta_i}(\mathbf{r_2}) \tilde{\psi}_{\gamma_i}(\mathbf{r_3}) \right|_{\text{SD}}, \tag{2.65}$$

where a_i is the coefficient for the ith product, and $\tilde{\psi}_{\alpha_i}(\mathbf{r_1})$ represents a modified hydrogenic orbital for electron 1 (similar for the other two factors). For example, $\tilde{\psi}_{\alpha_i}$ could be a modified ψ_{1s} orbital of hydrogen (as described earlier). For a judicious choice of orbitals, the ground state of lithium can be approximated as a *single* antisymmetrized product of one-electron orbitals. This turns out to give a reasonable description of the ground-state wave function and leads to a good first approximation of the energy. The sum in Equation 2.65 then reduces to a single term:

$$\Psi_{GS}(\mathbf{r_1}, \mathbf{r_2}, \mathbf{r_3}) = |\tilde{\psi}_{1s}\tilde{\psi}_{1s}\tilde{\psi}_{2s}|_{SD}, \qquad (2.66)$$

where the electron spins are taken to be up and down for the two electrons in the $1s$ orbital, and either up or down for the $2s$ orbital. We note that modification of hydrogenic orbitals to describe valence orbitals (the orbital for the outermost, unpaired electron) in multielectron states of atoms is well described by an approach known as quantum defect theory [3]. Especially successful for alkali atoms, quantum defect theory provides a simple correction to the energies of different angular momentum states that account for the different degrees to which they penetrate the core (and the subsequent changes in shielding by the inner electrons).

2.4 H$_2^+$ MOLECULE

We now consider the time-independent structure of molecules, and we again begin with the simplest case: H$_2^+$ with the nuclei fixed in place. Experimentally, H$_2^+$ is observed to be stable with a binding energy of about 2.65 eV in the ground state. As with helium, it is a three-body problem, although the situation is simplified considerably if we assume the nuclei are fixed at some distance R. Freezing the nuclei removes their kinetic energy from the Hamiltonian, and only the electron's motion remains. We also neglect both the Coulomb repulsion between the nuclei (the nuclear potential energy) and all spin. On the other hand, the problem is complicated by the fact that the potential seen by the single electron is no longer central; the electron experiences the combined Coulomb potentials of the two separate protons.

With these assumptions, the TISE is given entirely in terms of electron coordinates as

$$\left(-\frac{\nabla^2}{2} - \frac{1}{r_1} - \frac{1}{r_2}\right)\Psi = E\Psi, \qquad (2.67)$$

where r_1 and r_2 represent the displacement of the electron with respect to nuclei 1 and 2, respectively.[10] The coordinates are illustrated in Figure 2.1.

We begin by considering the electron as localized near either nucleus 1 or 2, with the other nucleus acting as a perturbation (this is essentially the limiting solution for large separation between the nuclei). Localization at either 1 or 2 are, of course, degenerate solutions (the degeneracy will be lifted by the perturbation of the other nucleus for finite distances between nuclei). Given that our beginning point involves degeneracy, we must use degenerate perturbation theory, and we choose the zeroth-order electronic wave function to be a sum of two atomic hydrogen orbitals centered around nuclei 1 and 2. Calling these two orbitals ψ_1 and ψ_2, we have for the spatial wave function

$$\Psi^{(0)}(\mathbf{r_1}, \mathbf{r_2}) = c_1\psi_1(\mathbf{r_1}) + c_2\psi_2(\mathbf{r_2}), \qquad (2.68)$$

Figure 2.1   Illustration of coordinates for H$_2^+$.

---

[10]Note that the 1 and 2 for H$_2^+$ refer to the nuclei (as opposed to helium where they labeled the electrons). In H$_2^+$ we use 1 and 2 to specify the nuclei instead of a and b to avoid future confusion with the notation for molecular potential energy surfaces.

where the c_i are coefficients and

$$\psi_1(\mathbf{r_1}) = R_{n_1 l_1}(r_1) Y_{l_1 m_1}(\theta_1, \phi_1) \tag{2.69a}$$

$$\psi_2(\mathbf{r_2}) = R_{n_2 l_2}(r_2) Y_{l_2 m_2}(\theta_2, \phi_2). \tag{2.69b}$$

Inserting this trial wave function into the TISE, we find[11]

$$\left(-\frac{\nabla^2}{2} - \frac{1}{r_1} - \frac{1}{r_2}\right) c_1 \psi_1 + \left(-\frac{\nabla^2}{2} - \frac{1}{r_2} - \frac{1}{r_1}\right) c_2 \psi_2 = E(c_1 \psi_1 + c_2 \psi_2). \tag{2.70}$$

The first two terms in each of the parentheses on the left side represent the Hamiltonian for atomic hydrogen acting on its eigenfunctions. These yield the (unperturbed) energies of hydrogen, and we can rewrite Equation 2.70 as

$$\left(E^{(0)} - E - \frac{1}{r_2}\right) c_1 \psi_1 + \left(E^{(0)} - E - \frac{1}{r_1}\right) c_2 \psi_2 = 0, \tag{2.71}$$

where $E^{(0)}$ is the unperturbed energy. If we now multiply this equation on the left by either ψ_1^* or ψ_2^* and integrate over all space (i.e. taking the inner product), we arrive at two equations for the coefficients, c_1 and c_2:

$$\left(E^{(0)} - E - \int d^3 r \psi_1^* \frac{1}{r_2} \psi_1\right) c_1$$
$$+ \left(\left(E^{(0)} - E\right) \int d^3 r \psi_1^* \psi_2 - \int d^3 r \psi_1^* \frac{1}{r_1} \psi_2\right) c_2 = 0 \tag{2.72a}$$

$$\left(\left(E^{(0)} - E\right) \int d^3 r \psi_2^* \psi_1 - \int d^3 r \psi_2^* \frac{1}{r_2} \psi_1\right) c_1$$
$$+ \left(E^{(0)} - E - \int d^3 r \psi_2^* \frac{1}{r_1} \psi_2\right) c_2 = 0. \tag{2.72b}$$

Using shorthand notation for the energy differences and integrals:

$$\Delta E \equiv E - E^{(0)} \tag{2.73a}$$

$$S \equiv \int d^3 r \psi_1^* \psi_2 \tag{2.73b}$$

$$V_{22} = V_{11} \equiv \int d^3 r \psi_2^* \frac{1}{r_1} \psi_2 = \int d^3 r \psi_1^* \frac{1}{r_2} \psi_1 \tag{2.73c}$$

$$V_{12} \equiv \int d^3 r \psi_1^* \frac{1}{r_1} \psi_2 = V_{21} \equiv \int d^3 r \psi_2^* \frac{1}{r_2} \psi_1, \tag{2.73d}$$

we can write Equation 2.72 in matrix form as (multiplying the entire expression by negative one):

$$\begin{bmatrix} \Delta E + V_{11} & \Delta E S + V_{12} \\ \Delta E S^* + V_{21} & \Delta E + V_{11} \end{bmatrix} \begin{bmatrix} c_1 \\ c_2 \end{bmatrix} = 0. \tag{2.74}$$

Solving this for ΔE, c_1, and c_2, we find

$$\Delta E_\pm = -\frac{V_{11} \pm V_{12}}{1 \pm S} \tag{2.75a}$$

---

[11] We note that $\mathbf{r_1}$ and $\mathbf{r_2}$ are not independent variables here.

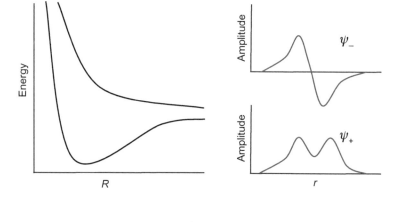

Figure 2.2 Energies and wave functions for H_2^+. The left panel shows the energies (potential energy curves) of the two lowest states as a function of proton separation (including the nuclear potential energy ($1/R$)), while the right two panels show the wave functions for the two solutions to Equation 2.75.

$$c_1 = \pm c_2. \qquad\qquad (2.75b)$$

All of the parameters in the expression for ΔE in Equation 2.75 depend implicitly on the separation between the two nuclei, R, and thus the correction to the energies will vary with R.

The integrals associated with evaluating $\Delta E(R)$ can be calculated explicitly using ellipsoidal coordinates, where the two nuclei lie at the foci of ellipsoids (see homework problem). The left panel of Figure 2.2 illustrates how the energies vary with R, where we have included the potential energy of nuclear repulsion ($1/R$). The right panel plots the shapes of the wave functions associated with the two states: $\psi_+ \implies c_1 = +c_2$, while $\psi_- \implies c_1 = -c_2$.

Figure 2.2 provides a qualitative picture of the energies and wave functions for H_2^+. For each nuclear separation, there are two possible energies, derived from the original two degenerate solutions at large R. Stitching these energies together and adding the nuclear repulsion yields the two "potential energy curves" seen in the left panel. We observe two characteristic shapes: the lower one is known as a "bonding" state and supports vibrational bound states of the two nuclei, while the upper one is known as an "antibonding" state and does not support any vibrational bound states. The associated wave functions, or molecular orbitals, are shown on the right. The electronic wave function is symmetric for the bonding curve and antisymmetric for the antibonding curve. The node in the middle of the antibonding orbital comes from destructive interference of the two localized atomic orbitals from which it was constructed. The node implies there is very little shielding of the positively charged repulsion between the two nuclei, and this leads to the repulsive potential energy curve. In contrast, the bonding orbital has no node, and the electron probability between the nuclei shields the two positive charges from each other. This creates a local minimum in the potential energy curve that can support bound states.

2.5 THE MOLECULAR HAMILTONIAN

In this section we aim to develop a general picture of electronic structure applicable to molecules. As in the earlier sections, we begin by considering systems isolated from the surrounding environment. While this is generally not the case for real-world systems, a complete description of the dynamics for any real object would require an unfeasible number of degrees of freedom and bring little additional insight to our particular focus on ultrafast dynamics. Nevertheless, interactions with the environment do play an important role in long-term behavior, and in Part IV, we discuss how to include coupling to such external degrees of freedom.

We wish to write down the general Hamiltonian operator for an isolated molecule with N atoms, taking into account all possible motions and interactions of the constituent electrons and nuclei except spin. Since the Hamiltonian for any many-body system is generally given by the sum of the kinetic and potential energies, it can be written as

$$\hat{H} = \sum_i \frac{-\nabla_{e,i}^2}{2} + \sum_{i,j>i} \frac{1}{|\mathbf{r}_i - \mathbf{r}_j|} + \sum_j \frac{-\nabla_{N,j}^2}{2M_j}$$
$$+ \sum_{i,j>i} \frac{Z_i Z_j}{|\mathbf{R}_i - \mathbf{R}_j|} - \sum_{i,j} \frac{Z_j}{|\mathbf{r}_i - \mathbf{R}_j|}, \tag{2.76}$$

where the $\mathbf{r}_i$ refers to the coordinate for the ith electron, $\mathbf{R}_j$ the coordinate for the jth nucleus, $\nabla_{e,i}^2$ is the Laplacian operator for the ith electronic coordinate, $\nabla_{N,j}^2$ is the Laplacian operator for the jth nuclear coordinate, M_j is the mass of the jth nuclei, and Z_j is the charge of the jth nuclei. Each of the five terms in the Hamiltonian above has a simple physical interpretation. The first two terms are entirely electronic in nature and represent the kinetic energies of individual electrons and the potential energy due to the Coulomb interaction between pairs of electrons. The third and fourth terms are the corresponding nuclear-only terms and represent the kinetic energy of the nuclei and the potential energy due to the Coulomb interaction between nuclei pairs. The final term represents the potential energy due to the Coulomb *attraction* between electrons and nuclei; it is this final term that is responsible for molecular bonding.

Analytically solving the TISE for the full molecular Hamiltonian of Equation 2.76 is not possible in general, and in fact requires approximations even to solve numerically for all molecules except H_2^+. In order to visualize the dynamics and gain an intuitive understanding, it is extremely useful to make what is known as the Born—Oppenheimer (BO) approximation.

2.5.1 The BO Approximation

The BO approximation allows one to treat the electrons and nuclei separately, greatly simplifying the problem. As it is usually described, the BO approximation relies on the fact that the timescale for electronic motion is generally much more rapid than that for nuclear motion due to the relative masses of the electrons and nuclei. Thus, one can think of the nuclei moving in a time-averaged potential generated by the rapidly moving electrons. (Alternatively, one can think of the electrons adiabatically following the slow-moving nuclei.) If the molecule is perturbed by something external, such as a laser field, the electrons respond rapidly while the nuclei react slowly to the new, time-averaged electronic distribution. In this section, we discuss the BO approximation in the time-independent picture and then address cases where the approximation breaks down.

Before proceeding with the mathematics of BO approximation, we briefly discuss the qualitative forms of motion available to a molecule. For an isolated molecule composed of N atoms, there are 3N total degrees of freedom available for motion of the *nuclei* in the three spatial dimensions (we consider only the nuclei, since the BO approximation decouples the electrons from the nuclear motion). Describing the system using the x, y, and z coordinates of each atom is cumbersome and rarely useful. As with classical mechanics of multibody problems, it is convenient to describe nuclear dynamics in terms of center-of-mass motion of the molecule and the relative motion of the constituent atoms. In this book, we are primarily interested in the *internal* degrees of freedom for a system, and so we ignore the center-of-mass motion, referring instead to the relative position of the nuclei. Alternatively, an isolated molecule is translationally invariant, and so translation of the center of mass can be ignored. In the absence of an applied field, an isolated molecule is also rotationally invariant.[12] Therefore, the three degrees of rotational freedom (e.g. rotation about the Euler angles) are also typically ignored, leaving 3N−6 degrees of freedom available for what is referred to as vibrational motion.[13] It is this vibrational motion that is most relevant in time-resolved spectroscopy, as it tends to mediate unimolecular reaction dynamics. For a one-dimensional diatomic molecule, the nuclear wave function $\chi(\mathbf{R})$ defined in Equation 2.77 describes vibrational motion along the internuclear coordinate.

Motivated by the separation of electronic and nuclear motion, we begin with an ansatz for the form of the total wave function $\Psi(\mathbf{r}_i, \mathbf{R}_j)$ in *any given electronic state*:

$$\Psi(\mathbf{r}_i, \mathbf{R}_j) = \psi(\mathbf{r}_i; \mathbf{R}_j)\chi(\mathbf{R}_j), \qquad (2.77)$$

where $\psi(\mathbf{r}_i; \mathbf{R}_j)$ is the *electronic* portion of the wave function and $\chi(\mathbf{R}_j)$ the *nuclear* portion.[14] Note that in $\psi(\mathbf{r}_i; \mathbf{R}_j)$, which describes the probability amplitude for the electrons as a function of the electronic coordinates $\mathbf{r}_i$, there is a parametric dependence on the nuclear coordinates $\mathbf{R}_j$, since the electrons are sensitive to changes in nuclear position. One can think of having a slightly different $\psi(\mathbf{r}_i)$ for each $\mathbf{R}_j$. For instance, one expects atomic-like wave functions in the limit of large nuclear separations, while at small nuclear distances, the wave functions of the electrons will be affected by the neighboring nuclei.

On the other hand, $\chi(\mathbf{R}_j)$ is a function of nuclear coordinates only, since in the BO approximation, we consider the nuclei as evolving in a time-averaged potential generated by the rapidly moving electrons. This time-averaged electronic potential, along with the nuclear repulsion, is known as the potential energy surface, or PES. It is on this PES that the nuclei move. Note that while $\chi(\mathbf{R}_j)$ does not depend on $\mathbf{r}_i$, it does depend quite sensitively on the nature of the particular electronic state of the molecule (i.e. different electronic states form different PESs).

Given our ansatz for the wave function, it is useful to split the Hamiltonian in Equation 2.76 into two parts: an "electronic" $\hat{H}^e$ that includes all the terms that go into the PES, and a nuclear $\hat{T}_N$ representing the nuclei kinetic energy.[15] For any given electronic state, $\psi(\mathbf{r}_i; \mathbf{R}_j)$ satisfies the TISE for the electronic portion of the Hamiltonian $\hat{H}^e$:

$$\hat{H}^e \psi(\mathbf{r}_i; \mathbf{R}_j) = E^e(\mathbf{R}_j)\psi(\mathbf{r}_i; \mathbf{R}_j), \qquad (2.78)$$

---

[12] We note there are some techniques that use aligned, gas-phase samples, where rotational dynamics are inherent in the alignment.

[13] For a diatomic molecule, the 3N−6 rule does not hold: rotation about the atom–atom bond is not considered a degree of freedom, since for point–particle nuclei, the molecule is unchanged for any amount of rotation. Thus, a diatomic molecule has only two degrees of rotational freedom and one degree of vibrational freedom (still consistent with 3N−3 total degrees of freedom neglecting translation).

[14] We note that $\chi(\mathbf{R}_j)$ in Equation 2.77 is the full vibrational wave function and not a particular vibrational eigenstate. It is sometimes useful to expand $\chi(\mathbf{R}_j)$ in a basis set of vibrational eigenstates for a given potential energy surface, but at this point, we stick with the full $\chi(\mathbf{R}_j)$.

[15] In some treatments of the BO approximation, the nuclear repulsion is treated separately from the other terms in $\hat{H}^e$, but here we include it such that $E_e(\mathbf{R}_j)$ represents the PES on which the nuclei move.

where

$$\hat{H}^e = \sum_i -\frac{\nabla_{e,i}^2}{2} + \sum_{i,j>i} \frac{1}{|\mathbf{r}_i - \mathbf{r}_j|} - \sum_{i,j} \frac{Z_j}{|\mathbf{r}_i - \mathbf{R}_j|} + \sum_{i,j>i} \frac{Z_i Z_j}{|\mathbf{R}_i - \mathbf{R}_j|}, \quad (2.79)$$

and $E^e(\mathbf{R}_j)$ is the $\mathbf{R}_j$-dependent energy associated with the electrons and nuclear repulsion. Substituting the trial wave function $\Psi(\mathbf{r}_i, \mathbf{R}_j)$ of Equation 2.77 into the full TISE (and making use of both Equation 2.78 and the chain rule for the Laplacian derivatives), we arrive at the following:

$$\hat{H}\psi(\mathbf{r}_i; \mathbf{R}_j)\chi(\mathbf{R}_j) = E\psi(\mathbf{r}_i; \mathbf{R}_j)\chi(\mathbf{R}_j)$$
$$= E^e(\mathbf{R}_j)\psi(\mathbf{r}_i; \mathbf{R}_j)\chi(\mathbf{R}_j) + \sum_j \frac{-1}{2M_j}\Big[\psi(\mathbf{r}_i; \mathbf{R}_j)\nabla_{N,j}^2\chi(\mathbf{R}_j)$$
$$+ 2\nabla_{N,j}\psi(\mathbf{r}_i; \mathbf{R}_j) \cdot \nabla_{N,j}\chi(\mathbf{R}_j) + \big(\nabla_{N,j}^2\psi(\mathbf{r}_i; \mathbf{R}_j)\big)\chi(\mathbf{R}_j)\Big].$$
$$(2.80)$$

Note that the three terms involving spatial derivatives in Equation 2.80 all have nuclear masses in the denominator. Given the large ratio between the electron and proton mass ($\sim$2000), it is a reasonable first approximation to ignore these terms altogether in solving the TISE for a molecule. This is equivalent to treating the nuclei as frozen, with no kinetic energy (like we did in Section 2.4 for H_2^+). Since the molecular dynamics we are interested in following explicitly involve nuclear motion, we clearly cannot ignore all three terms. The standard approach is to keep only the first of these terms when solving the molecular TDSE, because it is the only term that does not involve the $\nabla_{N,j}$ operator acting on the electronic wave function, $\psi(\mathbf{r}_i; \mathbf{R}_j)$ (corresponding to spatial derivatives of the *electronic* wave function with respect to *nuclear* coordinates). In other words, the latter two terms consider how the electronic wave function changes as the nuclei move closer or farther apart. As we will see later, these are generally small, and we therefore simplify Equation 2.80 by taking the limit where these derivatives vanish. Practically speaking, it is also convenient to drop these two terms, since they do not easily accommodate the separable solution proposed in Equation 2.77. Omitting the last two terms and factoring out $\psi(\mathbf{r}_i; \mathbf{R}_j)$, we arrive at the following simplified TISE for the nuclear wave function $\chi(\mathbf{R}_j)$ in the BO approximation:

$$\left(E^e(\mathbf{R}_j) + \sum_j \frac{-\nabla_{N,j}^2}{2M_j}\right)\chi(\mathbf{R}_j) = E\chi(\mathbf{R}_j). \quad (2.81)$$

Equation 2.81 allows us to solve for the (vibrational) dynamics of the nuclei in a given PES defined by $E^e(\mathbf{R}_j)$. The PES itself is calculated by solving the TISE for $\hat{H}^e$ (Equation 2.78) at various *fixed* nuclei positions; this is known as an electronic structure calculation. In this way, one can use the BO approximation in a time independent approach to calculate the electronic structure of the molecule, which subsequently drives time-dependent nuclear dynamics.

2.5.2 Breakdown of the BO Approximation

The validity of the BO approximation is typically excellent, since the electronic dynamics are generally much more rapid than any nuclear motion. However, there are cases where the BO approximation breaks down, and it is no longer possible to separate electronic and nuclear motions. In these cases, the timescales for electronic motion becomes comparable with those of the nuclei. One way to think about this is in terms of how the PESs change as a function of nuclear position. Timescales for quantum dynamics are generally dictated by the inverse of energy-level spacings, and the BO approximation is valid whenever the spacings between PESs (electronic spacings) are much larger than the spacings between vibrational levels (nuclear spacings). When two PESs come close together, the timescale for electronic dynamics in the vicinity of this near-degeneracy can be comparable to, or even longer than, the timescale for nuclear motion; therefore, the approximation breaks down near these "level crossings." An important example in multidimensional systems is the case of conical intersections (introduced in Section 2.5.3), which has implications for processes in physics, chemistry, and biology. For example, the photoprotection of DNA bases is understood in terms of molecular dynamics through conical intersections [4, 5].

Before considering what happens when the BO approximation breaks down, it is instructive to reconcile neglecting the last two terms in Equation 2.80 with our understanding of the BO approximation being valid when electronic dynamics are much more rapid than nuclear dynamics. Specifically, the last two terms in Equation 2.80 were

$$\sum_j \frac{-1}{2M_j} \left[2\nabla_{N,j} \psi(\mathbf{r}_i; \mathbf{R}_j) \cdot \nabla_{N,j} \chi(\mathbf{R}_j) + \left(\nabla_{N,j}^2 \psi(\mathbf{r}_i; \mathbf{R}_j) \right) \chi(\mathbf{R}_j) \right]. \qquad (2.82)$$

As mentioned, these terms involve gradients of the electronic wave function with respect to nuclear coordinates. The electronic states become strongly mixed in the vicinity of level crossings, implying there is a large change in electronic-state character (i.e. shape of the electronic wave function) with respect to any change in nuclear coordinate - cf. 2.6.1. Therefore, the gradients in Equation 2.82 become quite large exactly when electronic states come close together in energy (and the electronic dynamics evolve on the same timescale as nuclear dynamics).

We now examine what happens when the standard BO solutions of Equation 2.77 are no longer good eigenfunctions. Even when this occurs, it is still possible to construct an exact solution from a superposition of BO wave functions.[16] The BO solutions simply serve as a basis set, and in cases when the BO approximation breaks down, one must go beyond the first term in the expansion to adequately express the true eigenfunctions. We therefore begin by writing the full wave function as a sum over BO (or "adiabatic") solutions:

$$\Psi(\mathbf{r}_i, \mathbf{R}_j) = \sum_n \psi_n(\mathbf{r}_i; \mathbf{R}_j) \chi_n(\mathbf{R}_j), \qquad (2.83)$$

where $\psi_n(\mathbf{r}_i; \mathbf{R}_j)$ and $\chi_n(\mathbf{R}_j)$ correspond to the electronic and vibrational BO components of the wave function on the nth electronic state.[17] Figure 2.3 illustrates how one would compose a sum of two separate vibrational wave functions located on two different electronic states of H_2^+. Both the vibrational and electronic wave functions are shown, along with the adiabatic potential energy curves.

We insert the sum of Equation 2.83 into the full TISE, being sure to include the gradient terms we previously dropped in the BO approximation. This yields

---

[16]This assumes that the BO wave functions span the space. We do not attempt to prove this but take it as given.

[17]Note that here $\chi_n(\mathbf{R}_j)$ represents the full vibrational wave function – i.e. a superposition of vibrational eigenstates – on the nth electronic state.

Figure 2.3 Illustration of a total wave function consisting of a sum of BO product states for H_2^+. Lower: $\chi_a(\mathbf{R})$ is the vibrational wave function on the ground PES which is associated with the electronic wave function ψ_a. Upper: similar wave functions for the excited electronic PES. The full wave function is given by the sum of the products: $\Psi(\mathbf{r},\mathbf{R}) = \psi_a(\mathbf{r};\mathbf{R})\chi_a(\mathbf{R}) + \psi_b(\mathbf{r};\mathbf{R})\chi_b(\mathbf{R})$.

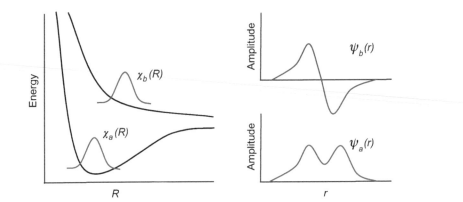

$$\hat{H}\sum_n \psi_n(\mathbf{r}_i;\mathbf{R}_j)\chi_n(\mathbf{R}_j) = \sum_n E_n^e(\mathbf{R}_j)\psi_n(\mathbf{r}_i;\mathbf{R}_j)\chi_n(\mathbf{R}_j)$$

$$+\sum_n \sum_j -\frac{1}{2M_j}\Big[\psi_n(\mathbf{r}_i;\mathbf{R}_j)\nabla_{N,j}^2\chi_n(\mathbf{R}_j)$$

$$+ 2\nabla_{N,j}\psi_n(\mathbf{r}_i;\mathbf{R}_j)\cdot\nabla_{N,j}\chi_n(\mathbf{R}_j) + \big(\nabla_{N,j}^2\psi_n(\mathbf{r}_i;\mathbf{R}_j)\big)\chi_n(\mathbf{R}_j)\Big]$$

$$= \sum_n E_n\psi_n(\mathbf{r}_i;\mathbf{R}_j)\chi_n(\mathbf{R}_j). \quad (2.84)$$

We project on the left with $\psi_m^*(\mathbf{r}_i;\mathbf{R}_j)$ and integrate over all space of the electronic coordinates, making use of a shorthand, $\hat{T}_N$, to represent the nuclear kinetic energy:

$$\hat{T}_N \equiv -\sum_j \frac{\nabla_{N,j}^2}{2M_j}. \quad (2.85)$$

In "braket notation," we have

$$\sum_n \big\langle \psi_m(\mathbf{r}_i;\mathbf{R}_j)\,\big|\,\hat{H}\,\big|\,\psi_n(\mathbf{r}_i;\mathbf{R}_j)\big\rangle \chi_n(\mathbf{R}_j) =$$

$$\sum_n \Big[\big(E_n^e(\mathbf{R}_j)+\hat{T}_N\big)\big\langle\psi_m(\mathbf{r}_i;\mathbf{R}_j)\,\big|\,\psi_n(\mathbf{r}_i;\mathbf{R}_j)\big\rangle$$

$$-\sum_j \frac{1}{M_j}\big\langle\psi_m(\mathbf{r}_i;\mathbf{R}_j)\,\big|\,\nabla_{N,j}\,\big|\,\psi_n(\mathbf{r}_i;\mathbf{R}_j)\big\rangle\cdot\nabla_{N,j}$$

$$-\sum_j \frac{1}{2M_j}\big\langle\psi_m(\mathbf{r}_i;\mathbf{R}_j)\,\big|\,\nabla_{N,j}^2\,\big|\,\psi_n(\mathbf{r}_i;\mathbf{R}_j)\big\rangle\Big]\chi_n(\mathbf{R}_j)$$

$$= \sum_n E_n\big\langle\psi_m(\mathbf{r}_i;\mathbf{R}_j)\,\big|\,\psi_n(\mathbf{r}_i;\mathbf{R}_j)\big\rangle\chi_n(\mathbf{R}_j). \quad (2.86)$$

Using the orthonormality of the BO eigenstates yields:

$$\sum_n \hat{H}_{mn}\chi_n(\mathbf{R}_j) = \sum_n \Big[\big(E_n^e(\mathbf{R}_j)+\hat{T}_N\big)\delta_{nm}$$

$$-\sum_j \frac{1}{M_j}\big\langle\psi_m(\mathbf{r}_i;\mathbf{R}_j)|\nabla_{N,j}\psi_n(\mathbf{r}_i;\mathbf{R}_j)\big\rangle\cdot\nabla_{N,j}$$

$$-\sum_j \frac{1}{2M_j} \langle \psi_m(\mathbf{r}_i;\mathbf{R}_j) | \nabla^2_{N,j} \psi_n(\mathbf{r}_i;\mathbf{R}_j) \rangle \Big] \chi_n(\mathbf{R}_j)$$

$$= E_m \chi_m(\mathbf{R}_j), \quad (2.87)$$

where $\hat{H}_{mn} \equiv \langle \psi_m(\mathbf{r}_i;\mathbf{R}_j) | \hat{H} | \psi_n(\mathbf{r}_i;\mathbf{R}_j) \rangle$ is the matrix element of the Hamiltonian between the electronic states ψ_m and ψ_n.

It is clear from Equation 2.87 that the Hamiltonian is nondiagonal in this basis, and the vibrational wave function $\chi_n(\mathbf{R}_j)$ on the nth BO electronic (adiabatic) state of the molecule is, in general, always coupled to the vibrational wave function on the mth BO electronic state, $\chi_m(\mathbf{R}_j)$. This "motional" coupling is proportional to derivatives of the electronic states with respect to vibrational coordinates — a result of the nuclear kinetic energy operator acting on the electronic wave function. For a vibrational wave function which is at rest on an adiabatic PES (e.g. near a minimum in the potential), there is very little motional coupling between adiabatic states, since the electronic wave function does not vary significantly over the extent of the vibrational wave function. However, motion of a wave packet on a given PES (i.e. a changing $\langle \mathbf{R}_j(t) \rangle$) can lead to significant coupling as the vibrational wave packet explores regions of the PES where the electronic states vary substantially with nuclear coordinate. Quasiclassical "surface hopping" calculations of wave packet dynamics on multiple PESs use the product of these gradients and the time derivative of $\langle \mathbf{R}_j(t) \rangle$ (i.e. $\langle \dot{\mathbf{R}}_j(t) \rangle \cdot \langle \psi_m | \nabla_N \psi_n \rangle$) to calculate the transition probability between adiabatic states [6].

2.5.2.1 First-Order Energy Corrections

If we are simply interested in the *energy* of a particular state, we can use first-order perturbation theory, which ignores the off-diagonal coupling terms. The term containing $\nabla_N \psi_n$ can be expressed using the product rule as

$$\langle \psi_m | \nabla_N \psi_n \rangle = \nabla_N \langle \psi_m | \psi_n \rangle - \langle \nabla_N \psi_m | \psi_n \rangle$$

$$= 0 - \langle \nabla_N \psi_m | \psi_n \rangle, \quad (2.88)$$

where we have again used the orthonormality of the eigenstates. If one chooses real wave functions, the only way for this last statement to be true for the diagonal terms ($n = m$) is if $\langle \psi_n | \nabla_N \psi_n \rangle = 0$. Therefore, the contribution to the energy vanishes for this term. Keeping only the diagonal contributions, we are left with the following expression for the energy of the mth adiabatic state:

$$\left(\hat{T}_N + E^e_m(\mathbf{R}_j) - \sum_j \frac{1}{2M_j} \langle \psi_m(\mathbf{r}_i;\mathbf{R}_j) | \nabla^2_{N,j} \psi_m(\mathbf{r}_i;\mathbf{R}_j) \rangle \right) \chi_m(\mathbf{R}_j)$$

$$= E_m \chi_m(\mathbf{R}_j), \quad (2.89)$$

where, as before, $\hat{T}_N$ represents the kinetic energy of the nuclei and $E^e_m(\mathbf{R}_j)$ is the solution to the *electronic* TISE at position $\mathbf{R}_j$, including the electrostatic repulsion between the nuclei.

2.5.2.2 Diabatic States

The full motional coupling between adiabatic states represented in Equation 2.87 is challenging to compute, as it involves multidimensional gradients of electronic wave functions with respect to nuclear coordinates (not single-point calculations). This motivates the choice of an alternative basis where, in particular, the $\langle \psi_m | \nabla_N \psi_n \rangle$ and

$\langle \psi_m | \nabla_N^2 \psi_n \rangle$ terms vanish. A trivial way to accomplish this is to choose states of the form $\psi(\mathbf{r}_i; \mathbf{R}_{j0})$, where $\mathbf{R}_{j0}$ is an appropriately chosen *fixed* "reference geometry" such that $\psi(\mathbf{r}_i; \mathbf{R}_{j0})$ no longer depends on $\mathbf{R}_j$. An example choice for the reference geometry is large nuclear separation, where the BO wave functions behave asymptotically (i.e. the wave function does not change with separation). With a trial solution based on these "diabatic states", we plug $\sum_n \psi_n(\mathbf{r}_i; \mathbf{R}_{j0}) \chi_{0n}(\mathbf{R}_j)$ into the TISE and multiply by $\langle \psi_m(\mathbf{r}_i; \mathbf{R}_{j0})|$ as earlier to arrive at

$$\sum_n \langle \psi_m(\mathbf{r}_i; \mathbf{R}_{j0}) | (\hat{H}_e + \hat{T}_N) | \psi_n(\mathbf{r}_i; \mathbf{R}_{j0}) \rangle \chi_{0n}(\mathbf{R}_j) = E_m \chi_{0m}(\mathbf{R}_j), \tag{2.90}$$

where $\chi_{0n}(\mathbf{R}_j)$ is the vibrational wave function on the nth diabatic state (note it is not the nth vibrational eigenstate). While $\chi_{0n}(\mathbf{R}_j)$ depends on $\mathbf{R}_j$, it is on a PES defined by electronic wave functions calculated at a fixed $\mathbf{R}_{j0}$. The operator $\hat{H}^e$ can be divided into the kinetic energy for the electrons, $\hat{T}_e \equiv \sum_i -\frac{\nabla_{e,i}^2}{2}$, plus the total electrostatic interaction energy:

$$\hat{H}^e = \hat{T}_e + V(\mathbf{r}_i; \mathbf{R}_j), \tag{2.91}$$

where

$$V(\mathbf{r}_i; \mathbf{R}_j) = \sum_{i,j>i} \frac{1}{|\mathbf{r}_i - \mathbf{r}_j|} + \sum_{i,j>i} \frac{Z_i Z_j}{|\mathbf{R}_i - \mathbf{R}_j|} - \sum_{i,j} \frac{Z_j}{|\mathbf{r}_i - \mathbf{R}_j|}. \tag{2.92}$$

Furthermore, we make use of the fact that $\psi_n(\mathbf{r}_i; \mathbf{R}_{j0})$ is an eigenstate of the electronic Hamiltonian:

$$\hat{H}^e \psi_n(\mathbf{r}_i; \mathbf{R}_{j0}) = E_n^e(\mathbf{R}_{j0}) \psi_n(\mathbf{r}_i; \mathbf{R}_{j0}), \tag{2.93}$$

and that the kinetic energy of the electrons does not depend on $\mathbf{R}_j$. This allows us to express the electronic kinetic energy $\hat{T}_e$ as

$$\hat{T}_e \psi_n(\mathbf{r}_i; \mathbf{R}_{j0}) = \left(E_n^e(\mathbf{R}_{j0}) - V(\mathbf{r}_i; \mathbf{R}_{j0}) \right) \psi_n(\mathbf{r}_i; \mathbf{R}_{j0}). \tag{2.94}$$

Substituting this into Equation 2.90 yields

$$\sum_n \langle \psi_m(\mathbf{r}_i; \mathbf{R}_{j0}) | (E_n^e(\mathbf{R}_{j0}) - V(\mathbf{r}_i; \mathbf{R}_{j0}) + V(\mathbf{r}_i; \mathbf{R}_j) + \hat{T}_N) |$$
$$\psi_n(\mathbf{r}_i; \mathbf{R}_{j0}) \rangle \chi_{0n}(\mathbf{R}_j) = E_m \chi_{0m}(\mathbf{R}_j). \tag{2.95}$$

Carrying out the spatial integral over r implicit in the braket notation (and using the fact that $\hat{T}_N$ does not operate on $\psi_n(\mathbf{r}_i; \mathbf{R}_{j0})$ at fixed $\mathbf{R}_{j0}$) leads to

$$\sum_n \left(\hat{T}_N \delta_{nm} + U_{mn} \right) \chi_{0n}(\mathbf{R}_j) = E_m' \chi_{0m}(\mathbf{R}_j), \tag{2.96}$$

where $E_m' \equiv E_m - E_m^e(\mathbf{R}_{j0})$ and

$$U_{mn} \equiv \langle \psi_m(\mathbf{r}_i; \mathbf{R}_{j0}) | (V(\mathbf{r}_i; \mathbf{R}_j) - V(\mathbf{r}_i; \mathbf{R}_{j0})) | \psi_n(\mathbf{r}_i; \mathbf{R}_{j0}) \rangle. \tag{2.97}$$

We now have a simple form for the nuclear TISE, where the couplings between diabatic basis states are known as "potential couplings," rather than the motional (derivative) couplings between the adiabatic states. Depending on the situation, either the diabatic or adiabatic basis set is more useful. Adiabatic states are generally preferred for relatively slow dynamics, where the motional coupling between states can be neglected. Diabatic states are better in the opposite limit, where rapid motion of a wave packet on a PES does not leave much time for potential coupling. Figure 2.4 illustrates both diabatic and adiabatic states for the molecule NaI.

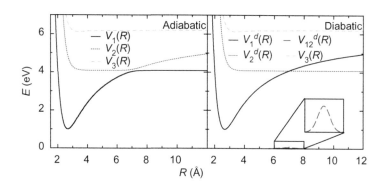

Figure 2.4 Adiabatic (left panel) and diabatic (right panel) potential energy curves for the first three states of NaI. Note that the diabatic curves cross, but have a coupling between them shown as a dashed line in the inset. Diagonalizing the Hamiltonian and transforming to the adiabatic states leads to a lifting of the degeneracy, and an avoided crossing of the curves. Figure courtesy of Philipp Marquetand.

2.5.3 Noncrossing Theorem

Based on our discussion of diabatic versus adiabatic states and the breakdown of BO approximation, we are in an excellent position to discuss the "noncrossing" theorem. We begin by considering two diabatic states in a diatomic molecule that happen to cross at position $R = R_d$. The coupling between them can be described by the following 2×2 matrix:

$$U = \begin{bmatrix} U_{11}(R) & U_{12}(R) \\ U_{21}(R) & U_{22}(R) \end{bmatrix}. \tag{2.98}$$

The diagonal elements $U_{11}(R)$ and $U_{22}(R)$ are the energies of the first and second states as functions of R; these are simply the diabatic potential energy curves for each state. The off-diagonal terms represent the couplings between the diabatic states, which, in general, are also functions of R. Diagonalizing this matrix removes the off-diagonal terms (and hence the coupling), yielding the uncoupled, adiabatic states. One speaks of the degeneracy (or crossing) between the diabatic states as being "lifted" once the matrix has been diagonalized.

The new adiabatic energies at R_d are given by setting the determinant of $(U - \lambda I)$ to zero and finding the new eigenvalues:

$$(U_{11}(R_d) - \lambda)(U_{22}(R_d) - \lambda) - U_{21}(R_d)U_{12}(R_d) = 0. \tag{2.99}$$

Solving for λ yields

$$\lambda = \frac{U_{11}(R_d) + U_{22}(R_d)}{2} \pm \frac{1}{2}\sqrt{(U_{11}(R_d) - U_{22}(R_d))^2 + 4|U_{12}(R_d)|^2}. \tag{2.100}$$

Relabeling the two eigenvalues as $V_\pm(R_d)$, we find the separation between adiabatic states at R_d to be

$$V_+ - V_- = \sqrt{(U_{11}(R_d) - U_{22}(R_d))^2 + 4|U_{12}(R_d)|^2}. \tag{2.101}$$

Since both elements under the square root in Equation 2.101 need to separately be equal to zero for degenerate eigenvalues (and hence a state crossing), one finds that adiabatic states of a diatomic molecule will, in general, not cross. This result is known as the noncrossing theorem.[18]

---

[18]We note that as it is derived here, the noncrossing theorem is more of a propensity rule, rather than an absolute one that completely rules out the possibility of curves crossing.

However, in a polyatomic molecule with more than one degree of vibrational free-dom $(R \rightarrow R_1, R_2 \ldots)$, it is possible for there to be positions in coordinate space with a degeneracy between adiabatic eigenvalues because one can move along multiple degrees of freedom to find positions where each of the terms in the square root on the right hand side of Equation 2.101 are zero independently. These degenerate points (or seams in general) correspond to intersections of multidimensional PESs. In two dimensions, the intersection of PESs is geometrically described by a cone near the intersection, leading to the name "conical intersections." As we shall see in Parts IV and V, the regions in coordinate space around conical intersections play an important role in polyatomic molecular dynamics.

2.6 MULTICONFIGURATIONAL WAVE FUNCTIONS

Much of the current understanding of molecular structure and dynamics is based on approximate solutions to the TISE in the BO approximation. Known as electronic structure theory, the results can be used to calculate the energies, geometries, and vibrational frequencies for different electronic states of a given molecule. In this section, we build on the discussions of Sections 2.2.2 and 2.4 to outline the general approach to constructing solutions for many-electron molecular systems (i.e. any molecule other than H_2^+).

For example, imagine we want calculate an excited-state PES of a molecule to simulate the evolution of a vibrational wave packet. Or perhaps we are interested in modeling an experimental measurement following dynamics on an excited state. These require calculating the excited-state energy at a series of different molecular geometries corresponding to different atomic positions (or vibrational coordinate values). Interpolating these calculated points creates a smooth potential surface.

In particular, for each point on the surface, one needs to solve the electronic TISE for the N-electron system:

$$\hat{H}^e \psi(\mathbf{r}_i; \mathbf{R}_j) = E^e(\mathbf{R}_j) \psi(\mathbf{r}_i; \mathbf{R}_j), \tag{2.102}$$

where $\hat{H}^e$ is given in Equation 2.79. A number of different approaches have been developed to numerically solve Equation 2.102, each with different levels of rigor and computation time. As with atomic helium (Section 2.2.2), it is common to start with the fact that in a system with identical, spin-$1/2$ particles, the total wave function must be antisymmetric with respect to exchange of any two electrons. This is ensured by constructing the total wave function in terms of Slater determinants of single-electron orbitals. The N-electron Slater determinant is a natural extension of the two-dimensional case and can be written as (including only the spatial portion)

$$\Psi(\mathbf{r_1}, \mathbf{r_2}, \cdots \mathbf{r_n}) = \frac{1}{\sqrt{N}} \begin{vmatrix} \phi_1(\mathbf{r_1}) & \phi_2(\mathbf{r_1}) & \cdots & \phi_N(\mathbf{r_1}) \\ \phi_1(\mathbf{r_2}) & \phi_2(\mathbf{r_2}) & \cdots & \phi_N(\mathbf{r_2}) \\ \vdots & \vdots & \ddots & \vdots \\ \phi_1(\mathbf{r_N}) & \phi_2(\mathbf{r_N}) & \cdots & \phi_N(\mathbf{r_N}) \end{vmatrix}. \tag{2.103}$$

With the appropriate choice of orbital functions, it is always possible to write the ground-state electronic wave function for a given set of nuclear coordinates in terms of a single Slater determinant. This is known as a "single configuration," and the orbitals that permit such a single-determinant wave function are known as "natural orbitals."

However, because of the interactions between electrons (electron–electron correlation), it is generally impossible to write all electronic states and all molecular geometries in terms of a single Slater determinant (which would amount to describing an N-body wave function in terms of N, one-body wave functions). For example, if one has chosen natural orbitals to describe the ground state in terms of a single Slater determinant, the excited states must typically be written as an expansion of additional Slater determinants (i.e. multiple configurations or "multiconfiguration"). Without the additional configurations, one cannot capture the correct electronic character and energy of the state. In addition, if two electronic states approach the same energy along a given nuclear coordinate, the BO approximation can be violated. In this case, individual BO solutions will not provide an accurate description of the molecular states, as the characters and configurations of the states become mixed. One again, multiple configurations are required. In the following section, we provide an example in H_2 demonstrating how these multiconfigurational wave functions are typically handled.

2.6.1 Wave Functions for H_2

The simplest example to consider is H_2. We start by constructing a molecular orbital based on atomic hydrogen orbitals:

$$\sigma = \psi_{1s_1} + \psi_{1s_2}, \tag{2.104}$$

where ψ_{1s_1} and ψ_{1s_2} represent $1s$ orbitals of atomic hydrogen localized on protons 1 and 2, respectively, and the label σ implies rotational symmetry about the internuclear axis. We assume a single-configuration wave function ansatz for the ground state of the molecule that requires only one Slater determinant of the *molecular* orbital σ:

$$\Psi(\mathbf{r_1}, \mathbf{r_2}, s_1, s_2) = \frac{1}{\sqrt{2}} \begin{vmatrix} \sigma(\mathbf{r_1})\chi_1(s_1) & \sigma(\mathbf{r_2})\chi_1(s_2) \\ \sigma(\mathbf{r_1})\chi_2(s_1) & \sigma(\mathbf{r_2})\chi_2(s_2) \end{vmatrix}, \tag{2.105}$$

where $\sigma(\mathbf{r}_i)$ represents the spatial portion of the orbital for the ith electron and $\chi_1(s_i)$ or $\chi_2(s_i)$ represents the spin portion of the orbital for the ith electron. Writing out the Slater determinant, we have

$$\begin{aligned} \Psi(\mathbf{r_1}, \mathbf{r_2}, s_1, s_2) &= \frac{1}{\sqrt{2}} \big(\sigma(\mathbf{r_1})\chi_1(s_1)\sigma(\mathbf{r_2})\chi_2(s_2) \\ &\quad - \sigma(\mathbf{r_1})\chi_2(s_1)\sigma(\mathbf{r_2})\chi_1(s_2) \big) \\ &= \frac{1}{\sqrt{2}} \big(\sigma(\mathbf{r_1})\sigma(\mathbf{r_2}) \big) \big(\chi_1(s_1)\chi_2(s_2) - \chi_2(s_1)\chi_1(s_2) \big). \end{aligned} \tag{2.106}$$

Note how this particular wave function is factorizable into a product of spatial and spin-dependent functions, since in our ansatz, both electrons occupy the same σ orbital (i.e. they can both be described by the same spatial wave function). Expressing the spatial dependence of the wave function in terms of the atomic orbitals, we arrive at

$$\Psi(\mathbf{r_1},\mathbf{r_2},s_1,s_2) = \frac{1}{\sqrt{2}}\left(\psi_{1s_1}(\mathbf{r_1}) + \psi_{1s_2}(\mathbf{r_1})\right)\left(\psi_{1s_1}(\mathbf{r_2}) + \psi_{1s_2}(\mathbf{r_2})\right)$$
$$\times \left(\chi_1(s_1)\chi_2(s_2) - \chi_2(s_1)\chi_1(s_2)\right)$$
$$= \frac{1}{\sqrt{2}}\left(\psi_{1s_1}(\mathbf{r_1})\psi_{1s_2}(\mathbf{r_2}) + \psi_{1s_2}(\mathbf{r_1})\psi_{1s_1}(\mathbf{r_2})\right. \tag{2.107}$$
$$\left. + \psi_{1s_1}(\mathbf{r_1})\psi_{1s_1}(\mathbf{r_2}) + \psi_{1s_2}(\mathbf{r_1})\psi_{1s_2}(\mathbf{r_2})\right)$$
$$\times \left(\chi_1(s_1)\chi_2(s_2) - \chi_2(s_1)\chi_1(s_2)\right).$$

The first two terms of the spatial portion of the wave function correspond to a covalent H–H bond (equal sharing of the electron probability between protons), whereas the last two terms correspond to an ionic H^+–H^- bond (electron probability localized on proton 1 or 2). Near the equilibrium bond length, this electron configuration provides a reasonably accurate description of the electronic ground state. However, as one moves toward dissociation, this no longer reflects the true character of the ground state wave function, as one expects a pure covalent bond with one electron localized around each proton. This covalent bond can be well-described by a sum of two different configurations: one in which the atomic orbitals are added to form a bonding molecular orbital (σ from Equation 2.104) and the one in which they are subtracted to form an "antibonding" molecular orbital, σ^*:

$$\sigma^* = \psi_{1s_1} - \psi_{1s_2}. \tag{2.108}$$

The full wave function is then written as a sum of these different configurations (or Slater determinants):

$$\Psi(\mathbf{r_1},\mathbf{r_2},s_1,s_2) = c_g \begin{vmatrix} \sigma(\mathbf{r_1})\chi_1(s_1) & \sigma(\mathbf{r_1})\chi_2(s_1) \\ \sigma(\mathbf{r_2})\chi_1(s_2) & \sigma(\mathbf{r_2})\chi_2(s_2) \end{vmatrix}$$
$$+ c_u \begin{vmatrix} \sigma^*(\mathbf{r_1})\chi_1(s_1) & \sigma^*(\mathbf{r_1})\chi_2(s_1) \\ \sigma^*(\mathbf{r_2})\chi_1(s_2) & \sigma^*(\mathbf{r_2})\chi_2(s_2) \end{vmatrix}, \tag{2.109}$$

where c_g and c_u are coefficients with the normalization requirement that $|c_g|^2 + |c_u|^2 = 1$.

Figure 2.5 illustrates how multiple configurations are required to describe the electronic ground state of molecular hydrogen (H_2) at internuclear separations much larger than the equilibrium ground-state distance. The figure shows the magnitudes of the two

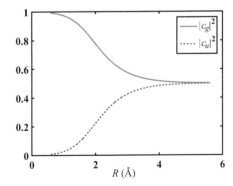

Figure 2.5 Magnitude squared of coefficients for ground-electronic-state configurations of H_2 as a function of bond length. Calculation courtesy of Spiridoula Matsika.

coefficients from Equation 2.109 as a function of bond length: the solid curve plots $|c_g|^2$ (corresponding to the bonding σ orbital), while the dotted curve plots $|c_u|^2$ (corresponding to the antibonding σ^* orbital). Near the equilibrium distance, the ground state is well described by a single configuration, as $|c_g|^2$ is the only nonzero coefficient. As the bond is stretched toward dissociation, the wave function contains equal contributions from both bonding and antibonding configurations (at large R, $c_g = -c_u$).

Notationally, since the bonding σ orbital is symmetric, it is frequently written as σ_g, with the g standing for *gerade* ("even" in German). Similarly, the σ^* orbital is often written as σ_u, with the u standing for *ungerade* ("uneven"). Both of the configurations correspond to doubly occupied orbitals (both electrons are described by the same spatial wave function, or orbital), and are thus often written with a superscript 2.

This concludes our introduction of the time-independent quantum mechanics used to describe molecular structure. While we clearly did not delve into significant detail in the final sections, we hope the discussion provides sufficient context for what is to come. In the next chapter we consider the basic principles governing how molecular systems interact with a time-dependent light field.

2.7 HOMEWORK PROBLEMS

1. Derive the recursion formula for the radial Schrödinger equation for the hydrogen atom:

$$A_k = -2A_{k-1}\left(\frac{z-(l+k)\sqrt{-2E}}{(l+k)(l+k+1)-l(l+1)}\right).$$

2. Use the programming language of your choice to numerically integrate the radial Schrödinger equation for hydrogen $(n,l) = (10,0)$, $(10,1)$ and $(10,5)$. Plot the calculated wave functions and compare with analytic expressions for the wave functions.

3. Numerically calculate the radial matrix element $\langle R_{n,l} | r | R_{n,l} \rangle$ (in atomic units) for the $(n,l) = (10,0)$, $(10,1)$, and $(10,5)$ states of atomic hydrogen. Compare these with values based on analytic solutions.

4. Consider adding together wave functions for $(n,l) = (10,0)$ and $(11,0)$. Plot their sum if they are added with phases of 0 and π. Calculate the expectation value $\langle r \rangle$ for the total wave function in both cases. Describe how this expectation value should vary as the superposition of two states evolves in time.

5. Calculate the spin–orbit coupling for the $n = 2, l = 1$ state in hydrogen.

6. Calculate the helium $2s$ energies $(1s^1 2s^1)$ for $S = 0, 1$ using first-order perturbation theory. Compare with measured values.

7. Based on our treatment of the H_2^+ molecule, calculate and plot the energies of the first two bonding and antibonding energy levels as a function of nuclear separation, R. You can calculate the integrals associated with S, V_{11} and V_{12} by making use of ellipsoidal coordinates, (μ, ν, ϕ), where each of the protons lies at one of the two foci. The coordinates are defined by: $\mu = \frac{r_1 + r_2}{R}$, $\nu = \frac{r_1 - r_2}{R}$ and ϕ corresponding to rotation about the internuclear axis. These can be visualized by considering planar elliptic coordinates, which are then rotated about

the internuclear axis. Lines of constant μ are given by ellipses with foci at the two nuclei, while lines of constant v are given by hyperbolas with foci at the two nuclei. The volume element in ellipsoidal coordinates is given by $dV = \frac{R^3}{8} \left(\mu^2 + v^2 \right) d\mu dv d\phi$, with $1 \leq \mu \leq \infty$, $-1 \leq v \leq 1$, and $0 \leq \phi \leq 2\pi$.

8. Estimate the vibrational period of H_2^+ based on the potential energy curve for the bonding state you calculated in Problem 7. Compare your calculated value with the experimentally measured one listed on the NIST chemistry webbook: http://webbook.nist.gov/chemistry/. Hint: Classically, the frequency of a harmonic oscillator is given by $\omega = \sqrt{\frac{k}{m}}$, where k is the second derivative of the potential. z

9. Find analytic expressions for the $X^1\Sigma^+$ and 0^+ diabatic molecular states of NaI, along with potential couplings, provided in the following reference: Wang, et al., *Chemical Physics Letters* v. 175, p. 225 (1990). Calculate and plot the adiabatic states based on the expressions for the diabatic states you find.

CHAPTER 3

Light–Matter Interaction

3.1 INTRODUCTION

For the quantum systems we wish to study, interaction with electromagnetic radiation (or "light"[1]) is the primary means for both initiating and probing dynamics. The electric field of light couples strongly to charged particles, and it is relatively easy to prepare and measure specific electromagnetic fields in the laboratory. In this chapter, we derive a detailed description of how light interacts with the quantum systems of interest.

The electric field of a light pulse causes charged particles to accelerate. Since the mass of a proton is about $2,000$ times the mass of an electron, the proton experiences roughly $2,000$ times smaller acceleration than an electron in the same electric field. Given this large difference, the nuclei play a far less important role than the electrons in light–matter interactions, and we therefore focus on the electrons. All the electrons in a molecule are identical, and the applied field interacts equally with all of them. This implies that the interaction of the molecule and light field is inherently a multielectron problem. While very challenging to solve in general, there are simplifications that make the problem tractable.

One of these involves the "single-active electron" (SAE) approximation (see Section A.7 for a more mathematical discussion of the SAE approximation). While all of the electrons respond equally to an applied field, the "portion of each electron occupying a given orbital" can have a dramatically different response than the portions residing in other orbitals.[2] For instance, if one applies a light field whose frequency is resonant with the $2s - 2p$ transition in lithium, it is an excellent approximation to consider only the $2s$ electron as interacting with the field. This electron is resonantly driven, while the other two electrons in 1s orbitals are far from resonance and experience very small excursions from equilibrium.

---

[1]We will generally refer to the interaction between the system and applied electromagnetic fields as the light–matter interaction, despite the fact that many of the electromagnetic fields fall outside the optical, or visible, portion of the spectrum.

[2]Because the electrons are indistinguishable, we cannot actually assign each electron in the system to a particular orbital. While the state is formally described by the Slater determinant, one can think about each electron having "a portion of it" in each of the accessible orbitals. Nevertheless, for simplicity we will refer to a "$2s$ electron" in lithium (for example), even though we technically mean the fractional probability of each of the three electrons that resides in the $2s$ orbital. Another way of thinking about this is in terms of a quasiparticle - the "$2s$ electron" is a quasiparticle which includes contributions from all of the identical electrons in the system.

On the other hand, if one applies a field resonant between the $1s^2$ and $1s2p$ states in helium, both electrons in the $1s^2$ orbital will have the same probability of making a transition to the $2p$ orbital. However, the probability for this transition can be calculated by considering only a single electron's response to the field. The reason this works so well is that the photon frequency required to promote *both* electrons to the $2p$ orbital is far greater than the $1s^2 \rightarrow 1s2p$ transition frequency, and thus far from resonance. The SAE approximation is generally valid when the probability of such multielectron excitation is small.

A nice tool for visualizing this basic physics is to consider a funnel shaped spiral wishing well, where one drops coins and watches them spiral into the center of the well (see Figure 3.1). The shape of the well approximates a Coulomb potential $(1/r)$, and two pennies spinning around the center represent two electrons in an atom. The penny closest to the center represents a core electron, while the one near the outer edge represents a loosely bound valence electron (perhaps in a highly excited Rydberg state). For such well-separated electrons, it is reasonable to neglect their interaction and consider them to be independent particles. The application of an external field is mimicked by tilting the wishing well. If one tilts the well slowly in one direction (comparable to a strong, DC electric field), the outer penny is affected more by the tilt and will leave the well more readily (just as the outer electron in a two-electron atom is affected more by a strong DC field). If the well is tilted back and forth, similar to an atom in an oscillating electric field, the frequency of the tilting determines which of the two pennies is more affected. Tilting the well at a frequency corresponding to the orbital frequency of either penny will result in that penny responding more to the field.

Another simplification to the multielectron problem comes from the discussion in Section 2.6, where we considered writing the full, N-dimensional electron wave function as a product of N, one-dimensional wave functions, or orbitals. We noted that due to electron correlation (the electron–electron interaction term in the molecular Hamiltonian), one cannot generally write the full N-electron wave function in terms of a single configuration (a single Slater determinant). Instead, one must express it as a sum over multiple configurations (multiple Slater determinants). However, in many cases the first term in this sum dominates, and the problem reduces to a single configuration.

Figure 3.1 Spiral "wishing well" with two pennies orbiting around the center. The shape of the wishing well is that of a large funnel. Coins launched along the outside of the well roll around and spiral toward the center, with their rotational frequency increasing as they approach the center. Their motion mimics that of classical electrons orbiting a nucleus.

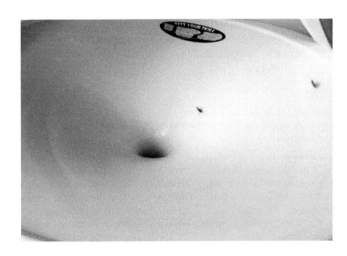

3.2 ELECTRON IN AN APPLIED FIELD

We began Chapter 2 by writing down the generic Hamiltonian for a molecule in a light field (see Equation 2.1). While Chapter 2 focused on the molecular portion of the Hamiltonian, in Chapter 3, we want to include terms related to the light–molecule interaction. We begin the process by rewriting the molecular Hamiltonian of Equation 2.76 in a way that focuses on a SAE. Since we are interested in the response of an *electron* in the molecule, we ignore the term corresponding to the kinetic energy of the nuclei (in this sense, it is really Equation 2.79 that we are rewriting). We group the potential energy terms (terms 2, 3, and 4 of Equation 2.79) into a general potential $V(r)$ that corresponds to the net Coulomb interaction between the electron of interest and all the other electrons and nuclei in the molecule. In addition, the electronic kinetic energy (term 1 of Equation 2.79) is only that of the SAE. With this rewriting, the Hamiltonian for a single electron in a molecule can be written as

$$H = \frac{\mathbf{p}^2}{2} + V(r), \tag{3.1}$$

where to simplify notation we do not write out the momentum operator $\mathbf{p}$ in terms of the Laplacian. Technically, H, $\mathbf{p}$, and V in Equation 3.1 are all operators, but we drop the normal "hat" notation for simplicity.

If the molecule is in the presence of an electromagnetic field, the Hamiltonian is modified by the substitution $\mathbf{p} \rightarrow (\mathbf{p} - q\mathbf{A})$, where q is the electron charge and $\mathbf{A}$ is the vector potential of the field ($\mathbf{B} = \nabla \times \mathbf{A}$, or $\mathbf{E} = -\partial \mathbf{A}/\partial t$ when $\nabla V(r) = 0$). We assume the vector potential can be expressed classically (i.e. without quantizing the radiation field). With this substitution (noting that $q = -1$ in atomic units), the Hamiltonian becomes

$$H = \frac{(\mathbf{p} + \mathbf{A})^2}{2} + V(r). \tag{3.2}$$

Equation 3.2 can be justified by noting that Hamilton's equations produce the correct Lorentz equation of motion for the electron in the field (see homework). In addition, modifying the Hamiltonian through $\mathbf{p} \rightarrow (\mathbf{p} + \mathbf{A})$ can be motivated by considering the momentum of the electron to be "boosted" (or shifted) by the vector potential of the field.[3]

In order to proceed, we choose to work in the Coulomb (or "transverse") gauge, where $\nabla \cdot \mathbf{A} = 0$.[4] In this case, $\mathbf{p}$ and $\mathbf{A}$ commute ($\mathbf{p} \cdot \mathbf{A} = \mathbf{A} \cdot \mathbf{p}$), and we can expand the quadratic term from the Hamiltonian in Equation 3.2 to yield

$$H = \frac{1}{2}(\mathbf{p}^2 + 2\mathbf{p} \cdot \mathbf{A} + \mathbf{A}^2) + V(r) \tag{3.3}$$

The term proportional to $\mathbf{A}^2$ is sometimes referred to as the pondermotive potential, and it represents the energy shift of a free electron oscillating in an electromagnetic field [8]. It can be neglected in most calculations for a couple of reasons. First, this term does not involve the position or momentum of the electron and is therefore not responsible for any *transitions* between eigenstates of the system Hamiltonian. The second is that for most experimental conditions, it is much smaller than the other terms

---

[3]A shift of the momentum corresponds to a linear phase variation in position space—i.e. $\Psi'(r) = e^{i\mathbf{r} \cdot \mathbf{A}}\Psi(r)$.

[4]Because the fields and physically measurable quantities are independent of gauge choice (e.g. Lorenz or Coulomb), one is free to choose the gauge that leads to the least cumbersome mathematical treatment.

in the Hamiltonian.[5] Although we ignore it in what follows, we note that in the case of strong-field ionization, this term must be considered since it can play an important role in determining the final energy of photoelectrons.

The other term in the Hamiltonian that contains the vector potential, $2\mathbf{p} \cdot \mathbf{A}$, corresponds to the interaction term. There are two different approaches to carrying out calculations resulting from this term. One works directly with $\mathbf{p} \cdot \mathbf{A}$, while the other implements a gauge transformation to produce an operator for which calculations are easier with the standard position-space wave function. We follow the gauge transformation approach to arrive at a simple and intuitive expression well suited for calculations in position space. For a detailed discussion of this approach, see [7].

3.2.1 Electric-Dipole Approximation

We assume a vector potential corresponding to an electromagnetic wave traveling in a direction $\mathbf{k}$ with a central frequency ω_0:

$$\mathbf{A} = -A_0(t)\hat{\varepsilon}\sin(\mathbf{k} \cdot \mathbf{r} - \omega_0 t) = \frac{i}{2}A_0(t)\hat{\varepsilon}\left(e^{+i(\mathbf{k} \cdot \mathbf{r} - \omega_0 t)} - e^{-i(\mathbf{k} \cdot \mathbf{r} - \omega_0 t)}\right), \qquad (3.4)$$

where $A_0(t)$ is the time-dependent envelope of the vector potential and $\hat{\varepsilon}$ its polarization. Inserting this vector potential into the Hamiltonian and neglecting the $\mathbf{A}^2$ term, we arrive at

$$H = \frac{1}{2}\left[\mathbf{p}^2 + i\mathbf{p} \cdot \hat{\varepsilon}A_0(t)\left(e^{+i(\mathbf{k} \cdot \mathbf{r} - \omega_0 t)} - e^{-i(\mathbf{k} \cdot \mathbf{r} - \omega_0 t)}\right)\right] + V(r). \qquad (3.5)$$

For a quantum system whose spatial extent is much smaller than the wavelength of the radiation described by the vector potential, the complex exponential term can be approximated as $e^{i(\mathbf{k} \cdot \mathbf{r} - \omega_0 t)} \sim e^{-i\omega_0 t}$, since the atomic coordinate r is generally small compared with $1/k$. This is the heart of the dipole approximation, which is generally quite good: the wavelength of visible light is ~ 500 nm, while the atomic unit of length is about 0.5 Å. Physically, the applied field is essentially uniform over the size of a typical atom or molecule, and any spatial dependence of the field can be neglected. Mathematically, a Taylor series expansion of the spatial component of the complex exponential about zero produces terms that give rise to separate couplings between states of the system:

$$e^{i\mathbf{k} \cdot \mathbf{r}} = 1 + i\mathbf{k} \cdot \mathbf{r} + \frac{(i\mathbf{k} \cdot \mathbf{r})^2}{2!} + \frac{(i\mathbf{k} \cdot \mathbf{r})^3}{3!} + \cdots \qquad (3.6)$$

The first (constant) term in Equation 3.6 is known as the electric dipole term, and this is the only one kept in the dipole approximation. The second term includes both the magnetic dipole and electric quadrupole interactions, while the third and higher terms contain higher order electric and magnetic multipoles. Already, the magnetic dipole and electric quadrupole contributions from the second term are generally much smaller than the electric dipole term for visible radiation interacting with ground-state atoms.[6] In the electric dipole approximation, the Hamiltonian becomes

$$H = \frac{1}{2}\left[\mathbf{p}^2 + i\mathbf{p} \cdot \hat{\varepsilon}A_0(t)\left(e^{-i\omega_0 t} - e^{+i\omega_0 t}\right)\right] + V(r). \qquad (3.7)$$

---

[5]For example, with a 10 mW visible laser, the ratio of the second to third term in the Hamiltonian is about 10^8, allowing us to safely neglect the pondermotive shift.

[6]These terms can become important for X-ray interactions, optical radiation interacting with molecules in highly excited states, or when the dipole term has a vanishing contribution due to symmetry.

3.2.2 Matrix Elements of the Interaction Hamiltonian

The part involving the field is known as the interaction Hamiltonian H_{int}:

$$H_{\text{int}} \equiv \frac{i}{2}\mathbf{p} \cdot \hat{\varepsilon} A_0(t) \left(e^{-i\omega_0 t} - e^{+i\omega_0 t} \right). \tag{3.8}$$

One typically represents the wave function in the position basis as opposed to the momentum basis, so it is useful to rewrite $\mathbf{p} \cdot \hat{\varepsilon} A_0(t)$ in terms of the spatial operator $\mathbf{r}$ (which is easier to evaluate on position-space wave functions than the differential operator $\mathbf{p}$). This is facilitated by transforming the wave function according to

$$\Psi(\mathbf{r},t) = e^{i\mathbf{A}(t)\cdot\mathbf{r}}\Psi'(\mathbf{r},t). \tag{3.9}$$

Applying the chain rule for the spatial and temporal derivatives, noting that $\mathbf{E}(t) = -\frac{\partial \mathbf{A}}{\partial t}$, and cancelling some terms, allows us to write the Schrödinger equation as

$$i\frac{\partial \Psi'(\mathbf{r},t)}{\partial t} = \left[\frac{\mathbf{p}^2}{2} + V(r) + \mathbf{r} \cdot \mathbf{E}(\mathbf{t}) \right] \Psi'(\mathbf{r},t), \tag{3.10}$$

where we have taken the vector potential to be quasimonochromatic, such that the envelope $A_0(t)$ varies slowly compared to the carrier frequency ω_0. $\Psi'(\mathbf{r},t)$ represents a "momentum boosted" version of the wave function, which is natural to work with in the presence of a vector potential. Therefore, for notational simplicity throughout the rest of the book, we drop the "prime" on the transformed wave function, with the understanding that $\Psi(\mathbf{r},t)$ is actually the transformed version. In the absence of a vector potential (e.g. before our time-dependent vector potential turns on), $\Psi(\mathbf{r},t) = \Psi'(\mathbf{r},t)$. With the new form of the Hamiltonian in the Schrödinger equation above, we find the matrix elements of H_{int} in the field-free basis:

$$\langle \psi_n | H_{\text{int}} | \psi_m \rangle = \left\langle \psi_n \left| \frac{1}{2}\mathbf{r} \cdot \hat{\varepsilon} E_0(t) \left(e^{-i\omega_0 t} + e^{+i\omega_0 t} \right) \right| \psi_m \right\rangle. \tag{3.11}$$

Finally, we define the transition dipole moment between the states as

$$\mu_{nm} \equiv \langle \psi_n | -\mathbf{r} | \psi_m \rangle, \tag{3.12}$$

with the projection of the dipole moment onto the polarization axis of the applied field given by

$$\mu_{nm} \equiv \langle \psi_n | -\mathbf{r} \cdot \hat{\varepsilon} | \psi_m \rangle = \mu_{nm} \cdot \hat{\varepsilon}. \tag{3.13}$$

In the case of molecular systems, which lack the central symmetry of atoms, it will frequently be useful to consider the matrix element of $\mathbf{r}$ only; in this case, the transition dipole moment will remain a vector as in Equation 3.12 (for example, see Section 3.3.1). For the simple two-level system in Section 3.2.3, we use the definition in Equation 3.13, and write the matrix elements of the interaction Hamiltonian as

$$\langle \psi_n | H_{\text{int}} | \psi_m \rangle = -\frac{1}{2} \left(e^{+i\omega_0 t} + e^{-i\omega_0 t} \right) \mu_{nm} E_0(t). \tag{3.14}$$

This form of the interaction Hamiltonian can be used to calculate the coupling between any two eigenstates of the system Hamiltonian in an applied field.

3.2.3 Two-Level System

Having established a useful form of the interaction Hamiltonian, we consider the limiting case where the dynamics can be described using a wave function that is a coherent sum of only two, field-free eigenstates. This is a good approximation for an atom or molecule interacting with an oscillating electromagnetic field close to resonance between two electronic states (at least one of which is populated), but sufficiently far from resonance with all other states so that their probability of being occupied is vanishingly small. The two-state wave function can be expressed in the Schrödinger picture as

$$\Psi(t) = c_n(t)e^{-i\omega_n t}|\psi_n\rangle + c_m(t)e^{-i\omega_m t}|\psi_m\rangle, \tag{3.15}$$

where c_i and ω_i are the amplitude and frequency, respectively, of the ith state.[7] We assume $|\psi_m\rangle$ to be the higher-energy state, such that $\omega_m - \omega_n$ is positive. The wave function $\Psi(t)$ evolves in time according the Schrödinger equation,

$$i\frac{\partial}{\partial t}\Psi(t) = \hat{\mathbf{H}}(t)\Psi(t), \tag{3.16}$$

where $\hat{H}(t) = \hat{H}_0 + \hat{H}_{\text{int}}(t)$. Inserting the wave function from Equation 3.15 into Equation 3.16 and projecting onto the eigenstates, we arrive at a set of coupled differential equations for the amplitudes:

$$i\dot{c}_n(t) = e^{-i\omega_{mn}t}\langle\psi_n|H_{\text{int}}|\psi_m\rangle c_m(t) + \langle\psi_n|H_{\text{int}}|\psi_n\rangle c_n(t) \tag{3.17a}$$

$$i\dot{c}_m(t) = e^{+i\omega_{mn}t}\langle\psi_m|H_{\text{int}}|\psi_n\rangle c_n(t) + \langle\psi_m|H_{\text{int}}|\psi_m\rangle c_m(t), \tag{3.17b}$$

where we have used the TISE in the form $H_0|\psi_n\rangle = \omega_n|\psi_n\rangle$ and the orthogonality of the electronic eigenstates.

For states of definite parity like atomic eigenstates (either even or odd), the diagonal elements of H_{int} are zero. Inserting the off-diagonal elements of H_{int} from Equation 3.14 into Equation 3.17, we have

$$\dot{c}_n(t) = -\frac{\mu_{nm}E_0(t)}{2i}e^{-i\omega_{mn}t}\left(e^{+i\omega_0 t} + e^{-i\omega_0 t}\right)c_m(t) \tag{3.18a}$$

$$\dot{c}_m(t) = -\frac{\mu_{mn}E_0(t)}{2i}e^{+i\omega_{mn}t}\left(e^{+i\omega_0 t} + e^{-i\omega_0 t}\right)c_n(t). \tag{3.18b}$$

Multiplying out the complex exponentials produces terms that evolve at rates of $\omega_{mn} - \omega_0$ and $\omega_{mn} + \omega_0$. If ω_0 is far from ω_{mn}, both of these terms will evolve rapidly compared with the electric field envelope $E_0(t)$. However, if one chooses ω_0 to be close to resonance, $\omega_0 \sim \omega_{mn}$, then $e^{i(\omega_{mn} - \omega_0)t}$ will vary slowly compared with $E_0(t)$ (while $e^{i(\omega_{mn} + \omega_0)t}$ will vary rapidly). The integral of a rapidly oscillating factor (which is equally positive and negative) multiplied by a slowly-varying one will generally be very small due to cancelation of the positive and negative portions of the integral. Since we are frequently interested in near-resonant cases where $\omega_0 \sim \omega_{mn}$, we neglect the

---

[7]An alternative way of defining the wave function involves absorbing the complex phase evolution of each eigenstate into the coefficient: $\Psi(t) = a_n(t)|\psi_n\rangle + a_m(t)|\psi_m\rangle$, where $a_n(t) = c_n(t)e^{-i\omega_n t}$. This is known as a "rotating frame transformation," since the linear, time-dependent phase in the a_n frame corresponds to the rotation of a vector in the complex plane.

contribution from the $e^{i(\omega_{mn}+\omega_0)t}$ term (ths step is known as the rotating-wave approximation, or RWA). We define $\Delta \equiv \omega_{mn} - \omega_0 = -(\omega_{nm} + \omega_0)$ and write

$$\dot{c}_n(t) = -\frac{\mu_{nm}E_0(t)}{2i}e^{-i\Delta t}c_m(t) \tag{3.19a}$$

$$\dot{c}_m(t) = -\frac{\mu_{mn}E_0(t)}{2i}e^{+i\Delta t}c_n(t). \tag{3.19b}$$

The factor $\mu_{nm}E_0(t)$ has units of frequency and controls the coupling between the atom and light field. It is known as the Rabi frequency and is often written as χ_{Rabi}[8]:

$$\chi_{\text{Rabi}}(t) \equiv \mu_{nm}E_0(t). \tag{3.20}$$

When the detuning Δ is close to zero (i.e. smaller than the spectral bandwidth of the applied field), the differential equations can be interpreted as an integral of a product of slowly oscillating terms that can have a nonzero value at long times. Physically, the probability (or "population") is transferred between states during the pulse, and the system is left in a coherent superposition of ground ($|\psi_n\rangle$) and excited ($|\psi_m\rangle$) states. This coherent superposition between two states with a nonzero transition dipole moment is known as a "coherence," and it naturally leads to an oscillating dipole at the end of the pulse. In this case, the expectation value of the electron position, $\langle \mathbf{r} \rangle$, oscillates in time at the frequency difference between states.

In the opposite extreme, the detuning is large compared with the bandwidth of the applied field, and the factor containing Δ is no longer slowly varying compared to the field envelope, $E_0(t)$. In this case the envelope doesn't change significantly during one detuning cycle, and the integral becomes a product of a rapidly-oscillating term times a slowly-varying one, resulting in very little population transfer for both intermediate and long times. This observation validates elimination of atomic and molecular states far from resonance with the applied field, as the wave function amplitude in those states remains negligible, except for the case of very strong applied fields, where χ_{Rabi} is comparable to Δ.[9]

The coupled, first-order differential equations of Equation 3.19 can be transformed into a set of uncoupled, second-order differential equations whose analytic solutions are well known. The probability of being in the two states is given by (see homework) [9]:

$$P_1(t) = \frac{1}{2}\left[1 + \left(\frac{\Delta}{\Omega}\right)^2\right] + \frac{1}{2}\left(\frac{\chi_{\text{Rabi}}}{\Omega}\right)^2\cos(\Omega t) \tag{3.21a}$$

$$P_2(t) = \frac{1}{2}\left(\frac{\chi_{\text{Rabi}}}{\Omega}\right)^2[1 - \cos(\Omega t)], \tag{3.21b}$$

where $\Omega = \sqrt{\chi_{\text{Rabi}}^2 + \Delta^2}$. Here, we show and discuss numerical solutions to the equations for a few different values of the key parameters Δ and $\chi_{\text{Rabi}}(t)$, assuming that the

---

[8]We include the subscript "Rabi" on χ due to the unfortunate fact that χ is also standard notation for the nuclear wave function in molecules.

[9]This two-level treatment can be extended to include an arbitrary number of states to account for multifrequency fields. Such calculations are known as "essential states calculations" and are relatively straightforward to implement numerically. We describe such calculations in Sections 3.2.4 and 6.5.2.

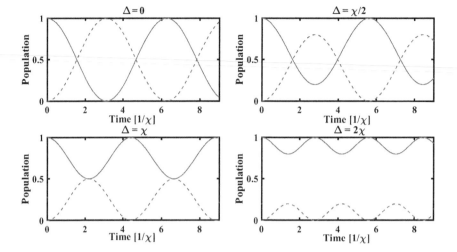

Figure 3.2 Probability of being in state one (solid curve) and state two (dashed curve) as a function of time for solutions to the two-level atom in a radiation field (Equation 3.21). Population starts in state one, and solutions are shown for different values of the detuning, Δ.

population starts entirely in one of the states. Figure 3.2 shows solutions to Equation 3.19 for different values of the detuning, Δ, plotted as a function of time in units of one over the Rabi frequency. There needs to be zero detuning between the field frequency and level separation to transfer 100% of the population from the ground to excited state. As the detuning increases, we note two primary effects in the two-level system: the amount of population transferred decreases, while at the same time, the frequency of oscillation of this transfer increases.

3.2.4 Multiphoton Absorption

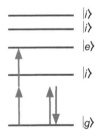

Figure 3.3 Illustration of two-photon absorption (left) and the AC-Stark shift (right). The horizontal lines represent the energies of the field-free eigenstates, while the arrows represent the photon energies. Although there is no one-photon resonance between any two eigenstates, there is a two-photon resonance between the ground state and one of the excited states (leading to two-photon absorption). The two-photon process on the right leads to an AC-Stark shift of the levels.

In this section, we build on the treatment of Section 3.2.3 to describe multiphoton coupling between different electronic states of an atom or molecule (ignoring any vibrational or rotational motion). Assuming the applied field is off-resonant between any pair of states, there is no single-photon resonance. However, if there is a *multiphoton* resonance, such as the two-photon resonance illustrated in Figure 3.3, we can derive an effective multiphoton Rabi frequency for the coupling using an approach known as "adiabatic elimination." This coupling is related to the molecular polarizability and affects the dynamic (AC) Stark shifts (DSSs) of the molecular eigenstates. Although we specifically consider a two-photon resonance, the approach is general for any number of photons ($E_2 - E_1 \approx n\omega_0$, n being an integer greater than one).

We begin by expanding the atomic or molecular wave function in terms of the field-free electronic eigenstates:

$$|\Psi(t)\rangle = \sum_{n=g,e,i} c_n(t) e^{-i\omega_n t} |\psi_n\rangle, \qquad (3.22)$$

where $c_n(t)$ are electronic eigenstate amplitudes. Here we have taken ψ_g to represent the ground state, ψ_e to represent an excited state that is two-photon resonant with the ground state, and ψ_i to represent all other electronic eigenstates that are coupled to ψ_g or ψ_e by the field. As with the two-level system of Section 3.2.3, the Schrödinger equation leads to a set of differential equations for the coefficients of the form:

$$i\dot{c}_m = \sum_{n=g,e,i} c_n(t) e^{+i\omega_{mn} t} \langle \psi_m | H_{\text{int}} | \psi_n \rangle, \qquad (3.23)$$

where $\omega_{mn} \equiv \omega_m - \omega_n$ and H_{int} is the molecule-field interaction Hamiltonian.

The sum over n contains all states i that have a nonzero transition dipole moment to either the g or e states. Since there is no one-photon resonance, the detuning for any transition to intermediate state i is large, and there is no significant population transfer to any of the intermediate states.[10] Dividing Equation 3.23 into ground ($n = g$), excited ($n = e$), and off-resonant ($n = i \neq e, g$) states, we can write the Schrödinger equation as

$$i\dot{c}_g = \sum_i c_i(t)e^{i\omega_{gi}t} \langle \psi_g | H_{\text{int}} | \psi_i \rangle \tag{3.24a}$$

$$i\dot{c}_e = \sum_i c_i(t)e^{i\omega_{ei}t} \langle \psi_e | H_{\text{int}} | \psi_i \rangle \tag{3.24b}$$

$$i\dot{c}_i = c_g(t)e^{-i\omega_{gi}t} \langle \psi_i | H_{\text{int}} | \psi_g \rangle + c_e(t)e^{-i\omega_{ei}t} \langle \psi_i | H_{\text{int}} | \psi_e \rangle. \tag{3.24c}$$

Adiabatic elimination of the rapidly oscillating, off-resonant amplitudes $c_i(t)$ begins by formally integrating the equation for $\dot{c}_i$. Writing out the interaction matrix elements from Equation 3.14, we have

$$c_i(t) = -\frac{1}{2i} \int_{-\infty}^{t} dt' \left[\mu_{ig}c_g(t')e^{-i\omega_{gi}t'} + \mu_{ie}c_e(t')e^{-i\omega_{ei}t'} \right] E_0(t') \left(e^{+i\omega_0 t'} + e^{-i\omega_0 t'} \right). \tag{3.25}$$

Equation 3.25 can be integrated by parts, where we will use an approximation similar to the slowly varying envelope approximation. We explicitly show the steps for one of the four terms in Equation 3.25 (the other three follow similarly). In particular, we look at the term of the form:

$$-\frac{\mu_{ig}}{2i} \int_{-\infty}^{t} dt' \left[c_g(t')E_0(t') \right] e^{-i(\omega_{gi}+\omega_0)t'}. \tag{3.26}$$

Integration by parts ($\int u\,dv = uv - \int v\,du$) is performed by letting $u = [c_g(t')E_0(t')]$ and $dv = e^{-i(\omega_{gi}+\omega_0)t'} dt'$. The expression given in 3.26 then becomes

$$-\frac{\mu_{ig}}{2i} \left(\left[c_g(t')E_0(t') \frac{e^{-i(\omega_{gi}+\omega_0)t'}}{-i(\omega_{gi}+\omega_0)} \right]_{-\infty}^{t} \right.$$
$$\left. - \int_{-\infty}^{t} \frac{d}{dt'} \left[c_g(t')E_0(t') \right] \frac{e^{-i(\omega_{gi}+\omega_0)t'}}{-i(\omega_{gi}+\omega_0)} dt' \right). \tag{3.27}$$

We first consider the remaining integral, which consists of the product of two factors: the derivative of $[c_g(t')E_0(t')]$ and a complex exponential. For nonresonant transitions, the exponent in the complex exponential will not be close to zero, and we expect the first factor to vary slowly when compared with the second. Specifically, both the state amplitude $c_g(t')$ and field envelope $E_0(t')$ must change slowly compared with the one-photon detuning frequency, as we would expect for a nonresonant transition. In this case, we have an integral of a slowly varying factor multiplied by a complex exponential, resulting in a vanishing or very small contribution.

---

[10]This can be checked by integrating the Schrödinger equation while explicitly including as many off-resonant levels as one likes.

When evaluating the boundary term, we note that $E_0(t \rightarrow -\infty) = 0$ (the field is a pulse that turns on). In this case, we are simply left with the following contribution from this term (canceling the $-i^2 = +1$):

$$-\frac{\mu_{ig}}{2}c_g(t)E_0(t)\frac{e^{-i(\omega_{gi}+\omega_0)t}}{(\omega_{gi}+\omega_0)}. \tag{3.28}$$

The other three terms in Equation 3.25 follow similarly, each resulting in a slightly different form of Equation 3.28. Combining all four terms yields the following expression for the intermediate-state amplitudes:

$$c_i(t) = -\frac{\mu_{ig}}{2}c_g(t)E_0(t)\left[\frac{e^{-i(\omega_{gi}+\omega_0)t}}{(\omega_{gi}+\omega_0)} + \frac{e^{-i(\omega_{gi}-\omega_0)t}}{(\omega_{gi}-\omega_0)}\right]$$
$$-\frac{\mu_{ie}}{2}c_e(t)E_0(t)\left[\frac{e^{-i(\omega_{ei}+\omega_0)t}}{(\omega_{ei}+\omega_0)} + \frac{e^{-i(\omega_{ei}-\omega_0)t}}{(\omega_{ei}-\omega_0)}\right]. \tag{3.29}$$

We next substitute this expression for $c_i(t)$ back into the expressions for $\dot{c}_g$ and $\dot{c}_e$ in Equation 3.24, where we again use Equation 3.14 for the interaction matrix elements. Starting with $\dot{c}_g$ we have

$$i\dot{c}_g = \sum_i \frac{\mu_{ig}}{2}\frac{\mu_{gi}}{2}c_g(t)E_0^2(t)$$
$$\times \left(e^{+i(\omega_{gi}+\omega_0)t} + e^{+i(\omega_{gi}-\omega_0)t}\right)\left[\frac{e^{-i(\omega_{gi}+\omega_0)t}}{(\omega_{gi}+\omega_0)} + \frac{e^{-i(\omega_{gi}-\omega_0)t}}{(\omega_{gi}-\omega_0)}\right]$$
$$+ \sum_i \frac{\mu_{ie}}{2}\frac{\mu_{gi}}{2}c_e(t)E_0^2(t)$$
$$\times \left(e^{+i(\omega_{gi}+\omega_0)t} + e^{+i(\omega_{gi}-\omega_0)t}\right)\left[\frac{e^{-i(\omega_{ei}+\omega_0)t}}{(\omega_{ei}+\omega_0)} + \frac{e^{-i(\omega_{ei}-\omega_0)t}}{(\omega_{ei}-\omega_0)}\right]. \tag{3.30}$$

We now multiply out the complex exponentials, noting that $\omega_{gi} - \omega_{ei} = \omega_g - \omega_e \equiv -\omega_{eg}$:

$$i\dot{c}_g = \sum_i \frac{\mu_{ig}}{2}\frac{\mu_{gi}}{2}c_g(t)E_0^2(t)\left[\frac{1+e^{-2i\omega_0 t}}{(\omega_{gi}+\omega_0)} + \frac{e^{+2i\omega_0 t}+1}{(\omega_{gi}-\omega_0)}\right]$$
$$+ \sum_i \frac{\mu_{ie}}{2}\frac{\mu_{gi}}{2}c_e(t)E_0^2(t)$$
$$\times \left[\frac{e^{-i\omega_{eg}t}+e^{-i(\omega_{eg}+2\omega_0)t}}{(\omega_{ei}+\omega_0)} + \frac{e^{-i(\omega_{eg}-2\omega_0)t}+e^{-i\omega_{eg}t}}{(\omega_{ei}-\omega_0)}\right]. \tag{3.31}$$

With an eye towards making the RWA, we define the *two-photon* detuning as $\Delta_2 \equiv \omega_e - \omega_g - 2\omega_0 = \omega_{eg} - 2\omega_0$. This allows us to rewrite Equation 3.31 as

$$i\dot{c}_g = \sum_i \frac{\mu_{ig}}{2}\frac{\mu_{gi}}{2}c_g(t)E_0^2(t)\left[\frac{1+e^{-2i\omega_0 t}}{(\omega_{gi}+\omega_0)} + \frac{e^{+2i\omega_0 t}+1}{(\omega_{gi}-\omega_0)}\right]$$
$$+ \sum_i \frac{\mu_{ie}}{2}\frac{\mu_{gi}}{2}c_e(t)E_0^2(t)$$
$$\times \left[\frac{e^{-i(\Delta_2+2\omega_0)t}+e^{-i(\Delta_2+4\omega_0)t}}{(\omega_{ei}+\omega_0)} + \frac{e^{-i\Delta_2 t}+e^{-i(\Delta_2+2\omega_0)t}}{(\omega_{ei}-\omega_0)}\right]. \tag{3.32}$$

This form of expression allows us to make the *two-photon* RWA, keeping only the three terms that oscillate at Δ_2 (or less). With a slight rearrangement of the leading factors, we have the final equation for $\dot{c}_g$, along with the corresponding one for $\dot{c}_e$:

$$i\dot{c}_g = \sum_i \frac{|\mu_{ig}|^2}{4} c_g(t) \left[\frac{E_0^2(t)}{(\omega_{gi} + \omega_0)} + \frac{E_0^2(t)}{(\omega_{gi} - \omega_0)} \right]$$

$$+ \sum_i \frac{\mu_{ie}\mu_{gi}}{4} c_e(t) \left[\frac{E_0^2(t)e^{-i\Delta_2 t}}{(\omega_{ei} - \omega_0)} \right] \tag{3.33a}$$

$$i\dot{c}_e = \sum_i \frac{|\mu_{ie}|^2}{4} c_e(t) \left[\frac{E_0^2(t)}{(\omega_{ei} + \omega_0)} + \frac{E_0^2(t)}{(\omega_{ei} - \omega_0)} \right]$$

$$+ \sum_i \frac{\mu_{ig}\mu_{ei}}{4} c_g(t) \left[\frac{E_0^2(t)e^{+i\Delta_2 t}}{(\omega_{gi} + \omega_0)} \right]. \tag{3.33b}$$

The term proportional to $c_g(t)$ in Equation 3.33a (and $c_e(t)$ in Equation 3.33b) is a Stark shift that scales with the field's time-dependent intensity envelope $E_0^2(t)$. For this reason, it is referred to as a *dynamic* Stark shift, which we define for either the ground ($\omega_g^S(t)$) or excited ($\omega_e^S(t)$) state to be:

$$\omega_g^S(t) \equiv -\sum_i \frac{|\mu_{ig}|^2}{2} E_0(t)^2 \frac{\omega_{ig}}{\omega_{ig}^2 - \omega_0^2} \tag{3.34a}$$

$$\omega_e^S(t) \equiv -\sum_i \frac{|\mu_{ie}|^2}{2} E_0(t)^2 \frac{\omega_{ie}}{\omega_{ie}^2 - \omega_0^2}, \tag{3.34b}$$

where we have combined terms using a common denominator and factored a minus sign out front (thereby reversing the order of ω_{gi} or ω_{ei}). For the other terms in Equation 3.33, one can similarly define a two-photon Rabi frequency, $\Omega(t)$:

$$\Omega(t) \equiv -\sum_i \frac{\mu_{ei}\mu_{ig}}{(2)^2} \frac{E_0^2(t)}{\omega_{ig} - \omega_0} \tag{3.35a}$$

$$\tilde{\Omega}(t) \equiv -\sum_i \frac{\mu_{gi}\mu_{ie}}{(2)^2} \frac{E_0^2(t)}{\omega_{ie} + \omega_0}, \tag{3.35b}$$

where we have again pulled out a minus sign from each to reverse the ordering in the denominators. Note that $\tilde{\Omega}(t) = \Omega(t)$ in the limit of two-photon resonance. With these definitions, Equation 3.33 becomes

$$i\dot{c}_g(t) = \omega_g^S(t)c_g(t) + \tilde{\Omega}(t)e^{-i\Delta_2 t}c_e(t),$$

$$i\dot{c}_e(t) = \omega_e^S(t)c_e(t) + \Omega(t)e^{i\Delta_2 t}c_g(t). \tag{3.36}$$

In matrix Hamiltonian form, this can be written compactly as

$$i\dot{\mathbf{c}} = \hat{\mathbf{H}}(t)\mathbf{c}, \tag{3.37}$$

where

$$\mathbf{c} = \begin{pmatrix} c_g \\ c_e \end{pmatrix} \quad \text{and} \quad \hat{\mathbf{H}}(t) = \begin{pmatrix} \omega_g^S(t) & \tilde{\Omega}(t)e^{-i\Delta_2 t} \\ \Omega(t)e^{i\Delta_2 t} & \omega_e^S(t) \end{pmatrix}. \tag{3.38}$$

Note that this Hamiltonian is strictly Hermitian only in the limit of zero two-photon detuning (i.e. two-photon resonance), when $\tilde{\Omega}(t) = \Omega(t)$.[11] An important distinction between this Hamiltonian and that for single-photon coupling between two levels is that the diagonal terms here are time dependent and scale with the field intensity. The energies of both levels follow the intensity envelope of the field, and the direction they move (closer or further apart) depends on the detuning of all the off-resonant states.

While this derivation is specific to two-photon coupling between states, it can easily be generalized to any order (although the mathematical description is tedious). Before moving on, we highlight two important points. First, this treatment is nonperturbative, meaning that it is valid even for nonnegligible depletion of the ground state (or other initial state) and in the presence of substantial Stark shifts. The key assumption is that one must be able to neglect population in the adiabatically eliminated, off-resonant states. This can be checked by integration of the TDSE with explicit inclusion of the off-resonant states. Second, all strong-field coupling between atomic and molecular states can be put in this framework, where the coupling between initial and final states is mediated by intermediate, off-resonant states. Without the intermediate states and their transition dipole moments, there would be no multiphoton coupling between initial and final states.

3.2.5 "Classical" Time-Domain Description

Before moving on, it is worth pointing out that one can also describe an electron in a field without referring to the eigenstate picture (as one would hope, since we're interested in what a single electron does in time). This can be approached using the Lorentz model of an atom, which assumes that the electron responds to an applied field as if it were a mass on a spring. An electron in an oscillating electric field is then simply a damped, driven oscillator. The electron oscillates at the frequency of the field, with the amplitude of oscillation depending on the frequency of the drive field relative to the natural oscillation frequencies (or resonances) of the electron.

In the limit where the pulse is far from resonance, the electron starts at rest, oscillates with low amplitude when the field is on, and then ends essentially at rest.[12] The phase of the oscillation when compared with the drive field when far from resonance is either zero (when $\omega_{\text{drive}} \ll \omega_0$) or π (when $\omega_{\text{drive}} \gg \omega_0$). In the resonant case, the electron cannot follow the field, but rather is partially out of phase with respect to the drive field (by $\pi/2$ when exactly on resonance). This excitation generates a time-dependent dipole that persists after the electric field has turned off. The phase difference between the electron displacement and the drive field is what leads to energy transfer from the field to the system. For a more detailed discussion of this process, including how the drive field is affected, see Section A.8.2.

---

[11] For a non-Hermitian Hamiltonian, the total probability is not necessarily conserved.

[12] Note this assumes the field turns off slowly compared with the damping rate so that the electron amplitude can follow the field adiabatically as it turns off. In the frequency domain, this implies that the spectrum of the drive field does not overlap with the resonance. If the drive field turns off too rapidly, it would contain frequency components on resonance.

3.3 MOLECULES IN RADIATION FIELDS

3.3.1 Electronic Transitions and the Condon Approximation

In Section 3.2.3, we considered a two-level, electronic system in a light field and found that the coupling between two electronic states is given by the (electronic) transition dipole moment, μ_{nm}. Here we generalize this to include nuclear degrees of freedom and multiple electrons, as would be typical for a molecular system. Taking our earlier Born Oppenheimer molecular wave functions, and accounting for the contributions from all of the electrons in the molecule, the molecular transition dipole moment can be written as a sum and factored into electronic and nuclear components:

$$
\begin{aligned}
\mu_{nm}^{\mathrm{mol}}(\mathbf{R}_j) &= \left\langle \Psi_n(\mathbf{r}_i, \mathbf{R}_j) \left| \sum_k -\mathbf{r}_k \right| \Psi_m(\mathbf{r}_i, \mathbf{R}_j) \right\rangle \\
&= \left\langle \psi_n(\mathbf{r}_i; \mathbf{R}_j)\chi_n(\mathbf{R}_j) \left| \sum_k -\mathbf{r}_k \right| \psi_m(\mathbf{r}_i; \mathbf{R}_j)\chi_m(\mathbf{R}_j) \right\rangle \\
&= \left\langle \psi_n(\mathbf{r}_i; \mathbf{R}_j) \left| \sum_k -\mathbf{r}_k \right| \psi_m(\mathbf{r}_i; \mathbf{R}_j) \right\rangle \left\langle \chi_n(\mathbf{R}_j) \left| \chi_m(\mathbf{R}_j) \right\rangle \right. \\
&\equiv \mu_{nm}(\mathbf{R}_j) \left\langle \chi_n(\mathbf{R}_j) \left| \chi_m(\mathbf{R}_j) \right\rangle \right.
\end{aligned}
\tag{3.39}
$$

where μ_{nm}^{mol} is the full, molecular transition dipole moment, including the vibrational wave function overlap between the two states, while μ_{nm} is the electronic portion of the transition dipole moment. If the variation of the electronic states with nuclear coordinate is small, then one can neglect the $\mathbf{R}_j$ dependence of the the dipole moment: $\mu_{nm}(\mathbf{R}_j) \to \mu_{nm}$. This is known as the Condon approximation.

Equation 3.39 nicely shows how the transition dipole moment, and therefore the likelihood of absorption or emission of radiation between these two states, depends on both an electronic component and a nuclear component. The last term in this expression—the vibrational wave function overlap between the two states—is known as the Franck–Condon factor. To the extent that the Condon approximation is valid, the vibrational wave function in the final electronic state is simply a projection of the initial vibrational wave function onto the vibrational eigenstates of the final potential energy surface (PES). This leads to the notion of a "vertical" transition, where the vibrational wave function is simply transferred vertically in energy (but not laterally in position) to the new PES.

Even when one cannot ignore the spatial variation of the transition dipole moment (i.e. when the Condon approximation is not valid), the transition will still be approximately vertical. In this case, the varying transition dipole moment slightly shifts the vibrational wave function in the excited PES to larger or smaller R. However, an electronic transition can never generate a vibrational wave function in the excited PES that is completely displaced from the ground-state wave function, since there is no transition probability to R values where the ground-state vibrational wave function is zero.

This result is simple to interpret physically: the nuclei do not have time to move during an electronic oscillation period (the inverse of the electronic transition frequency). We note that this interpretation is valid independent of the applied field pulse duration. At any given instant during the pulse, the ground-state vibrational wave function is transferred vertically to the excited state, at which point it begins to evolve. For a pulse whose duration is much shorter than the timescale for appreciable vibrational

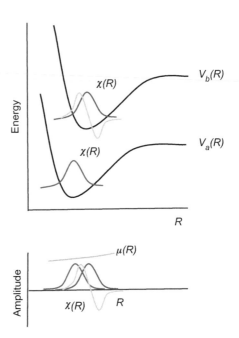

Figure 3.4 (See color insert.) Illustration of the vibrational wave functions, $\chi(R)$, and transition dipole moment, $\mu_{ba}(R)$, for a diatomic molecule. The upper portion of the diagram shows the ground vibrational eigenstate in the lower PES, along with both the ground and first-excited eigenstates in the upper PES. The bottom portion of the figure shows all three wave functions projected down onto a single coordinate axis to visualize the Franck–Condon factors $\langle \chi_b(R) | \chi_a(R) \rangle$. The raised line shows how the dipole moment, $\mu_{ba}(R)$, might vary with R.

motion, the entire electronic transition can be thought of as taking place with "frozen nuclei"; a portion of the wave function on the initial PES is impulsively transferred to the final PES without any motion until after the pulse is gone. On the other hand, when the laser pulse duration is long compared with vibrational motion timescales, the vibrational wave function has time to evolve on the excited PES during the pulse. Here one can think of the pulse as continuously promoting portions of the ground-state wave function vertically to the excited state. The portions that arrive early will have already evolved on the excited PES before the portions arriving later in the pulse.

In fact, for a long pulse these different portions interfere with each other on the excited-state surface, leading to a final vibrational wave function after the pulse that generally does not resemble the initial, ground-state wave function. This time-domain description can be connected to the frequency-domain picture: a long duration pulse has a small spectral bandwidth (assuming the pulses are bandwidth-limited), and in the limit of a very long pulse, the wave function on the excited PES continues to evolve and interfere until it eventually resembles the single vibrational eigenstate accessible by the long-duration, narrow-band pulse.[13]

Figure 3.4 illustrates this idea, depicting the vibrational wave function in the ground electronic state, as well as the first two vibrational eigenstates in an excited-state potential. All three vibrational wave functions, along with the transition dipole moment, are shown as a function of R on the lower axis. The transition dipole moment $\mu_{ba}(R)$ is assumed to vary slowly with R, in which case this dependence can be ignored. Assuming the spectral bandwidth of the pulse is sufficient to excite both vibrational eigenstates in the upper PES, they will be populated according to their wave function overlap in the transition dipole moment (the Franck–Condon factor).

---

[13]We note that in the case when the laser frequency and bandwidth are such that only the ground vibrational eigenstate in the excited PES is populated, the vibrational wave function on the excited PES may once again resemble that on the ground PES (just like in the case of short-pulse excitation).

Finally, we note that the *electronic* portion of the transition dipole moment connecting any two electronic states, μ_{nm}, can vary greatly in magnitude depending on the symmetry of the states and where the electron is likely to be found. For instance, in the case of atomic hydrogen, the definite parity of the eigenstates results in the well-known "selection rules" that govern optical transitions: states with the same parity (e.g. two *s* states) have a vanishing transition dipole moment because the integral associated with the electronic component of Equation 3.39 is zero due to the odd parity of the position operator **r**. Molecular systems tend to have less symmetry than atomic ones, and the selection rules are generally not as strict. However, there are frequently "dark" states that have very small transition dipole moments with the ground state and are therefore unlikely to be accessed by a light-induced transition.

3.3.2 Infrared and Raman Transitions

In addition to nuclear motion occurring as a result of an electronic transition in a molecule, an applied field can also directly drive transitions between vibrational states in the same PES. Since a major goal of time-resolved spectroscopy is to excite and detect vibrational motion in molecules, it is important to have a good physical picture of how the laser-molecule interaction both generates and detects vibrations. In this section we consider driving molecular vibrations through two different mechanisms: infrared (IR) or Raman transitions.

3.3.2.1 IR Transitions

The electric field of the laser exerts a classical force on both the electrons and protons in the molecule of the form $\mathbf{F} = q\mathbf{E}$, where we ignore the small magnetic component of the Lorentz force for nonrelativistic velocities. For a diatomic molecule with a *permanent* dipole moment (or polyatomic molecules with a permanent dipole moment that has a nonzero projection along a given vibrational coordinate R), an oscillating electric field can directly excite vibrations in the molecule by simply driving the charges back and forth. The applied field can resonantly drive vibrational transitions whose natural frequency matches the frequency of the field. Molecular vibrations are typically between 1–100 THz (e.g. the C–H stretching frequency is approximately 90 THz), corresponding to IR excitation wavelengths. Because of this, transitions between vibrational eigenstates driven by an R-dependent, permanent dipole moment are known as IR transitions.

Classically, the potential energy for a dipole in an electric field can be written as

$$U = -\mu(R) \cdot E(t), \tag{3.40}$$

where, in order to be concise, we have written the permanent dipole moment of the molecule in the nth electronic state, $\mu_{nn}(R)$, as $\mu(R)$. This is equivalent to the interaction term we derived above in Equation 3.14. In the dipole approximation (where the electric field does not vary significantly over the size of the molecule), the magnitude of the force can be written as

$$F = |-\nabla U| = \mathbf{E}(t) \cdot \frac{d\mu(R)}{dR}. \tag{3.41}$$

This expression provides a simple way to see how laser fields can drive vibrations in molecules. However, the anharmonicity typical in most bonding potentials implies that driving large-amplitude vibrational motion via IR transitions is challenging, since the anharmonic potential leads to a vibrational frequency that changes with the amplitude of vibration. Thus, one needs to correspondingly vary the excitation frequency as a function of time to drive large amplitude vibrations. This difficulty of "climbing an anharmonic ladder" is illustrated (in the frequency domain) in Figure 3.5. If the vibrational anharmonicity of a given potential is sufficiently large, an applied field on-resonance between a particular pair of vibrational eigenstates will be off-resonance for all other pairs. In this case, it is possible to consider the laser-molecule interaction in terms of just two states, making it straightforward to connect the classical picture of the IR transition implied in Equation 3.41 to the quantum eigenstate picture using an approach similar to the two-level system of Section 3.2.3.

In analogy to our treatment of the two state electronic system in section 3.2.3, we begin by expressing the total vibrational wave function of the system in terms of two eigenstates $|\chi_1(R)\rangle$ and $|\chi_2(R)\rangle$:

$$\chi(R,t) = c_1(t)e^{-i\omega_1 t}|\chi_1(R)\rangle + c_2(t)e^{-i\omega_2 t}|\chi_2(R)\rangle, \quad (3.42)$$

where as usual c_i is the amplitude and ω_i the frequency of the ith state. We assume this vibrational wave packet is in a single electronic state $\psi_n(\mathbf{r}_i; R)$, in which case the full-wave function is given in the Born–Oppenheimer approximation by

$$\Psi(\mathbf{r}_i, R, t) = |\psi_n(\mathbf{r}_i; R)\rangle \left(c_1(t)e^{-i\omega_1 t}|\chi_1(R)\rangle + c_2(t)e^{-i\omega_2 t}|\chi_2(R)\rangle \right). \quad (3.43)$$

This wave function evolves according the time dependent Schrödinger equation,

$$i\frac{\partial}{\partial t}\Psi(\mathbf{r}_i, R, t) = \mathbf{H}(t)\Psi(\mathbf{r}_i, R, t). \quad (3.44)$$

In the presence of an external field, $\mathbf{H}(t) = H_0 + H_{\text{int}}(t)$, where H_0 is the molecular Hamiltonian and the interaction matrix elements of H_{int} come from Section 3.2.2 (with a few, minor modifications):

$$\langle \psi_n(\mathbf{r}_i; R)\chi_1(R) | H_{\text{int}} | \psi_n(\mathbf{r}_i; R)\chi_2(R) \rangle$$
$$= -\frac{1}{2} \left(e^{+i\omega_0 t} + e^{-i\omega_0 t} \right) E_0(t) \langle \chi_1(R) | \mu(R) | \chi_2(R) \rangle \cdot \hat{\varepsilon}, \quad (3.45)$$

where $E_0(t)$ is the electric field envelope and $\hat{\varepsilon}$ is the polarization vector.

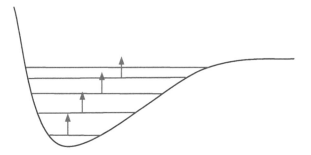

Figure 3.5 Illustration of the difficulty of driving large amplitude vibrational motion. For an anharmonic potential, the spacing between vibrational eigenstates (shown with horizontal lines) is not uniform, generally decreasing with increasing quantum number. This means that light resonant for the transition between the ground and first-excited states will generally not be resonant for transitions between higher lying levels (as indicated by the vertical arrows).

If the molecule has no permanent dipole moment (e.g. a homonuclear diatomic molecule), this interaction Hamiltonian vanishes, and there is no coupling of the vibrational states by the field. Furthermore, if the permanent dipole moment of the molecule is completely independent of vibrational coordinate R (the "zeroth-order" Condon approximation of Section 3.3), $\mu(R) \rightarrow \mu_0$ can be pulled out of the inner product in Equation 3.45, and orthogonality of the vibrational eigenstates ensures there is no field-driven coupling. In either case, there are no IR transitions.

However, it is typically the case that $\mu(R)$ depends at least somewhat on the displacement R. The dependence is driven by the fact that $\psi_n(\mathbf{r}_i; R)$ depends parametrically on R. Since the electronic wave function varies with R, the dipole moment (which depends on electron displacement) generally also varies with R; it is this dependence that is neglected in the Condon approximation. Assuming a Taylor series expansion for small displacement near the equilibrium separation R_0, expanding the dipole moment to first order yields a linear dependence on R:

$$\mu(R) \approx \mu_0 + R \frac{d\mu(R)}{dR}. \tag{3.46}$$

With this approximation the dipole-moment factor in Equation 3.45 becomes:

$$\langle \chi_1(R) | \mu(R) | \chi_2(R) \rangle \approx \left\langle \chi_1(R) \left| R \frac{d\mu(R)}{dR} \right| \chi_2(R) \right\rangle \equiv \mu_{12}^v, \tag{3.47}$$

where the superscript v indicates that this dipole moment connects *vibrational* states 1 and 2.

After applying the RWA and projecting onto an eigenstate basis (as in the case of the two-level system), Equation 3.44 becomes an expression for the state amplitudes:

$$\dot{c}_1(t) = -\frac{\mu_{12}^v \cdot \varepsilon E_0(t)}{2i} e^{-i\Delta t} c_2(t) \tag{3.48a}$$

$$\dot{c}_2(t) = -\frac{\mu_{21}^v \cdot \varepsilon E_0(t)}{2i} e^{+i\Delta t} c_1(t), \tag{3.48b}$$

where $\Delta \equiv (\omega_2 - \omega_1) - \omega_0 = -(\omega_{12} + \omega_0)$ is the detuning of the applied field with respect to the field-free vibrational energy level difference. This set of differential equations describes IR transitions between two vibrational eigenstates in a single electronic PES. Consistent with our intuition based on Equation 3.41, it is the change in the dipole moment as a function of vibrational coordinate shown in Equation 3.47 that drives the transition.

3.3.2.2 Raman Transitions

In systems where there is no permanent dipole moment (or the dipole moment does not vary with coordinate), one cannot directly drive vibrations with an applied field as described earlier.[14] However, one can generate an *induced* dipole moment with an applied electric field via the molecular polarizability α. The polarizability is defined as the proportionality constant relating the induced dipole moment in a molecule, μ, to an applied field $\mathbf{E}(t)$:

$$\mu(t) \equiv \alpha \mathbf{E}(t). \tag{3.49}$$

---

[14]In multidimensional systems, we note that this is also the case for vibrational coordinates where the projection of the dipole moment along that particular coordinate is zero.

We note that, in general, a molecule has different polarizabilities along different axes, and thus α is a tensor. For simplicity, we will ignore the tensor nature of the polarizability and treat α as a scalar. Similar to the case of the permanent dipole, if the induced dipole moment varies with R, the molecule will experience a force given by the gradient of the potential energy of the induced dipole in the electric field. Since $\mathbf{E}(t)$ is assumed to be spatially uniform in the dipole approximation, it is $\alpha = \alpha(R)$ that introduces the R-dependence to μ. In analogy to Equation 3.46, we expand the polarizability in a Taylor series expansion for the vibrational coordinate:

$$\alpha(R) \approx \alpha_0 + R\frac{d\alpha(R)}{dR}, \tag{3.50}$$

where α_0 is the polarizability of the molecule at equilibrium. The magnitude of the force is calculated from the gradient of the potential energy of the oscillating dipole in the electric field:

$$|\mathbf{F}(t)| = \frac{d}{dR}\left(\mu(R,t)\cdot\mathbf{E}(t)\right) \approx \frac{d\alpha}{dR}E^2(t), \tag{3.51}$$

where μ and $\mathbf{E}$ are parallel when α is a scalar (Equation 3.49), and we have only kept the first-order derivatives of $\alpha(R)$ with respect to R. If we consider an oscillating field, we can write the magnitude of the time-averaged force as

$$\langle F(t)\rangle = \frac{1}{2}\frac{d\alpha}{dR}E_0^2(t), \tag{3.52}$$

where $E_0(t)$ is the electric field envelope. Note that in contrast to IR transitions where the force is proportional to the strength of the laser field, the force is now proportional to the field squared. One can think of the first order of the field creating the dipole, with the second order interacting with the generated dipole.

Absorption of energy by a molecule via such second-order interactions comes from Raman scattering of the applied field. Raman scattering is an inelastic process, where the scattered light is at a different frequency (the Stokes frequency) than the input. The energy difference between the input and scattered light is deposited in the molecule. *Stimulated* Raman scattering occurs when there is sufficient light at the Raman-shifted Stokes frequency to interfere with the input light. This creates a modulation of the light field at the vibrational frequency of the molecule, thereby enhancing the Raman process. If the bandwidth of the excitation source is broad enough, there may be photons at the Stokes frequency present in the input light. In this case of *impulsive* stimulated Raman scattering, the *second-order* power spectrum of the pulse (corresponding to $E_0^2(t)$) contains a component at the vibrational frequency of the molecule that drives the Raman scattering, and the Raman process is immediately stimulated without having to build up from the spontaneous scattering.

Finally, we saw that direct IR excitation requires the laser frequency match the vibrational frequency of the molecule to generate significant motion. In contrast, Raman scattering can drive vibrations using frequencies far from the vibrational frequency of the molecule (e.g. optical). If the frequency of the laser matches a difference in energy between *electronic* states of the molecule, the polarizability is greatly enhanced, and the force applied with such radiation is far greater than the force for nonresonant light. This is known as *resonance* Raman scattering.

Figure 3.6 summarizes the different approaches for coupling two vibrational states in a molecule. An IR driving field directly couples the two states by matching the frequency of the transition, while in a Raman process, the pump and Stokes fields

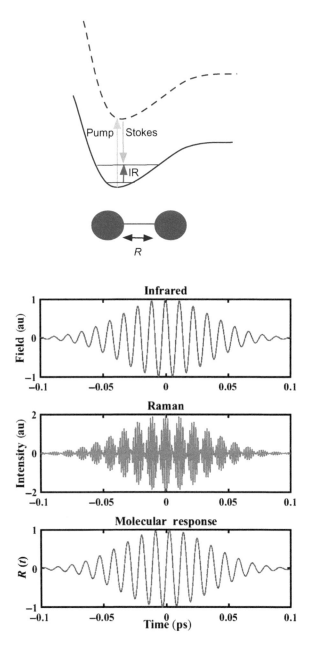

Figure 3.6 Illustration of both IR and Raman driving of molecular vibrations. Top portion shows the energetics of the two processes, while the lower portion shows the corresponding fields and molecule response as functions of time. The low-frequency, IR field connecting two vibrational states directly drives the molecular vibrations, while in the Raman process the pump and Stokes fields together interfere to drive molecular vibrations at the appropriate frequency.

combine to drive the molecular vibration. Note that the molecular response follows the carrier frequency in the case of IR transitions and the envelope of the beat frequency for Raman transitions.

The polarizability that drives Raman transitions can be calculated similar to the multiphoton absorption derivation of Section 3.2.4. Two-photon absorption and impulsive Raman scattering are both second-order processes proportional to the square of the applied field. In two-photon absorption, the interaction can be thought of as "up-up" (cf. Figure 3.3), whereas for Raman transitions, the interaction is "up-down" (c.f. Figure 3.6). Since both processes are second order in the applied field, a key connection between the two is the spectrum of $E^2(t)$, where $E(t)$ is the applied laser field. This spectrum contains two peaks: one at twice the carrier frequency and one near zero frequency. In two-photon absorption, the energy difference between electronic states must fall within the bandwidth of the spectrum near twice the carrier frequency. For impulsive Raman scattering, the energy difference between the initial and final states must fall within the portion of the spectrum of $E^2(t)$ near zero frequency [10].

In conclusion, the simplest and most direct light–matter interaction is a linear interaction, proportional to the field of the pulse (linear with E). However, as we saw with IR transitions in molecules, there may be symmetry or other considerations that lead to a vanishing coupling between states at first order. Raman scattering is a second-order light–matter interaction that allows one to excite vibrations not possible with a linear interaction. As we will see, it will be useful to consider the general nth-order interaction between the applied field and the system of interest, either for the pump or probe separately, or in terms of the total field (pump and probe together).

3.4 HOMEWORK PROBLEMS

1. Calculate the dipole matrix element for the $1s \rightarrow 2p$ transition in atomic hydrogen.

2. Show that one arrives at the correct equations of motion for a charged particle in an electromagnetic field if you transform the momentum according to: $\mathbf{p} \rightarrow \mathbf{p} - q\mathbf{A}$.

3. Transform the set of coupled differential equations of Equation 3.19 into a set of uncoupled equations. Show that the solutions to these uncoupled equations are given by Equation 3.21.

4. Design a laser pulse that perfectly inverts the population in a two-level atom. In other words, specify the detuning and intensity envelope such that an atom initially in the ground state will have 100% probability of being found in the excited state at the end of the pulse. Consider a separation of ground and excited states of 0.1 atomic units, and a transition dipole moment of 0.5 atomic units. Make a graph of the pulse intensity as a function of time, along with the probability of being in the excited and ground states as a function of time.

5. In Section 3.2.3, we considered a two-level system in an AC radiation field. Similar to the multiphoton case (see Equation 3.37), the Schrödinger equation for the two-level system can be written in matrix form as

$$i\dot{\tilde{\mathbf{c}}} = \hat{\mathbf{H}}(t)\tilde{\mathbf{c}}, \quad \tilde{\mathbf{c}} = \begin{pmatrix} \tilde{c}_g \\ \tilde{c}_e \end{pmatrix}, \tag{3.53}$$

where $\tilde{\mathbf{c}}$ is the state vector in a particular basis. The Hamiltonian matrix in this basis is given as

$$\hat{\mathbf{H}}(t) = \frac{1}{2} \begin{pmatrix} -\Delta & -\chi_{\text{Rabi}}(t) \\ -\chi_{\text{Rabi}}(t) & \Delta \end{pmatrix}, \tag{3.54}$$

where, like earlier, Δ is the detuning and $\chi_{\text{Rabi}}(t) = \mu_{mn}E_0(t)$ is the Rabi frequency coupling the two levels. Now consider the basis transformation defined by the equations:

$$\tilde{c}_g(t) = c_g(t)e^{+i\frac{\Delta}{2}t} \tag{3.55a}$$

$$\tilde{c}_e(t) = c_e(t)e^{-i\frac{\Delta}{2}t}. \tag{3.55b}$$

Use Equation 3.55 to rewrite the Schrödinger equation of Equations 3.53 and 3.54 in the c_g and c_e basis (no tilde). Verify your result agrees with what we found in Equation 3.19.

6. In Problem 9 of Chapter 2, you started with the diabatic states for the NaI molecule and calculated the adiabatic states by transforming to a basis where the potential was diagonal. This problem involves performing a similar calculation of adiabatic atomic eigenstates for a two-level atom in a laser field. Calculate the eigenstates of atom and field (the "dressed states") for a two-level atom in a strong, continuous-wave laser field as a function of detuning, from -0.01 to 0.01 atomic units with a Rabi frequency of 0.001 atomic units and a separation between atomic levels of 0.1 atomic units. Plot the dressed-state energies as a function of detuning and discuss the nature of the eigenvectors for the limits of large positive and negative detuning. How do these eigenvectors compare with the bare (field-free) states for large positive/negative detuning? For zero detuning? How do these eigenvectors behave as the field is turned off for positive and negative detuning? What can you infer about what happens to an atom in a strong laser field whose frequency is swept through resonance and which is turned on slowly before the frequency sweep and off slowly after the frequency sweep?

7. The one-photon dipole matrix elements and transition wavelengths for atomic lithium are given in Table 3.1. In this problem, we consider a *two-photon* transition from the 2s ground state to the 3s excited state.

   (a) Using adiabatic elimination of the off-resonant states, calculate an effective two-level Hamiltonian for the atom when exposed to an intense, two-photon-resonant laser pulse with a central wavelength of 735 nm. Only consider contributions from states shown in the table.

   (b) Assume the laser pulse has a Gaussian-shaped intensity profile with 300 fs duration (full-width, half-max) and an area of π. In other words, the integral of the two-photon Rabi frequency is equal to π, and the pulse would

Table 3.1 One-Photon Wavelengths and Transition Dipole Moments for Relevant Transitions in Atomic Lithium

Transition	Wavelength (nm)	Dipole Moment $[10^{-29}$ Cm]
$2s - 2p$	670.8	1.99
$2s - 3p$	323.3	0.118
$2s - 4p$	274.1	0.102
$3s - 2p$	812.6	1.20
$3s - 3p$	2688	5.10
$3s - 4p$	1079	0.041

invert the population were it not for the "DSS" terms in the Hamiltonian. Calculate and plot the $3s$ excited-state population as a function of time. Briefly discuss your result.

8. In this problem, you will calculate the coupling strength for a Raman transition. Consider applying a laser field to drive a two-photon transition in a molecule between two different vibrational eigenstates in the same electronic state. This situation is analogous to the treatment of multiphoton transitions discussed in Section 3.2.4. However, in the Raman case, both the initial and final states are below the intermediate, off-resonant state (see Figure 3.7), leading to the description of a Raman transition as "up-down" instead of "up-up." We assume the bandwidth of the laser is such that both photon frequencies in Figure 3.7 are present in the pulse (the Raman excitation is "impulsive"); in this situation, the process is two-photon resonant.

Calculate the *two-photon* Rabi frequency of a Raman transition between the ground and first-excited vibrational eigenstates of the lowest electronic state in diatomic sodium. Assume the laser has a field strength of 0.01 atomic units and a central wavelength of 800 nm. Take the electronic component of the transition dipole moment (do not worry about Franck–Condon factors or rotational levels) between the X and A states to be roughly 10 Debye (see *The Journal of Chemical Physics* **66**, p. 1477 (1977)); you can look up the transition frequency between the X and A electronic states. Note that the coupling strength will be proportional to E_0^2 as in Equation 3.52, and that the proportionality constant is contains $d\alpha/dR$ from the classical description.

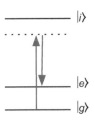

Figure 3.7 Diagram representing a (two-photon) Raman transition between two vibrational eigenstates ($|g\rangle$ and $|e\rangle$) in a molecule. The central laser frequency is one-photon detuned from the excited electronic state $|i\rangle$.

CHAPTER 4

Introduction to Experimental Techniques

I n Chapter 2 we considered time-independent quantum mechanics in simple atoms and molecules to understand the basics of atomic and molecular structure. Chapter 3 presented a discussion of the light–matter interaction relevant for interpreting the various spectroscopies covered in the book. In this chapter, we outline four archetypal approaches to time-resolved spectroscopy that serve as a framework for the rest of the book. These four approaches, along with their extensions and modifications, are developed in detail in later chapters.

4.1 THE PUMP–PROBE FRAMEWORK

While there are an amazing number of techniques that follow quantum dynamics using an array of interactions and pulse sequences, we aim to discuss all the various approaches within the framework of a simple "pump–probe" analysis. What we mean by this are two, local-in-time interactions with two separate pulses: a pump pulse creates an initial, time-dependent state of the system, and a probe pulse subsequently interacts with the evolving system. After the probe interaction, light and/or charged particles are collected as the measurement observable. Variation in the yield is monitored as a function of the time delay between pump and probe interactions; it is this variation with pump–probe delay that constitutes the time-resolved measurement. While the pump pulse is almost always a short, coherent electromagnetic field (e.g. a laser pulse), the probe can be either a second laser pulse or a pulse consisting of electrons.

In some cases, the pump or probe process may involve multiple interactions with more than one field. Nevertheless, we argue it is still useful to conceptually separate the experiment into two distinct interactions: one that prepares a time-dependent quantum state, and one that probes it. Almost all approaches for performing time-resolved measurements can be understood in terms of this pump–probe framework, even those involving pump or probe pulses composed of multiple fields. The main point is that a time-dependent state is prepared in a quantum system or ensemble of systems, and an observable of this time-dependent state is measured as a function of time delay between the pump and probe interactions.

We will consider molecular interactions with the applied fields in both the perturbative ("weak-field") and nonperturbative ("strong-field") regimes. When the pump or probe field is sufficiently weak, the process can be described using the lowest order of time-dependent perturbation theory (see Section A.3). In this case, the interaction with each field is linear: the amplitude of the excited state wave function is linearly proportional to the field strength. However, as the field strength grows, the interaction may depend nonlinearly on the field, and require either higher-order perturbation theory or a nonperturbative treatment to describe adequately. As mentioned in Chapter 1, the entire pump–probe sequence is an inherently nonlinear process with respect to the *total* applied field.

A useful starting point for discussion of the various experimental pump–probe approaches is to divide them into two groups: coherent and incoherent detection techniques. Incoherent processes result from the incoherent (or independent) addition of signal amplitudes from each source. These sources can be different molecules in an ensemble, or different atoms in a single molecule. On the other hand, for coherent processes, the measured signal results from the coherent addition of scattering or emission from multiple sources. Whether a detection technique is coherent or not has a major influence on how the measurement is carried out and what details require attention.

4.2 INCOHERENT PROCESSES

In this section, we focus on two prototypical examples of incoherent time-resolved techniques: laser-induced fluorescence (LIF), and time-resolved photoelectron spectroscopy (TRPES). While we briefly mention some related methods, we focus on LIF and TRPES because they serve as clear archetypal illustrations of incoherent time-resolved spectroscopies, and provide examples of both light and charged particle detection.

4.2.1 Laser-Induced Fluorescence

Laser-induced fluorescence, along with the related technique of fluorescence gating, are standard incoherent, time-resolved techniques. The general idea of an LIF measurement is shown in Figure 4.1. The top panel illustrates the typical experimental arrangement, where a pump pulse incident on the molecular sample initiates the dynamics. A probe pulse, whose time delay relative to the pump can be varied, then interacts with the same molecular sample. Fluorescence light emitted by the sample after interaction with the probe is collected and measured with a photodetector (typically, a photomultiplier tube or PMT).

The middle panel depicts the interaction energetically, using model potential energy surfaces, where the horizontal axis represents some molecular coordinate. In a simple diatomic, this would be the internuclear separation R, but in a larger molecule, it could represent a more complicated, coordinated motion of several atoms. In this view of the interaction, the molecule starts in the ground-electronic state before the pump pulse promotes a portion of the ground-state wave function to an excited state (here assumed to be a bound state, but not necessarily so). In general, the excited state will differ in curvature and equilibrium location from the ground state, and so the vibrational wave function will evolve in time on this PES. Sometime later, the probe pulse arrives and, depending on the location and spread of the wave packet in the intermediate state, may transfer population to a third, higher-lying state. Population in this state can be measured via fluorescence decay; the emitted light is collected and serves as the signal. Typically, excitation to the third state is reasonably localized along the molecular coordinate, as the probe photon energy is only resonant over a small range

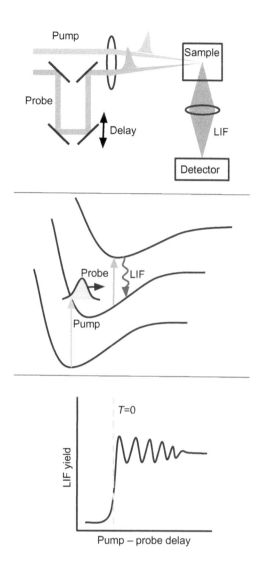

Figure 4.1   (See color insert.) Cartoon diagram illustrating LIF. The top panel illustrates the experimental apparatus; the middle panel shows the interaction with the pump and probe pulses, along with evolution of the time-dependent molecular wave function, on molecular potential energy surfaces; and the bottom panel shows the LIF yield as a function of pump–probe delay.

of positions crossed by the wave packet on the intermediate state. Since the wave packet is oscillating on the intermediate state, the intensity of the fluorescence signal will vary in time as the pump–probe delay is changed.

The oscillating fluorescence yield is reflected in the bottom panel of Figure 4.1, where the LIF signal is plotted as a function of the delay time between the pump and probe pulses. Time $T = 0$ indicates the position of temporal overlap of the two pulses. Before time zero, there is no population in the intermediate state, and therefore no excitation by the probe to the upper fluorescent state. The fluorescence signal turns on after time zero and oscillates with a period determined by the shape of the intermediate state PES (for a bound state). Due to the generally anharmonic nature of any molecular PES, the vibrational wave packet will tend to dephase, or delocalize, over time (see Section 5.5.2 for a discussion). This appears as a gradual decay of the oscillation amplitude in the LIF signal. Note that in LIF, the dynamics on the upper state are not captured since the fluorescence signal is not time resolved. The temporal information encoded in the signal reflects dynamics on the intermediate state where the wave function evolves between pump and probe pulses.

As we shall see, although the details vary for different techniques, the framework illustrated in Figure 4.1 provides a qualitative understanding of the basic approach of incoherent time-resolved spectroscopy. While the types of pulses and detected signals vary, all of the techniques monitor a time-dependent signal that reflects dynamics initiated by the pump pulse and is proportional to the number of molecules in the ensemble.

4.2.2 Time-Resolved Photoelectron Spectroscopy

As a second example of an incoherent technique, we consider TRPES. As shown in the top panel of Figure 4.2, the experimental arrangement is quite similar to LIF. The primary difference is that instead of fluorescent light, the detected signal comes from electrons created when the probe pulse ionizes the sample. Experiments are typically performed in a vacuum chamber so that the emitted electrons are easily collected and

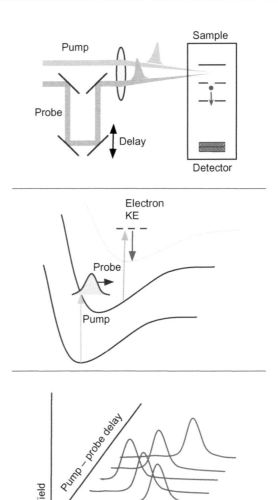

Figure 4.2   (See color insert.) Cartoon diagram illustrating TRPES. Top: Pump and probe pulses are focused into a vacuum chamber where the atomic or molecular sample is ionized. The photoelectron energy can be measured via time-of-flight or velocity-map imaging, where the time/position of the electrons encodes their energy/momentum, respectively. Middle: After the pump pulse excites a vibrational wave packet on an excited state, the time-delayed probe pulse ionizes the molecule, resulting in the ejection of an electron. Bottom: The electron yield is plotted as a function of both pump–probe delay and the KE of the emitted electron.

detected using an electron multiplier. The electron kinetic energy (KE) can be determined based on the detector geometry and the time taken for the electrons to arrive. The detector could be a simple single-channel device, although microchannel plates are more commonly used. Microchannel plates allow for spatial resolution in the perpendicular plane, a fact that is utilized in more complicated detection schemes such as velocity-map imaging (see Section 7.3.2).

In the middle panel of Figure 4.2, we see that the initial pump process is identical to that of LIF: a pump pulse promotes a portion of the ground-state wave function to an intermediate state where it begins to evolve. However, in TRPES, the final state of the molecule is now ionic, and the emitted electron is in a continuum of energy states. The energy of the photoelectron depends on the difference between the intermediate and final PESs; by measuring this energy, one can track the location of the wave packet on the intermediate surface as a function of time. TRPES is particularly well suited for studying complicated excited-state dynamics that involve nonadiabatic processes such as internal conversion (especially relevant in polyatomic molecules and discussed in detail in later chapters).

This energy-resolved measurement provides an important difference when compared with traditional LIF: in TRPES, the KE of the electron provides a "second dimension" to the data.[1] As seen in the bottom panel of Figure 4.2, the electron yield is plotted as a function of not only pump–probe delay but also the KE release of the electron. At early times, the probe comes before the pump and no electron signal is recorded. After time zero, we see a peak turn on in the photoelectron spectrum, with the position of the peak corresponding to the KE of the emitted electron. The position of this peak as a function of pump–probe delay provides a measurement of molecular dynamics occurring on the intermediate state. As will be discussed in detail in Chapter 7, there is a wealth of information contained in the photoelectrons, as their energy and angular distributions can provide details of the wave packet evolution and character of the intermediate-state surface (shape of the electronic state wave function).

4.2.2.1 Time-Resolved Ion Spectroscopy

Before moving on to coherent processes, we note that a simple change to the experimental setup allows one to perform time-resolved, time-of-flight mass spectroscopy on the molecular *ions* produced after the pump–probe process. By switching the polarity of the voltages on the extraction plates in the top panel of Figure 4.2, one directs the positively charged ions to the detector instead of the negatively charged electrons. (As you might expect, there are more complicated approaches that collect both ions and electrons, either by having detectors at both ends or by reversing the polarity of the plates during the experiment.) The ions arrive at the detector at different times depending on their mass-to-charge ratio. Through an appropriate calibration of the detector, one can label the peaks in a time-of-flight signal as coming from certain mass-to-charge ratios, providing the ability to identify particular molecular fragments. The data look similar to the bottom panel of Figure 4.2, with the electron KE axis replaced by the mass-to-charge ratio of the ionic fragments. We note however, that the information contained in the fragment ion yields is not as directly related to the excited state wave packet dynamics as the information contained in a photoelectron spectrum.

---

[1] The additional, second dimension in the TRPES data is akin to spectrally resolving the fluorescence signal in LIF. This arrangement is typically implemented using coherent optical techniques, and the possibilities it provides are discussed in Section 4.3.1.

4.3 COHERENT PROCESSES

In the previous section, we discussed two techniques where the detected signal resulted from the *incoherent* addition of signal amplitudes from many sources. For example, in LIF, each excited molecule in the sample was assumed to fluoresce independently of the others. In this section, we consider spectroscopies where the detected signal arises from the *coherent* addition of signal amplitudes from multiple sources. Since coherent measurements are sensitive to the relative phase between the scattering sources, our treatment will necessitate a discussion of "phase matching" (as mentioned in Section 8.3, phase matching is related to conservation of momentum during the interaction). The two prototypical coherent spectroscopies we consider in this chapter are transient absorption (TA) and ultrafast electron diffraction (UED).

4.3.1 Transient Absorption

The basic experimental arrangement for TA is shown in the top panel of Figure 4.3. Up until the sample, the setup is identical to that of LIF; experimentally at least, it is only how the light is collected and analyzed that distinguishes the two approaches. In TA, the probe beam itself is collected and spectrally resolved in a spectrometer. It is the pump-induced changes to the probe beam that are of interest.

Typical data for a TA experiment are plotted in the bottom panel of Figure 4.3. Similar to TRPES, TA offers two dimensions for the data: pump–probe delay and wavelength of the detected probe light. The vertical scale shows the *differential change in absorption* of the probe light, or how much probe light is present at each wavelength when the pump pulse is present as compared to without the pump pulse. Positive signals indicate enhanced absorption due to the pump, while negative signals indicate less absorption (enhanced transmission, or "bleach"). The differential absorption signal is plotted as a function of wavelength for different pump–probe delays.

Energetically, the process looks similar to LIF, as seen in the middle panel of Figure 4.3. The pump pulse promotes a portion of the ground-state wave function to an intermediate state, upon which the wave function evolves in time. The time-delayed probe then interacts with the sample. The probe arrow pointing up in Figure 4.3 is the coherent analog to LIF, now measured via changes in the probe light instead of fluorescence. In particular, evolution of the wave function on the intermediate state leads to a time-dependent, enhanced absorption of the probe as the wave packet comes into the resonance region for the probe photon energy (manifested in a positive differential absorption signal). The probe arrow pointing down in Figure 4.3 does not have a corresponding analog in LIF. This process represents less absorption of the probe due to wave function evolution out of a resonance region (manifested in a negative differential absorption signal, or bleach). The subtle distinction between the possible pathways in TA and LIF comes about due to the coherent nature of the interaction in TA, where the probe field interacts with the pump-induced, time-dependent polarization of the molecular sample. This polarization acts as a source of radiation that mixes coherently with the probe field. Details of the TA process are addressed in Chapter 8. In addition, Chapter 8 also discusses a number of techniques related to TA, including coherent anti-Stokes Raman scattering.

4.3.2 Ultrafast Electron Diffraction

We now come to the final example in our introduction to experimental, time-resolved techniques: UED. In one sense, UED fits nicely into the set of three techniques

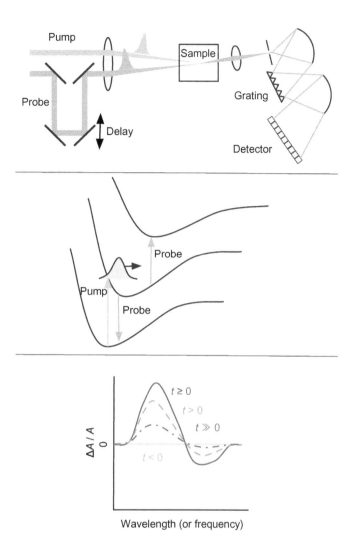

Figure 4.3   (See color insert.) Cartoon diagram illustrating TA. Top: As with both LIF and TRPES, the pump and probe pulses are focused into the molecular sample with a variable time delay. In TA, the coherently scattered probe light is collected and spectrally resolved in a spectrometer. Middle: Once again the pump pulse promotes a portion of the ground-state wave function to an excited state, where it subsequently evolves in time. In TA, the time-delayed probe pulse interacts with the induced polarization of the molecule, resulting in either enhanced or suppressed absorption of the probe. Bottom: The change in absorption of the probe pulse is plotted as a function of wavelength for different time delays.

already described, as we have considered incoherent light detection (LIF), coherent light detection (TA), and incoherent electron detection (TRPES). UED, which involves coherent electron detection, is a natural final choice. However, UED is different than any of the previous examples in that the probe is now an ultrafast pulse of electrons (composed of charged particles) instead of an ultrafast light pulse (composed of photons). Nevertheless, UED can be described as a coherent scattering process similar to TA, and it provides a new perspective on what can be measured using time-resolved techniques.

The experimental arrangement is shown in the top panel of Figure 4.4. The primary difference when compared with the previous techniques is that instead of interacting directly with the molecular sample, a time-delayed probe laser pulse impinges on a photocathode to generate an ultrashort burst of electrons. These electrons are accelerated and focused into the sample, where they serve as the probe in the usual pump–probe framework. Since the deBroglie wavelength of the incident electrons is

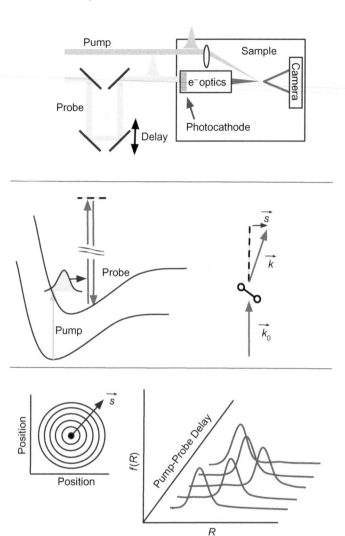

Figure 4.4   (See color insert.) Cartoon diagram illustrating UED. Top: A pump pulse excites the molecules and a time-delayed probe generates an ultrafast bunch of electrons that diffracts off the molecules. The spatial diffraction pattern as a function of pump–probe delay is recorded on a camera. Middle: Both energy (left) and momentum (right) pictures of the interaction are shown. In the case of diffractive imaging, the momentum viewpoint is typically more useful since the scattering is elastic (illustrated by the arrows of the same length on the left). Bottom: Typical data for a UED experiment, including the raw image on the 2D detector (left), and the radial distribution function, $f(R)$, as a function of pump–probe delay.

comparable to or shorter than the typical atom–atom distance in a molecule, the structure of the molecule is reflected in the coherent scattering process (if the deBroglie wavelength were significantly larger than the molecule, the electrons would scatter off the molecule as a point particle with no information about the internal structure). The diffracted electrons are collected on a spatially resolved detector such as a 2D camera.

The energetics of the interaction are shown on the left side of the middle panel in Figure 4.4. As usual, the pump pulse initiates wave function dynamics on an excited state. However, the energy picture for the probe process is not particularly relevant or informative here, as the electrons are elastically scattered from the molecules during the interaction. A description in momentum space is much more useful for depicting this type of interaction. The right side of the middle panel shows a ball-and-stick diagram of a diatomic molecule. An incoming electron beam interacts with the molecules via a diffraction process that involves a *momentum transfer* to the electrons that depends on the coherent addition of scattering from all atomic centers in the molecule. The particulars of the diffraction process depend on the molecular geometry at any given time, resulting in a diffraction pattern that changes with changing molecular structure.

We note that while both momentum and energy are conserved for all of the experiments we discuss, the representation of a particular measurement in an energy or momentum picture is not always relevant. In this case of electron diffraction, an energy diagram doesn't offer much insight into the measurement process, and a ball-and-stick figure or momentum diagram is more useful. On the other hand, in measurements that do not involve phase matching (e.g. LIF), the fact that momentum is conserved does not yield any additional insight into the measurement process, and an energy picture is more relevant.

The bottom panel of Figure 4.4 shows typical data for a UED experiment. The left image depicts the raw data collected on the camera at a particular pump–probe time; the intensity of the signal at any point (R, θ) represents the number of electrons diffracted to that particular location (the undiffracted electrons would arrive at the center of the detector but are typically blocked). For gas-phase samples, the random orientation of the molecules within the ensemble leads to azimuthal symmetry in the signal for an unpumped sample. The ring-like structure contains information about the spatial arrangement and bond lengths of the molecules. The right image in the bottom panel shows the radial distribution function, $f(R)$, for different pump–probe delays. The radial distribution function is a Fourier transform of the processed, angularly averaged intensity pattern measured by the camera, and it has peaks at radial values that correspond to bond lengths present in the sample (see Chapters 9 and 14).

In contrast to the other three approaches that use optical probes, the diffractive nature of UED provides the potential for a more direct measurement of changing molecular structure (a related technique utilizes the diffraction of X-rays instead of electrons as the probe). Diffractive measurements are discussed in detail in Chapter 9, including how one can extract molecular structure from the diffraction patterns.

4.4 CONCLUDING REMARKS

This chapter presented four experiments that highlight prototypical approaches to time-resolved spectroscopy. Additional details about the different techniques are presented in later chapters, where we discuss a variety of experimental examples. As we shall see, some of the most powerful results (from both a physical and pedagogical perspective) come about when comparing multiple techniques in the same molecular system. We also note that Appendix B discusses a number of general experimental considerations that arise in time-resolved spectroscopies, including detection methods and measurements of molecular ensembles. Although some of the factors are more pertinent in certain spectroscopies than others, it is important to have a basic understanding of the issues at play when discussing implementations of the various techniques.

Before concluding this chapter, it is worth commenting on a basic feature of the data shown in all the figures in this chapter. In general, time-resolved spectroscopy measurements cannot distinguish between a lack of motion in a given degree of freedom and delocalization of the wave function along that coordinate. This is due to the fact that time-resolved spectroscopies are typically sensitive to variations in the *first moment* of the wave function along a given coordinate, such as $\langle R(t) \rangle$ in a diatomic molecule. If a vibrational wave packet has sufficient displacement to experience the anharmonicity of the potential, the wave function will spread such that it extends over the entire range of motion after multiple vibrational periods in a bound state potential. This leads to a decay in the amplitude of modulation in the signal, which is our measure of $\langle R(t) \rangle$. In the sense that quantum evolution of the system does not lead to a variation in expectation values, the dynamics have "stopped" for the delocalized state. This actually agrees quite well with our classical intuition, since a delocalized wave function is not "moving" in any classical sense. Time-resolved spectroscopies are sensitive to the expectation values of operators, which by Ehrenfest's theorem is what we expect to agree most closely with classical laws [11].

For many potentials, the wave function can relocalize (rephase) before coupling to other degrees of freedom (see Chapter 10), leading to the phenomenon of wave packet revivals. In the eigenstate picture, where each time-independent state oscillates with its own phase, the revival time is the time it takes for all the eigenstates to (approximately) come back into phase with each other (this is seen in some of the experiments discussed in Chapter 7). During the time when the wave function is delocalized, there is little or no modulation of the signal, and one is not sensitive to the dynamics.

PART II

Quantum Dynamics in One Dimension

CHAPTER 5

Field-Free Dynamics

5.1 INTRODUCTION

In Part I we presented a basic theoretical description of atomic/molecular structure and light–matter interaction. In Part II we move on to explicitly consider *dynamics* in quantum systems. This chapter focuses on how molecular systems evolve in time without any applied field ("field-free dynamics"), while Chapter 6 presents the basic theoretical framework required for field-driven dynamics. We restrict ourselves to dynamics in one dimension in Part II, saving multidimensional dynamics until Part IV. While most systems of experimental interest have multiple degrees of freedom, beginning in one dimension keeps the focus on the fundamental dynamics without unnecessary complications. We will find that the one-dimensional treatment transfers over readily to systems with more degrees of freedom.

As discussed in Chapter 1, our primary quantity of interest in time-domain spectroscopy is $\Psi(\mathbf{r}, t)$, the wave function describing the quantum system. For even the simplest example of the hydrogen atom, $\Psi(\mathbf{r}, t)$ has six spatial degrees of freedom (three each for the proton and electron) as well as two spin degrees of freedom. As we did in Chapter 2, one often begins by considering the full-wave function to be a product of spin and spatial portions, focusing on the spatial part initially while ignoring the spin. One further fixes the proton at the origin, implying that for the hydrogen atom without spin, $\Psi(\mathbf{r}, t)$ describes only the position of the electron. As the complexity of the system grows, the possible degrees of freedom that can be included in $\Psi(\mathbf{r}, t)$ increases. For practical purposes, throughout the book we take the wave function to include only those aspects of the system we are currently interested in examining; in this chapter, that is a one-dimensional, spatial wave function.

By experimentally measuring $\Psi(\mathbf{r}, t)$ at many different times, one can map out how the wave function evolves. In some sense, for experimental time-domain spectroscopy, this is everything there is to know about a quantum system, and our job is done. However, one cannot directly measure the complete wave function, and much of the work in experimental spectroscopy is determining exactly what is being measured and how additional measurements might provide access to different aspects of $\Psi(\mathbf{r}, t)$. This is especially true for multidimensional systems, and the "art" of experimental science is often putting together as complete a picture as possible of the time-dependent wave function from the pieces that one captures in a collection of measurements.

On the theoretical side, the focus is on developing approximations that solve the time-dependent Schrödinger equation (TDSE) with model Hamiltonians that capture the essential physics (e.g. see [11], Chapter 9). Comparison between experiment and theory allows one to test the accuracy of the model Hamiltonian.

The primary goal of this chapter is to develop an intuitive understanding of field-free quantum dynamics in one dimension. We accomplish this by interpreting *numerical* solutions to the TDSE. While it is important to develop analytic approaches for solving the Schrödinger equation, the number of problems that can be solved exactly is very limited. In addition, many analytic solutions rely on a time-independent approach, which is not always the most useful basis for understanding the dynamics. On the other hand, in the absence of an applied field, it is relatively straightforward to solve the TDSE on a computer, and the solutions can be illuminating. Thus, we begin this chapter outlining an approach for solving the TDSE numerically. We make use of the split-operator [12, 13] method since it is intuitive, physical, and simple to implement, but note there are other approaches (e.g. Crank–Nicolson [14]) that can be more efficient or stable.

5.2 SPLIT-OPERATOR APPROACH

We are interested in quantum dynamics without any applied field, and so we begin with the one-dimensional TDSE for a time-independent Hamiltonian:

$$i\frac{\partial \psi(x,t)}{\partial t} = \hat{H}\psi(x,t). \tag{5.1}$$

This equation is separable and can be written as

$$\frac{\partial \psi(x,t)}{\psi(x,t)} = -i\hat{H}\partial t. \tag{5.2}$$

Integrating both sides from $t = 0$ and using the fact that $\hat{H}$ does not depend on time, we have

$$\ln \frac{\psi(x,t)}{\psi(x,0)} = -i\hat{H}(t - 0). \tag{5.3}$$

Finally, exponentiating both sides yields

$$\psi(x,t) = e^{-i\hat{H}t}\psi(x,0), \tag{5.4}$$

where $e^{-i\hat{H}t}$ is the *time-evolution operator* (or "propagator") for the quantum state under the Hamiltonian $\hat{H}$.[1] The time-evolution operator is often called $U(t,0)$, which carries an initial wave function at $t = 0$ to a final wave function at time t:

$$\psi(x,t) = U(t,0)\psi(x,0). \tag{5.5}$$

If one knows the initial wave function and the Hamiltonian, it is possible to calculate the wave function at any other time. However, in most cases, applying the evolution operator for a finite time interval is a nontrivial task. This can be seen by noting that the Hamiltonian is a sum of kinetic and potential terms:

$$\hat{H} = \hat{T} + \hat{V} = \frac{\hat{p}^2}{2m} + V(x), \tag{5.6}$$

---

[1]The exponential of an operator is assumed to be carried out through its Taylor series expansion: $e^{-i\hat{H}t} = 1 + (-i\hat{H}t) + (-i\hat{H}t)^2/2 + (-i\hat{H}t)^3/3! + \ldots$.

where $\hat{p}$ is the momentum operator, which expressed in the position basis is $\hat{p} = -i\hbar\partial/\partial x$. In quantum mechanics, the kinetic and potential operators do not generally commute:

$$[\hat{T}, \hat{V}] \neq 0, \tag{5.7}$$

which implies that $e^{-i\hat{H}t} \neq e^{-i\hat{T}t}e^{-i\hat{V}t}$ (see calculation later). While this factorization is not possible for finite times t, it would greatly aid in calculating the time evolution of ψ. In particular, one could first implement the potential portion of the time-evolution operator in position space (acting on $\psi(x)$), Fourier transform the wave function to momentum space, and then implement the kinetic portion in momentum space (acting on $\psi(p)$). This avoids the complexity of operating with $\hat{p}^2$ in position space, where it is a differential operator.

Since we can perform the calculations using a computer, a straightforward solution is to break up the propagator into a series of small intervals Δt (for N total intervals with $N \cdot \Delta t = t$). Dropping the "hats" for simplicity, we have

$$e^{-iHt} = e^{-iH\Delta t}e^{-iH\Delta t}\ldots e^{-iH\Delta t}. \tag{5.8}$$

Now the implementation of the factorized time-evolution operator can be carried out for each small interval as described earlier, with an error that scales favorably with Δt. The order of the error is calculated by considering the Taylor series expansions for both $e^{-iH\Delta t} = e^{-i(T+V)\Delta t}$ and $e^{-iT\Delta t}e^{-iV\Delta t}$:

$$e^{-i(T+V)\Delta t} = 1 - i(T+V)\Delta t - \frac{1}{2}(T+V)^2\Delta t^2 + \ldots$$

$$= 1 - i(T+V)\Delta t - \frac{1}{2}\left(T^2 + TV + VT + V^2\right)\Delta t^2 + \ldots$$

$$e^{-iT\Delta t}e^{-iV\Delta t} = \left(1 - iT\Delta t - \frac{1}{2}T^2\Delta t^2 + \ldots\right)$$

$$\times \left(1 - iV\Delta t - \frac{1}{2}V^2\Delta t^2 + \ldots\right)$$

$$= 1 - i(T+V)\Delta t - \frac{1}{2}\left(T^2 + V^2\right)\Delta t^2 - TV\Delta t^2 + \ldots$$

To second order the difference between these two is

$$\frac{(2TV - (TV + VT))\Delta t^2}{2} = \frac{(TV - VT)\Delta t^2}{2} = \frac{1}{2}[T,V]\Delta t^2.$$

Since the calculation time is proportional to $1/\Delta t$ but the error proportional to Δt^2, it is possible to find a Δt value for which the error is tolerable, while the time for the calculation is manageable. The recipe for implementing the propagator from $t = 0$ to a final time t is

$$\psi(x,t) = U(t,0)\psi(x,0)$$

$$= \prod_n U(n\Delta t, (n-1)\Delta t)\psi(x,0)$$

$$= \prod_n e^{-iT\Delta t}e^{-iV\Delta t}\psi(x,0), \tag{5.9}$$

where $\prod$ implies the product.

For each term in the product, we carry out the potential part of the propagator in position space, Fourier transform the wave function to momentum space, carry out the kinetic part of the propagator in momentum space, and finally Fourier transform back to position space. This four-step process amounts to

$$\psi'(x,t+\Delta t) = e^{-iV\Delta t}\psi(x,t) \tag{5.10a}$$

$$\psi'(p,t+\Delta t) = \sqrt{\frac{1}{2\pi}}\int_{-\infty}^{\infty}\psi'(x,t+\Delta t)e^{-ipx}dx \tag{5.10b}$$

$$\psi''(p,t+\Delta t) = e^{-iT\Delta t}\psi'(p,t+\Delta t) \tag{5.10c}$$

$$\psi''(x,t+\Delta t) = \sqrt{\frac{1}{2\pi}}\int_{-\infty}^{\infty}\psi''(p,t+\Delta t)e^{+ixp}dp, \tag{5.10d}$$

where $\psi(x,t)$ is the wave function at the beginning of the interval and $\psi''(x,t+\Delta t)$ is the new wave function at time $t+\Delta t$. This sequence is repeated for each Δt, with $\psi''(x,t+\Delta t)$ serving as the new initial wave function for the next step. While one must be careful to ensure that the step size is sufficiently small for accuracy, the approach is inherently unitary and does not lead to a loss of population as perturbative approaches do.

To visualize how each part of the propagator modifies the wave function, we plot it in both the position and momentum representations before and after propagation of a given time step. Figure 5.1 shows snapshots of the wave function in a harmonic potential before and after applying $e^{-iV\Delta t}$ and $e^{-iT\Delta t}$. The top-left panel shows the initial wave function in position space at time t: $\psi(x,t)$. The top-right panel plots the wave function after operating with $e^{-iV\Delta t}$: $\psi'(x,t+\Delta t)$. Application of the potential part of the propagator corresponds to giving the wave function a position-dependent phase advance, with the amount of phase proportional to the position-dependent potential. Similarly, the bottom-left panel shows the initial wave function in momentum space: $\psi(p,t)$. The bottom-right panel shows the wave function after operating with $e^{-iT\Delta t}$: $\psi''(p,t+\Delta t)$. As with the case of the potential portion, application of the kinetic part of the propagator leads to a momentum-dependent phase advance. To make the phase

Figure 5.1   Illustration of the split-operator approach to wave function propagation. Left panels show the initial wave function in both position (top) and momentum (bottom). Right panels show the wave functions after application of the potential and kinetic energy portions of the propagator (the first and third lines of Equation 5.10 correspond to the top and bottom panels, respectively). The gray curves in the right panels correspond to the wave function after application of the potential and kinetic energy portions of the propagator, while the black curves show the original wave functions for comparison.

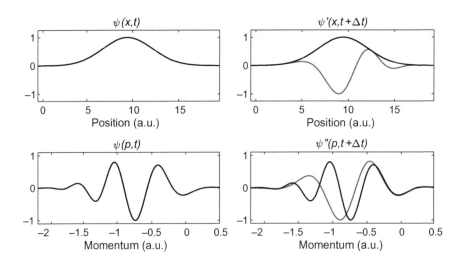

advance of the wave function clear, a much larger time step was chosen than would be used in an actual calculation.

If the phase advance as a function of position and momentum were constant, there would be no change in the expectation value of any operator, since the absolute phase of the wave function does not carry a measurable physical significance. In fact, if the phase advance were simply linear in either domain, it would correspond to a simple shift or displacement in the complementary domain by virtue of the Fourier-shift theorem. The quadratic phase advance in both domains due to the harmonic potential leads to both shifting and spreading of the wave packet in the complementary domains. In the sections that follow, we use the split-operator approach to numerically calculate wave function dynamics in various one-dimensional potentials.

5.3 THE FREE PARTICLE

As a first example, we consider a particle with one spatial degree of freedom where $V(x,t) \equiv 0$ everywhere (the particle is unbound or "free"). Even though the particle may have a spin degree of freedom, we ignore this. While the resulting dynamics are not surprising, it is worth building up our approach starting with a simple system before moving on to more complicated examples. In addition, since the three-dimensional problem is separable into different coordinates, consideration of one dimension easily generalizes to three.

Conceptually, the probability of finding the particle at any given position and time is determined by the modulus squared of the wave function. For concreteness, we consider an initial Gaussian wave function of the form:

$$\Psi(x, t = 0) = A e^{-a(x-x_o)^2}, \tag{5.11}$$

where A is an amplitude determined by normalization, x_o is the center position, and a is related to the width of the wave function: the full-width, half-maximum, or FWHM, is $2\sqrt{\ln 2/a}$.[2] The wave function of Equation 5.11 evolves according to the TDSE with $V(x) = 0$, which in atomic units is

$$i\frac{\partial \Psi}{\partial t} = -\frac{1}{2m}\frac{\partial^2 \Psi}{\partial x^2}, \tag{5.12}$$

where we have left the mass as m for generality (if the particle is a single electron, $m = 1$ in atomic units).

The typical analytic approach is to separate the spatial and temporal dependencies and solve for the time-independent eigenstates; these time-independent states are then combined to form a time-dependent "wave packet." As it turns out, the free particle is a nice starting point for a fully time-dependent treatment, as the time-independent solution to the free particle is somewhat awkward due to the fact that, unlike other textbook examples, the particle cannot actually exist in one of the stationary states. Instead, it must always be in a superposition of them for the wave function to be normalizable.

---

[2]Note that this wave packet could be constructed from a coherent superposition of sine waves (Fourier synthesis), corresponding to the eigenstates of the free particle. The analogous decomposition of a wave packet in terms of bound eigenstates for a harmonic oscillator potential is illustrated in Figure A.2.

The question we wish to answer in the time-dependent approach is, given an initial wave function $\Psi(x, t = 0)$ in the potential $V(x)$, how does the wave function $\Psi(x,t)$ evolve in time? There are three pieces of information needed, all of which we have: (1) the differential equation governing the dynamics (the TDSE), (2) the potential $V(x)$ (in this case, zero), and (3) the initial condition, or wave function, $\Psi(x, t = 0)$. Given this information, we are ready to propagate $\Psi(x, t = 0)$ forward in time.

We make use of the split-operator method described in Section 5.2 to numerically calculate the evolution of $\Psi(x,t)$. The first three panels of Figure 5.2 show snapshots of the wave function at three different times. Each panel shows both the magnitude (blue) and real part (red) of $\Psi(x)$, as well as the potential $V(x)$ (green). The last panel shows the position, $\langle x \rangle$, and width as measured by the standard deviation, $\sigma_x = \sqrt{\langle (x - \langle x \rangle)^2 \rangle}$, of the wave function as a function of time. While the dynamics are obviously simple for this particular case, it is worth highlighting general features.

The wave function has no net translational momentum, which is not surprising, given the initial conditions and the flat potential:

$$\langle x(t) \rangle = \int_{-\infty}^{+\infty} \Psi^*(x,t) x \Psi(x,t) dx = 0 \implies \langle p(t) \rangle = m \frac{d \langle x(t) \rangle}{dt} = 0. \tag{5.13}$$

While the expectation value of the position remains zero, the width of the distribution increases monotonically with time. In other words, the resulting "motion" is simply a spreading of the wave function such that the probability of finding the particle becomes distributed over a larger range, as seen in the increasing width of $|\Psi(x)|$ as time progresses (blue curves in Figure 5.2). This spreading is a purely quantum-mechanical behavior that results from the wave mechanics underlying quantum dynamics.

5.3.1 Momentum in Quantum Mechanics

The free-particle problem connects nicely with the Heisenberg uncertainty principle and the concept of momentum in quantum mechanics. The fact that the initial position of the quantum object is constrained (the initial wave function has a finite extent) implies there must be an uncertainty, or spread, in the object's momentum. This is typically expressed in terms of operator commutators:

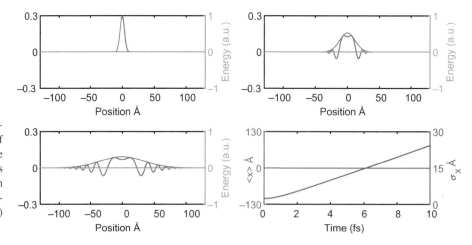

Figure 5.2 First three panels: magnitude (blue) and real part (red) of $\Psi(x)$ for the free particle at three different times. Right axis depicts the potential (green), which is flat in this case. Bottom-right panel: position ($\langle x \rangle$) and standard deviation (σ_x) of the wave function versus time.

$$\sigma_x^2 \sigma_p^2 \geq \left(\frac{1}{2i} \langle [\hat{x}, \hat{p}] \rangle \right)^2 \implies \sigma_x \sigma_p \geq \frac{1}{2}. \quad (5.14)$$

The spread in momentum leads to different "portions" of the wave function moving outward at different rates, such that the wave function covers a larger spatial extent than it did originally. Like the spatial spread, the spread in momentum can also be seen directly in the figure. The red curves plot the real portion of $\Psi(x)$, and the frequency of the oscillations in these curves represents the momentum of the particle: $\hat{p} = -i\frac{\partial}{\partial x}$. As time progresses, the lower momenta portions of the wave function (slower oscillations) stay near the origin, while the higher momenta portions (faster oscillations) spread outward left or right, depending on the sign. As you would expect, the more well known the initial position (the narrower the initial wave function), the faster it spreads in time due to its higher momentum components. This is a direct consequence of wave mechanics and the Fourier relationship between momentum and position (see Section A.1).

Net translational momentum can be explicitly introduced by modifying the initial wave function to contain a spatially varying phase:

$$\Psi(x, t = 0) = Ae^{-a(x-x_o)^2} e^{ikx}, \quad (5.15)$$

where k is a constant with units of one over distance. This modification of Ψ makes the wave function a traveling wave: e^{ikx} is simply a plane wave with wave number k at time $t = 0$. Calculation of the expectation value of x for this wave function yields

$$\langle x \rangle = x_o + \frac{k}{m} t. \quad (5.16)$$

As expected, the position of the wave packet is no longer fixed, and the wave function moves to the right as it spreads due to the spatially varying phase. The quantity k clearly relates to the momentum of the particle, since

$$\langle p \rangle = m \frac{d\langle x \rangle}{dt} = k. \quad (5.17)$$

Given this behavior, one must be careful when discussing a particle's location. In particular, while the peak of the distribution represents the most *likely* location of the particle, it is not necessarily obvious what one means when considering the question of how *long* it takes the particle to pass between two points. After all, the probability distribution can extend out significantly from the center. One way to think about this question is to consider the time Δt it takes the expectation value of the particle's position to move by one standard deviation [15]:

$$\Delta t = \frac{\sigma_x}{\left| \frac{d\langle x \rangle}{dt} \right|}. \quad (5.18)$$

Equation 5.18 can be used to form an uncertainly relation in terms of energy and time. For any operator Q that does not depend explicitly on time, Equation 5.14 can be rewritten as

$$\sigma_H^2 \sigma_Q^2 \geq \left(\frac{1}{2i} \langle [\hat{H}, \hat{Q}] \rangle \right)^2 = \left(\frac{1}{2} \right)^2 \left(\frac{d\langle Q \rangle}{dt} \right)^2, \quad (5.19)$$

where $\hat{H}$ is the Hamiltonian operator. Noting that σ_H is a measure of the uncertainly of the energy ($\sigma_H \equiv \Delta E$), and using Equation 5.18 with $x \rightarrow Q$, we can rewrite Equation 5.19 as

$$\Delta E \Delta t \geq \frac{1}{2}. \tag{5.20}$$

This energy-time uncertainly relation connects nicely with classical optical pulses: a pulse of light, or one whose intensity turns on and off, must be composed of more than one frequency, a fact that can be seen directly from Equation 5.20 (alternatively, it can be shown using the Fourier representation of the field).

5.3.2 Wave Packet Dispersion

We have seen that the wave function of the free particle will spread, or disperse, in time. Formally, the dispersion of a system is determined by the relationship between the angular frequency, ω, of the wave, and its wavenumber, k. For example, consider an electromagnetic plane wave propagating in vacuum according to Maxwell's wave equation for the electric field:

$$\frac{\partial^2 \mathbf{E}}{dx^2} = \frac{1}{c^2} \frac{\partial^2 \mathbf{E}}{dt^2}. \tag{5.21}$$

A plane-wave solution is given by

$$\mathbf{E}(x,t) = \hat{\varepsilon} A e^{i(kx - \omega t)}, \tag{5.22}$$

which upon substitution into the wave equation yields the following constraint on ω and k (assuming positive values for both):

$$\omega^2 = c^2 k^2$$
$$\omega = ck. \tag{5.23}$$

In a nondispersive system such as this, the phase velocity (ω/k) is equal to the group velocity $(d\omega/dk)$, and a pulse of light travels unaffected, maintaining its shape as it propagates. However, in a dispersive system where there is a *nonlinear* relationship between ω and k, the group and phase velocities are unequal, and a pulse will spread or compress as it travels. Whether a medium is dispersive or nondispersive for light depends on the response of the electrons in the material to the light field. Most transparent dielectric materials are moderately dispersive in the visible portion of the spectrum, so only short pulses (with broad frequency bandwidths) are affected.

In quantum mechanics, we need to look at the dispersion generated by the Schrödinger equation instead of Maxwell's wave equation. A basic difference between the two equations is that Maxwell's wave equation is second order in both time and space, whereas the Schrödinger equation is second order in space but only first order in time. Thus, if we consider a plane-wave solution for a flat ($V = 0$) potential in Schrödinger's equation (see Equation 5.12), we find the following dispersion relation:

$$\omega = \frac{k^2}{2m}. \tag{5.24}$$

The fact that the relationship between ω and k is nonlinear implies that a localized wave function will disperse (or spread out) in time for a flat potential, despite the absence of any outside classical forces. Actually, this is generally true for any potential: the wave function will spread due a difference between group and phase velocities. We note there is the special case of the harmonic oscillator, for which a particular wave function can propagate without spreading due to a delicate balance between the natural dispersion and the action of the potential (see Section 5.5.1). We will return to the idea of dispersion after considering the linear potential in the following section.

5.4 LINEAR POTENTIAL

As a natural extension of the free particle, we consider a linear potential with a constant, nonzero slope:

$$V(x) = c(x - x_o), \tag{5.25}$$

where c is a constant representing the slope of the potential and x_o the position where $V(x) = 0$. Our procedure is essentially the same as for the free particle of Section 5.3, except that the TDSE contains a nonzero $V(x)$:

$$i\frac{\partial \Psi}{\partial t} = -\frac{1}{2m}\frac{\partial^2 \Psi}{\partial x^2} + c(x - x_o)\Psi. \tag{5.26}$$

This potential is included in the numerical integration (again assuming a Gaussian initial wave function centered at x_o with no spatial phase), and we arrive at the results shown in Figure 5.3.

Similar to Figure 5.2, the wave function spreads in time. However, the center of the wave packet also moves to the left, as one would expect classically. It is easiest to see these effects by plotting the expectation value of the position, $\langle x \rangle$, and standard deviation, σ_x, as functions of time (lower-right panel). As expected, the probability distribution as a whole gradually moves off to the left as the wave function continues to spread. It can also be seen in Figure 5.3 that the larger momentum components again move out more quickly. In contrast to the free particle, note that components of the wave function with initial positive momentum (present even for the wave packet initially at rest because of its finite initial size) will be continually slowed to lower momentum, eventually turning around due to the potential. As the wave function moves off to the left, the amplitude of the positive momentum components will go to zero. In fact, we see here that it is somewhat risky talking about the advance or retardation of different momentum components relative to one another for an arbitrary potential. This is because a gradient in the potential leads to changes in the momentum distribution; not only are the components advancing or being delayed with respect to one another, but their amplitudes are also changing.

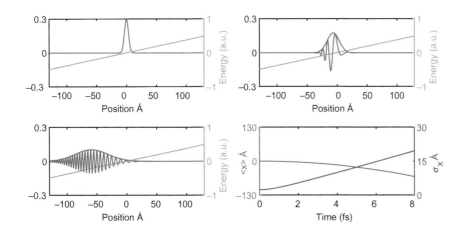

Figure 5.3 First three panels: magnitude (blue) and real part (red) of $\Psi(x)$ for a linear potential at three different times. Right axis depicts the potential (green). Bottom-right panel: position ($\langle x \rangle$) and standard deviation (σ_x) of the wave function versus time.

5.5 BOUND SYSTEMS

In Sections 5.3 and 5.4, we considered a single particle free to move away to large distances in at least one direction. Perhaps the next obvious step up in complexity would be to confine the single particle to a certain region of space. This is the approach in most undergraduate books, presenting a time-independent treatment of the infinite and/or finite potential well. The main result is that the imposition of boundary conditions on the wave function due to the confining potential results in a discrete set of normalizable eigenstates that are sinusoidal between the well edges (and decaying exponentials outside the well in the finite case). While relevant in some spectroscopy experiments, this situation is not essential for most of the time-domain experiments we will consider. We briefly discuss the time-domain, finite-well problem at the end of this chapter (Section 5.5.3), where it is used to demonstrate particular subtleties in the numerical simulations. Instead, we consider here two particles bound together.

Given the important role that vibrational dynamics play in time-resolved spectroscopy, we consider a two-atom, diatomic molecule as our bound system. If one is primarily interested in the relative motion of the two atoms, the problem can be treated in one dimension by defining a reduced mass and using the internuclear separation as the relevant degree of freedom. As a simple example, two hydrogen atoms can form a bond, essentially sharing the two valence electrons between them. This bond establishes a minimum potential energy separation about which the two hydrogen nuclei reside. The motion of the nuclei in this potential is what gives rise to molecular vibrations. Near the minimum of the potential, the curve is nearly harmonic, and like all continuous functions near an equilibrium, it is well approximated by a parabola. We therefore begin our treatment of bound systems with the quantum harmonic oscillator.

5.5.1 The Harmonic Oscillator

We consider the one-dimensional quantum harmonic oscillator with $V(x)$ of the form:

$$V(x) = \frac{1}{2}m\omega^2 x^2, \tag{5.27}$$

where m is the mass of the particle and ω represents the classical angular frequency of oscillation. As a reminder, the time-independent, stationary-state solutions for this potential are calculated in most undergraduate physics textbooks using either an algebraic approach with raising and lowering operators, or through a series solution to the differential equation. The primary result is that the eigenstates can be written in terms of the Hermite polynomials and have evenly spaced energies:

$$E_n = \left(n + \frac{1}{2}\right)\hbar\omega, \quad n = 0, 1, 2, \ldots \tag{5.28}$$

where n is the quantum number of the state.

Instead of following the eigenstate approach, we again choose to directly calculate how a wave function $\Psi(x,t)$ evolves in the potential. We use a Gaussian initial wave function, offsetting the Gaussian from the center of the well by twice its FWHM width. The evolution of the wave function and the corresponding width and position are shown in Figure 5.4. The upper-left panel plots the initial wave function, while the upper-right panel shows the wave function after four periods of oscillation. Note that both the central position and spread of the wave function have returned to their original values.

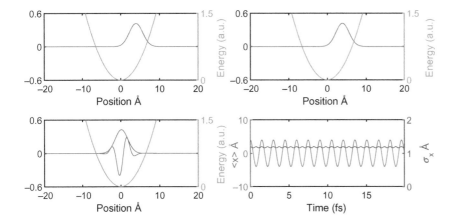

Figure 5.4 First three panels: magnitude (blue) and real part (red) of $\Psi(x)$ for the harmonic oscillator at three different times. Right axis depicts the potential (green), which is a parabola in this case. Bottom-right panel: position ($\langle x \rangle$) and standard deviation (σ_x) of the wave function versus time.

The lower-left panel plots the wave function at the end of the simulation, when it is roughly in the middle of the well. At this location in its period, the wave function is at its widest.

The lower-right panel of Figure 5.4 plots the position and width of $\Psi(x,t)$ as functions of time. While the oscillation in $\langle x \rangle$ is characteristic of any bound potential, note that in contrast to previous examples, the width of the wave function does not continue to increase. Instead, σ_x also oscillates in time (at twice the frequency). For this special case of the harmonic oscillator, the natural dispersion of the wave function in the TDSE is balanced by the parabolic shape of the potential. Conceptually, higher-momentum components in the wave function have to travel farther to the parabolic barrier before the classical turning point, giving the lower-momentum components a chance to catch up on each half-round-trip. The quadratic dependence of the potential on distance means that the extra distance traveled by the higher-momentum components is proportional to the momentum. Since the time for each half round trip is inversely proportional to the momentum, this dependence cancels, preserving the width of the wave function over long times.[3]

5.5.2 Anharmonic Potential

While most physically relevant potentials can be described as harmonic near an equilibrium point, any realistic potential will be anharmonic overall, with the importance of the anharmonicity increasing as the wave function moves farther from equilibrium. We therefore consider wave function dynamics in an anharmonic Morse potential, which serves as a good approximation for many molecular bonds.[4] As we shall see, the resulting motion combines the periodic oscillations seen in the harmonic oscillator with the spreading seen in the free particle.

The Morse potential is given as

$$V(R) = D_e \left(1 - e^{a(R-R_e)} \right)^2, \tag{5.29}$$

---

[3]In fact, if the initial wave function corresponds to the ground eigenstate simply displaced from the minimum, the width stays exactly the same with no oscillations (this is known as a "coherent state").

[4]The general features of the potential are similar to that of the two-body gravitational problem in classical mechanics, with an angular momentum barrier at low separation and an asymptotic decay out at large distances.

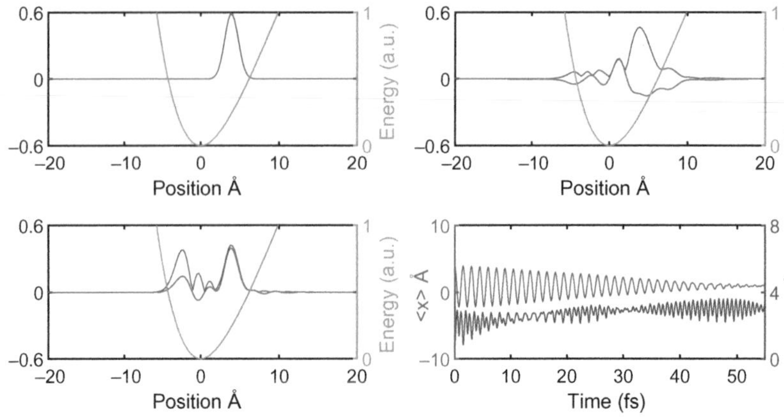

Figure 5.5 First three panels: magnitude (blue) and real part (red) of $\Psi(x)$ at three different times. Right axis depicts $V(x)$ (green), which is a Morse potential in this case. Bottom-right panel: position ($\langle x \rangle$) and standard deviation (σ_x) of the wave function versus time.

where D_e is the well depth, R_e is the equilibrium position, and a is a parameter controlling the width of the well. Figure 5.5 shows snapshots of a wave function evolving in a Morse potential. The top-left panel shows the initial wave packet, displaced from zero with no initial momentum. The top-right panel shows the wave packet four periods later, where it is again localized near the outer turning point of the potential. Note that in contrast to the harmonic oscillator, the wave packet no longer retains its exact shape; due to the anharmonicity of the potential, it has started to disperse. The bottom-left panel shows the wave packet at the end of the simulation when it is substantially delocalized.

The bottom-right panel of Figure 5.5 plots the position and width of the wave packet as a function of time. As in the case of the harmonic oscillator, the position of the wave packet oscillates back and forth with a period determined largely by the harmonic term in the potential. However, these oscillations decay with time as the wave packet spreads due to the anharmonicity of the potential. In addition, the perfect cancelation of wave packet spreading and rephasing we saw in the harmonic potential does not occur here, and the width increases over time. Since the wave packet is composed of a finite number of eigenstates, it can undergo a revival as the constituent eigenstates of the wave packet come back into phase. Similar to the way in which a Fourier series composed of a finite sum of sine waves repeats itself, the wave packet will relocalize and again display oscillations in the position. For the wave packet shown in Figure 5.5, a revival occurs at approximately 110 fs (the wave packet at $t = 55$ fs is delocalized such that the position essentially does not oscillate). In the next section, we mathematically connect the time-dependent wave packet to its constituent eigenstates.

5.5.2.1 Spectrum of the Wave Packet

It is possible to determine the eigenvalues and eigenfunctions for any one-dimensional potential directly from numerical solutions to the time-*dependent* Schrödinger equation. After solving the TDSE for the time-dependent wave function, $\Psi(x,t)$, one can construct the wave packet autocorrelation function defined as

$$C(t) \equiv \langle \Psi(x,0) | \Psi(x,t) \rangle. \tag{5.30}$$

We next take the Fourier transform of the autocorrelation function, writing both the initial and time-dependent wave functions as coherent superpositions of eigenstates of the potential:[5]

$$\mathscr{F}\{C(t)\} = \sqrt{\frac{1}{2\pi}} \int_{-\infty}^{\infty} \langle \Psi(x,0) \,|\, \Psi(x,t) \rangle \, e^{i\omega t} \, dt$$
$$= \sqrt{\frac{1}{2\pi}} \int_{-\infty}^{\infty} \int_{-\infty}^{\infty} \sum_{n,m} c_n^* \psi_n^*(x) c_m \psi_m(x) e^{-i\omega_m t} e^{i\omega t} \, dx \, dt, \tag{5.31}$$

where $\omega_m = E_m$ in atomic units and in the second line we have written the inner product as an integral over all space. Next, we make use of the fact that the eigenfunctions $\psi_n(x)$ are orthonormal, using the spatial integral to eliminate one of the sums:

$$\mathscr{F}\{C(t)\} = \sqrt{\frac{1}{2\pi}} \int_{-\infty}^{\infty} \sum_{n,m} c_n^* c_m \delta_{nm} e^{-i\omega_m t} e^{i\omega t} \, dt$$
$$= \sqrt{\frac{1}{2\pi}} \sum_{n} |c_n|^2 \int_{-\infty}^{\infty} e^{-i\omega_n t} e^{i\omega t} \, dt$$
$$= \sum_{n} |c_n|^2 \delta(\omega - \omega_n), \tag{5.32}$$

where ω_n is the frequency of the nth eigenstate.

This demonstrates that the Fourier transform of the wave packet autocorrelation function corresponds to the eigenspectrum (spectrum of eigenstates) of the wave packet. Using this approach, Figure 5.6 shows the calculated spectrum of the wave packet from Figure 5.5. Of course, the spectrum generated in this manner contains only those states required to construct the initial wave function, and depending on the displacement, shape, and width of the initial wave function, eigenstates of interest may not appear in its spectrum. Nevertheless, the spectrum can be calculated including states of interest by moving the initial wave function in the potential.

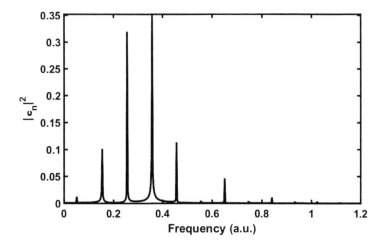

Figure 5.6 Spectrum of the Morse-potential wave packet from Figure 5.5.

---

[5] In writing the wave functions this way, we use the fact that the eigenstates are assumed to form a complete basis. An arbitrary wave function can be expanded in any complete basis, and the eigenstate basis is chosen for convenience.

Once one has calculated the eigenspectrum, it is straightforward to calculate the wave function for any eigenstate represented in the spectrum. The basic idea is to find the Fourier component of the space and time-dependent wave function that oscillates at the frequency corresponding to the eigenenergy of interest. We label this component $f_n(x)$:

$$
\begin{aligned}
f_n(x) &\equiv \int_{-\infty}^{\infty} \Psi(x,t) e^{iE_n t} dt \\
&= \int_{-\infty}^{\infty} e^{-i\hat{H}t} \Psi(x,0) e^{iE_n t} dt \\
&= \int_{-\infty}^{\infty} e^{-i\hat{H}t} \left(\sum_m c_m \psi_m(x) \right) e^{iE_n t} dt \\
&= \int_{-\infty}^{\infty} \sum_m e^{-iE_m t} c_m \psi_m(x) e^{iE_n t} dt \\
&= \sum_m c_m \psi_m(x) \delta(E_n - E_m) \sqrt{2\pi} \\
&= \sqrt{2\pi} c_n \psi_n(x),
\end{aligned}
\tag{5.33}
$$

where we have used that fact that the Hamiltonian propagator acting on the mth eigenstate returns $e^{-iE_m t}$. Thus, we see that the Fourier component of the space and time-dependent wave function $\Psi(x,t)$ oscillating at frequency E_n is proportional to the eigenstate $\psi_n(x)$. Therefore, knowledge of $\Psi(x,t)$ also allows one to calculate $\psi_n(x)$ for any n by numerically taking the Fourier component at frequency E_n (see Problem 4 at the end of this chapter).

5.5.2.2 Comparison to Classical Dynamics

We saw that an initially localized wave function in the anharmonic oscillator will delocalize over time. In the eigenstate picture, this corresponds to a dephasing of the constituent states in the wave packet. An interesting question arises when considering quantum dynamics of the anharmonic oscillator: are the dephasing dynamics a result of uniquely quantum evolution under the TDSE, or can they be captured by classical calculations which sample different initial conditions on the anharmonic potential?

To address this question, we compare classical and quantum calculations of the dynamics in the same anharmonic potential (i.e. under the influence of the same Hamiltonian). Figure 5.7 shows the expectation value of the position operator as a function of time, $\langle x(t) \rangle$, for both a quantum wave packet (left panel) and an ensemble of classical particles (right panel). Note that there are many rapid oscillations underneath the envelopes in both panels which are not resolved because of the finite line thickness used in the plots. The quantum calculation solves the TDSE as before, using a Gaussian wave packet displaced from equilibrium as the initial wave function. The classical calculation solves Hamilton's equations for an ensemble of particle trajectories, and then calculates $\langle x(t) \rangle$ averaging over the ensemble. The particles in the ensemble are given initial positions and momenta with probabilities that correspond to the amplitude of a quantum-mechanical, phase-space distribution (e.g. a Husimi distribution [11]).

The calculations show quantitative agreement for the behavior of $\langle x(t) \rangle$ during early times as the quantum wave packet begins to oscillate and then delocalize. This

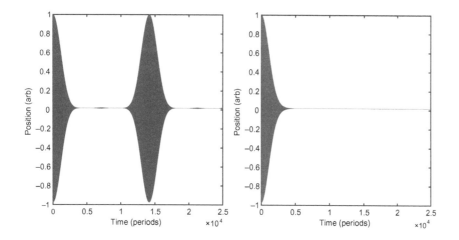

Figure 5.7 Expectation value $\langle x(t) \rangle$ for an anharmonic oscillator calculated using the quantum TDSE (left panel) and classical Hamilton's equations using an ensemble of initial conditions (right panel). Note the time scale is such that the fast oscillations under the envelopes are not resolved.

observation illustrates the fact that delocalization, or spreading of the quantum wave function, is not a unique feature of the TDSE; it is also reflected in the broad spread of final positions determined by the classical equations of motion starting from a distribution of initial conditions. Delocalization is a result of the anharmonic potential, independent of whether the evolution is classical or quantum.

However, while the quantum wave packet rephases and undergoes a revival of localized probability density, the classical ensemble does not over the same time scale. Although the quantum wave function in the potential well can be described in terms of a discrete sum of eigenstates, the ensemble of classical particles in this potential cannot. The smooth (essentially continuous) sum of initial conditions for the classical ensemble washes out any revival structure, and the wave packet revivals are a purely quantum feature of the dynamics.[6] Indeed, if one does not sample the initial conditions for the classical ensemble sufficiently densely, one sees "artificial" revivals in the classical case that occur at times related to the sampling of the initial conditions (as opposed to the parameters of the potential).

These observations motivate the possibility of using classical calculations to model quantum molecular dynamics for cases where the quantum calculations may be prohibitively expensive in computation time (e.g. some of the multidimensional dynamics discussed in Part IV). Since the revivals seen in the one-dimensional case do not generally occur in large-dimensional systems, the discrepancies between the quantum and classical cases due to revivals are typically unimportant. The two primary effects that cannot be captured by classical calculations are interference and nonadiabatic coupling to other potential energy surfaces (introduced in Section 2.5.2). Interference effects can generally be neglected because different portions of an initially localized wave packet will rarely come to the same point at the same time in an N-dimensional space. Calculations that aim to capture the effects of nonadiabatic couplings between potentials combine solving Hamilton's equations classically with "surface hopping" calculations that take the quantum nonadiabatic coupling into account separately [6].

---

[6]We note that the defining feature is whether the energy distribution is discrete or continuous. The quantum energy distribution in a bound potential well is discrete, while the classical energy distribution is continuous (as it is in most cases).

5.5.3 Finite Well

We finish this chapter with a brief discussion of wave packet dynamics in a finite, square well (given the discontinuities in the gradient of the potential at the boundaries, the square well is in some sense the ultimate anharmonic potential). Similar to earlier examples in this chapter, we consider an initial Gaussian wave function localized near the center of the well. As opposed to the case of the harmonic or Morse oscillator, we now start the wave function with nonzero initial momentum (since the gradient of the potential at the starting point is zero, in the absence of initial momentum the wave function would simply spread until the edges reach the walls of the potential). The upper-left panel of Figure 5.8 shows the initial wave packet centered in the well and moving to the right with a positive momentum (the momentum is indicated by the wiggles in the wave function). The upper-right panel shows the wave function the first time it encounters the left barrier (three-quarters of a period later). At the hard edge, the wave function is compressed as it turns around. The bottom-left panel shows the wave function at the final time, where the wave function is once again centered in the well but now significantly dispersed. The wave packet oscillations, as well as the overall spreading with compression at the turning points, can be seen in the plots of $\langle x \rangle$ and $\langle x^2 \rangle$ in the lower-right panel.

Qualitatively, the dynamics are similar to the other bound systems we treated. The primary difference comes at the edges. Unlike the other cases, oscillations in the wave function do not decrease in frequency as the wave packet approaches the barrier (since the potential is essentially unchanged right until the edge). One has to be careful with a numerical treatment of the propagation because of the infinite slope at the edges of the potential. There is a maximum slope in the potential $\left(\frac{dE}{dx}_{\max} \right)$ that can occur given a fixed momentum grid and time step in the numerical integration. If the wave packet encounters a region with a slope greater than $\frac{dE}{dx}_{\max}$, it will acquire momentum components larger than the maximum momentum allowable ($p_{\max}$). This problem can be mitigated by sampling the potential more densely or using smaller time steps, but both of these solutions are computationally expensive. For a given amount of computing resources, there is a maximum slope in the potential that can be treated numerically. In other words, one can never treat the finite-well exactly, since the potential must change over a finite distance.

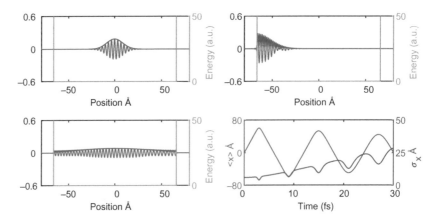

Figure 5.8 First three panels: magnitude (blue) and real part (red) of $\Psi(x)$ at three different times. Right axis depicts $V(x)$ (green), which is a finite-square-well potential in this case. Bottom-right panel: position ($\langle x \rangle$) and standard deviation (σ_x) of the wave function versus time.

5.6 HOMEWORK PROBLEMS

1. Numerically calculate the evolution of a free-particle wave packet as in Section 5.3. Aim to reproduce the results shown in Figure 5.2 using a mass of one atomic unit.

2. Consider the linear potential discussed in Section 5.4. Numerically, calculate the evolution of a wave packet that has an initial net translational momentum directed uphill. Be sure that the magnitudes of the initial momentum and the linear slope are such that the wave packet does not run off the spatial grid on the way uphill.

3. Similarly, numerically calculate the evolution of a wave packet in a harmonic potential as in Section 5.5.1. Aim to reproduce the results shown in Figure 5.4 using a mass of one atomic unit.

4. Making use of wave packet propagation, determine the first five eigenenergies and eigenstates for the Morse potential: $V(R) = D_e(1 - e^{-a(R-Re)})^2$, with $D_e = 1.5$ eV, $a = 1$ Å^{-1}, and $R_e = 2$ Å. Hint: You will need to propagate a wave packet whose projection onto the first five eigenstates is significant. Start with a Gaussian wave packet centered near the minimum of the potential and gradually increase the displacement until the spectrum of the wave packet contains the first five eigenvalues.

5. Consider the Morse potential described in Problem 4. Calculate the evolution of a wave packet with an energy just slightly above the dissociation energy. Start with a Gaussian wavepacket (or displaced ground eigenstate of the Morse) and start the evolution of the wave packet at the inner turning point. Make use of the impedance-matched imaginary potential approach developed by Kosloff and Kosloff (*Journal of Computational Physics* **63**(2):363–376, 1986) to avoid reflections of your wave packet from the edge of your grid as it moves toward dissociation.

6. Consider the potential given by

$$V(R) = \frac{1}{20}R^2 + \frac{0.03(R + 10)^3}{1 + e^{-0.5(R+40)}} - \frac{10}{1 + e^{0.5(R+40)}}. \qquad (5.34)$$

This could correspond to a "predissociative" potential, which does not support truly bound states since the barrier is finite. Such predissociative potentials can arise from the coupling between two diabatic potential energy surfaces (one bound and one dissociative).

  (a) Calculate the evolution of a wave packet on this potential for an initial Gaussian wave packet with FWHM = 2.36 a.u. and $\langle R(t = 0) \rangle = 7$ a.u. Use an x-axis extending from -700 a.u. to 20 a.u., a mass of 1, and propagate your wave packet for a total time of 80 a.u. Make a movie of the evolution from 0 to 80 a.u.

  (b) Calculate $\langle R(t) \rangle$ and the probability of the molecule remaining bound (i.e. within the well) as a function of time.

  (c) Describe the motion of the wave packet. Discuss the dispersion of the wave packet and tunneling through the barrier.

  (d) Why is the first reflection from the barrier different from subsequent ones?

CHAPTER 6

Field-Driven Dynamics

6.1 INTRODUCTION

In Chapter 5, we examined the evolution of quantum systems that takes place independent of any external field. Since time-resolved spectroscopy experiments involve applied electromagnetic fields, developing our understanding of field-driven dynamics is clearly important as well. This chapter builds up the mathematical structure necessary to describe the experimental results shown in Part III. We do this in stages, assuming that both the Born–Oppenheimer (BO) (Section 2.5.1) and dipole (Section 3.2.1) approximations are valid. Our ultimate goal is two-fold: to develop analytic models describing the dynamics, as well as the tools needed for numerical simulations.

The analytic models, applicable in the perturbative regime, provide a clear interpretation of measurements. One can typically identify a contributing term in an expression and relate it to the physical picture. On the other hand, the numerical calculations (valid in both the perturbative and strong-field regimes) offer the ability to visualize wave function dynamics in "real-time" on a computer. The combination of these two is helpful for developing an understanding of the physics underlying the dynamics probed in time-resolved spectroscopy experiments.

Developing a mathematical framework will allow us to calculate our basic quantity of interest in time-resolved spectroscopy: the time-dependent wave function on multiple electronic states of a molecule. From this result, we can construct measurement observables such as ion or fluorescence yields, or macroscopic polarizations and absorptions. We will find that interpreting experimental results is nontrivial, and typically requires a mathematical structure that parallels the physical apparatus used to carry out the experiment. This is best accomplished by presenting the associated mathematical models for the different measurements along with the experimental data and its interpretation. These models provide a framework for understanding the measurements, and it is hard to fully appreciate one without the other. We therefore use this chapter to build up the basic mathematical background of field-driven dynamics, saving the description of particular observables until the corresponding experiments in later chapters.

6.2 MOLECULAR WAVE FUNCTION EVOLUTION

In Section 2.5.1, the BO approximation led to a separation of the molecular wave function into electronic and nuclear components. The full wave function, $\Psi_n(\mathbf{r}_i, \mathbf{R}_j)$, for a vibrational wave packet on the nth electronic state can be written as

$$\Psi_n(\mathbf{r}_i, \mathbf{R}_j) = \psi_n(\mathbf{r}_i; \mathbf{R}_j)\chi_n(\mathbf{R}_j), \tag{6.1}$$

where $\psi_n(\mathbf{r}_i; \mathbf{R}_j)$ represents the electronic wave function that depends explicitly on the electronic coordinates $\mathbf{r}_i$ and parametrically on the nuclear coordinates $\mathbf{R}_j$, while $\chi_n(\mathbf{R}_j)$ represents the nuclear wave function and depends only on the nuclear coordinates. The subscript n indicates the specific electronic PES on which the nuclear dynamics take place (e.g. the B state of I_2).

In the BO approximation, $\psi_n(\mathbf{r}_i; \mathbf{R}_j)$ is an eigenstate of the electronic Hamiltonian: $\hat{H}^e = \hat{T}_e + \hat{V}_{ee} + \hat{V}_{eN} + \hat{V}_{NN}$ (see Equation 2.79). In general, $\chi_n(\mathbf{R}_j)$ is not a vibrational eigenstate, but instead represents an arbitrary vibrational wave function on the nth electronic PES. Therefore, unlike the stationary state $\psi_n(\mathbf{r}_i; \mathbf{R}_j)$, $\chi_n(\mathbf{R}_j)$ evolves in time even when no light fields are present. With the time dependence made explicit, the state becomes

$$\Psi_n(\mathbf{r}_i, \mathbf{R}_j, t) = \psi_n(\mathbf{r}_i; \mathbf{R}_j)\chi_n(\mathbf{R}_j, t). \tag{6.2}$$

The most general state of the system is a superposition of different $\Psi_n(\mathbf{r}_i, \mathbf{R}_j, t)$

$$\Psi(\mathbf{r}_i, \mathbf{R}_j, t) = \sum_n \psi_n(\mathbf{r}_i; \mathbf{R}_j)\chi_n(\mathbf{R}_j, t). \tag{6.3}$$

The total wave function $\Psi(\mathbf{r}_i, \mathbf{R}_j, t)$, as well as each electronic eigenstate $\psi_n(\mathbf{r}_i; \mathbf{R}_j)$, is normalized. However, note that the amplitude coefficient for each electronic eigenstate $\psi_n(\mathbf{r}_i; \mathbf{R}_j)$ is contained in $\chi_n(\mathbf{R}_j, t)$, the full vibrational wave function on the nth electronic state. If desired, $\chi_n(\mathbf{R}_j, t)$ could be expressed in terms of a superposition of vibrational eigenstates, with each eigenstate having an amplitude coefficient associated with it:

$$\chi_n(\mathbf{R}_j, t) = \sum_m c_m(t)e^{-i\omega_m t}\chi_{nm}(\mathbf{R}_j), \tag{6.4}$$

where c_m is the coefficient and $\chi_{nm}(\mathbf{R}_j)$ is the mth vibrational eigenstate on the nth electronic state. The sum here would naturally be replaced by an integral for an unbound potential. While $\chi_n(\mathbf{R}_j, t)$ is not normalized, each of the m vibrational eigenstates is:

$$\int |\chi_{nm}(\mathbf{R}_j)|^2 d\mathbf{R}_j = 1. \tag{6.5}$$

6.2.1 Evolution with No Applied Field

Although the basic idea was presented in Chapter 2, it is worth repeating here using our notation for the full, molecular wave function. In the absence of an applied field, the time-dependent Schrödinger equation (TDSE) for this state reads:

$$i\frac{\partial}{\partial t}\Psi(\mathbf{r}_i, \mathbf{R}_j, t) = \hat{H}\Psi(\mathbf{r}_i, \mathbf{R}_j, t), \tag{6.6}$$

where the Hamiltonian $\hat{H}$ is separated into electronic and nuclear components:

$$\hat{H} = \hat{H}^e + \hat{T}_N = \left(\hat{T}_e + \hat{V}_{ee} + \hat{V}_{eN} + \hat{V}_{NN}\right) + \hat{T}_N. \tag{6.7}$$

In terms of $\hat{H}^e$ and $\hat{T}_N$, the TDSE for the general state becomes

$$i\frac{\partial}{\partial t}\left(\sum_n \psi_n(\mathbf{r}_i;\mathbf{R}_j)\chi_n(\mathbf{R}_j,t)\right) = (\hat{H}^e + \hat{T}_N)\sum_n \psi_n(\mathbf{r}_i;\mathbf{R}_j)\chi_n(\mathbf{R}_j,t). \qquad (6.8)$$

Since $\psi_n(\mathbf{r}_i;\mathbf{R}_j)$ is assumed to be an eigenstate of the electronic Hamiltonian, we can write $\hat{H}^e \psi_n(\mathbf{r}_i;\mathbf{R}_j) = E_n^e(\mathbf{R}_j)\psi_n(\mathbf{r}_i;\mathbf{R}_j)$ on the right-hand side, where $E_n^e(\mathbf{R}_j)$ is the $\mathbf{R}_j$-dependent eigenenergy of the nth electronic state. $\hat{H}^e$ and $\hat{T}_N$ only operate on ψ and χ, respectively, and we can write

$$i\sum_n \psi_n(\mathbf{r}_i;\mathbf{R}_j)\frac{\partial}{\partial t}\chi_n(\mathbf{R}_j,t)$$
$$= \sum_n \left(E_n^e(\mathbf{R}_j)\psi_n(\mathbf{r}_i;\mathbf{R}_j)\chi_n(\mathbf{R}_j,t) + \psi_n(\mathbf{r}_i;\mathbf{R}_j)\hat{T}_N\chi_n(\mathbf{R}_j,t)\right). \qquad (6.9)$$

Since the factor $\psi_n(\mathbf{r}_i;\mathbf{R}_j)$ appears in each term, this equation can be simplified by exploiting the orthonormality of the BO electronic eigenstates. Specifically, multiplying on the left by $\psi_m^*(\mathbf{r}_i;\mathbf{R}_j)$ and integrating over all space (in the electronic coordinates) picks out the $n = m$ term from the sum: $\int \psi_m^*(\mathbf{r}_i;\mathbf{R}_j)\psi_n(\mathbf{r}_i;\mathbf{R}_j)d\mathbf{r}_i = \delta_{mn}$. This results in the following expression:

$$i\frac{\partial}{\partial t}\chi_m(\mathbf{R}_j,t) = \left(\hat{T}_N + E_m^e(\mathbf{R}_j)\right)\chi_m(\mathbf{R}_j,t). \qquad (6.10)$$

Note that the electronic states are no longer present and that only the mth vibrational wave function appears. As expected, in the absence of an externally applied field or nonadiabatic coupling (i.e. in the BO approximation), population stays in the initial electronic eigenstate. Equation 6.10 shows that the vibrational wave function evolves in time according to an effective Hamiltonian $\hat{H}_N^{\text{eff}} \equiv \hat{T}_N + E_m^e(\mathbf{R}_j)$, with an effective potential energy that includes both the nuclear repulsion and attraction between the electrons and nuclei. As in Section 5.2, Equation 6.10 is separable and can be written as

$$\frac{\partial \chi_m(\mathbf{R}_j,t)}{\chi_m(\mathbf{R}_j,t)} = -i\hat{H}_N^{\text{eff}}\partial t \qquad (6.11)$$

Integrating both sides from $t = 0$ and using the fact that $\hat{H}_N^{\text{eff}}$ does not depend on time, we have

$$\ln\frac{\chi_m(\mathbf{R}_j,t)}{\chi_m(\mathbf{R}_j,0)} = -i\hat{H}_N^{\text{eff}}(t-0). \qquad (6.12)$$

Finally, exponentiating both sides yields

$$\chi_m(\mathbf{R}_j,t) = e^{-i\hat{H}_N^{\text{eff}}t}\chi_m(\mathbf{R}_j,0), \qquad (6.13)$$

where $e^{-i\hat{H}_N^{\text{eff}}t}$ is the "time propagator" for the nuclear wave function under $\hat{H}_N^{\text{eff}}$. Note that the $E_m^e(\mathbf{R}_j)$ term in $\hat{H}_N^{\text{eff}}$ contains both the variation of the electronic state energy with $\mathbf{R}_j$, as well as the energy offset above the ground state (the difference in the minimum energy between an excited electronic state and ground state). This vertical offset energy leads to an overall phase advance of the vibrational wave function on the higher PES when compared with the lower. It is this phase advance one aims to match with a resonant applied field coupling two electronic states.

6.2.2 Evolution with an Applied Field

In order to model a time-resolved experiment, we next consider the TDSE for the general state $\Psi(\mathbf{r}_i, \mathbf{R}_j, t)$ in the presence of a time-dependent, externally applied electric field such as an optical laser pulse. We are primarily interested in the interaction in the dipole approximation (see Section 3.2.1), in which case the Hamiltonian picks up a new term of the form:

$$\hat{H} = \hat{H}^e + \hat{T}_N + \hat{H}_{\text{int}} = \hat{H}^e + \hat{T}_N - \mu E(t), \tag{6.14}$$

where $\mu = -\sum_k \mathbf{r}_k \cdot \hat{\varepsilon}$ is electronic component of the transition dipole moment (containing contributions from all electrons), and $E(t)$ is the time-dependent electric field of the laser of the form $E(t) = E_0(t)\frac{1}{2}\left(e^{+i\omega_0 t} + e^{-i\omega_0 t}\right)$, where $E_0(t)$ is the slowly varying envelope and ω_o the angular frequency. For simplicity, we have incorporated the polarization vector of the field ($\hat{\varepsilon}$) into the dipole moment so that it is a scalar quantity. The addition of the field affects only the right side of the TDSE (Equation 6.6), and as before, we use the orthonormality of the electronic eigenstates to rewrite Equation 6.10 with the interaction term as

$$i\frac{\partial}{\partial t}\chi_m(\mathbf{R}_j, t) = E_m^e(\mathbf{R}_j)\chi_m(\mathbf{R}_j, t) + \hat{T}_N\chi_m(\mathbf{R}_j, t)$$
$$- \sum_n \mu_{mn}E(t)\chi_n(\mathbf{R}_j, t), \tag{6.15}$$

where

$$\mu_{mn} \equiv \int \psi_m^*(\mathbf{r}_i; \mathbf{R}_j)\left(-\sum_k \mathbf{r}_k \cdot \hat{\varepsilon}\right)\psi_n(\mathbf{r}_i; \mathbf{R}_j)d\mathbf{r}_i, \tag{6.16}$$

is the transition dipole moment connecting electronic states m and n.[1]

Equation 6.15 describes the time dependence of each $\chi_m(\mathbf{R}_j, t)$ in terms of its own evolution on the mth electronic state (first two terms), as well as changes due to field-induced coupling with population on other electronic states (last term). In general, the sum is over all other possible states. If only one state is near resonance with the frequency of the applied field, it typically suffices to ignore coupling to all the other states. For concreteness, we consider the specific case of only two electronic states ($m, n = a, b$) coupled by a quasimonochromatic laser field. We also consider only one degree of vibrational freedom so that $\mathbf{R}_j \to R$ and define $V_m(R) \equiv E_m^e(R)$ to emphasize the fact that $E_m^e(R)$ serves as the effective potential for the nuclei. With these substitutions, the coupled equations for the two vibrational wave functions (Equation 6.15) become

$$i\frac{\partial}{\partial t}\begin{pmatrix} \chi_a(R, t) \\ \chi_b(R, t) \end{pmatrix} = \begin{pmatrix} \hat{T}_N + V_a(R) & -\mu_{ab}E(t) \\ -\mu_{ba}E(t) & \hat{T}_N + V_b(R) \end{pmatrix}\begin{pmatrix} \chi_a(R, t) \\ \chi_b(R, t) \end{pmatrix}. \tag{6.17}$$

Equation 6.17 analytically describes the evolution of a molecule along one vibrational degree of freedom with an applied field $E(t)$ resonantly coupling two electronic states. Like in Section 3.2.3, we can make a rotating-wave approximation (RWA) that removes the rapid time dependence from the Hamiltonian brought in by the oscillating field $E(t)$ coupling the two states. As we shall see in Section 6.3, this has the added benefit of favorably changing the required time step for numerical implementation.

---

[1] For homonuclear diatomic molecules, $\mu_{mm} = 0$ due to the symmetry of the electronic eigenstates.

In the molecular case where there is nontrivial evolution of the vibrational wave function independent of the applied field, the RWA becomes a three-step process. We first perform a transformation where the potential energy contributions to the diagonal elements of the Hamiltonian become phases in the off-diagonal terms. Then, just as in Section 3.2.3, we can carry out the RWA by dropping terms with phases that vary rapidly (compared with the more slowly varying phases). Finally, we transform back to a frame that has only slow time variation in the off-diagonal terms (they depend on the envelope of the field, not the carrier).

The initial transformation is defined by

$$\chi_a'(R,t) \equiv \chi_a(R,t)e^{-i(\omega_b-\omega_a)t/2} \tag{6.18a}$$

$$\chi_b'(R,t) \equiv \chi_b(R,t)e^{+i(\omega_b-\omega_a)t/2}, \tag{6.18b}$$

where we have used the shorthand of ω_a to represent the R-dependent potential $V_a(R)$ (similarly for ω_b). The quantity $\omega_b - \omega_a = V_b(R) - V_a(R)$ is the R-dependent energy difference between the two PESs. With this transformation, the left side of Equation 6.17 involves a product rule for the derivative. Keeping the χ_i' term on the left side and multiplying both sides by $e^{\mp i(\omega_b-\omega_a)t/2}$ yields Equation 6.19, where we have suppressed the (R,t) dependence in the wave functions for compactness:

$$i\frac{\partial \chi_a'}{\partial t} = \left(\hat{T}_N + V_a(R) + \frac{(\omega_b - \omega_a)}{2} \right) \chi_a'$$
$$- \mu_{ab}E_0(t)\chi_b' \left(\frac{e^{+i\omega_0 t} + e^{-i\omega_0 t}}{2} \right) e^{-i(\omega_b-\omega_a)t} \tag{6.19a}$$

$$i\frac{\partial \chi_b'}{\partial t} = \left(\hat{T}_N + V_b(R) - \frac{(\omega_b - \omega_a)}{2} \right) \chi_b'$$
$$- \mu_{ba}E_0(t)\chi_a' \left(\frac{e^{+i\omega_0 t} + e^{-i\omega_0 t}}{2} \right) e^{+i(\omega_b-\omega_a)t}. \tag{6.19b}$$

We simplify these expressions by noting that since $\omega_{a/b}$ is simply a shorthand for $V_{a/b}(R)$, the terms $V_a(R) + (\omega_b - \omega_a)/2$ and $V_b(R) - (\omega_b - \omega_a)/2$ combine to yield $(\omega_a + \omega_b)/2$. In addition, the field-dependent terms in the expressions multiply the trailing exponential to produce factors of the form:

$$e^{\pm i(\omega_0 - (\omega_b - \omega_a))t} \qquad \text{and} \qquad e^{\mp i(\omega_0 + (\omega_b - \omega_a))t}. \tag{6.20}$$

The first one of these oscillates at the frequency $\Delta \equiv \omega_0 - (\omega_b - \omega_a)$. For a near-resonance laser field, $\omega_0 \sim (\omega_b - \omega_a)$, and the phase in the exponent evolves slowly. On the other hand, the second of these factors oscillates at approximately $2\omega_0$ for a near-resonant field. Making the RWA involves dropping the term with the rapidly oscillating phase, since it will integrate to a much smaller value than the other term. This allows us to rewrite the TDSE in the RWA as

$$i\frac{\partial \chi_a'}{\partial t} = \left(\hat{T}_N + \frac{(\omega_a + \omega_b)}{2} \right) \chi_a' - \frac{\mu_{ab}E_0(t)}{2} e^{+i\Delta t} \chi_b' \tag{6.21a}$$

$$i\frac{\partial \chi_b'}{\partial t} = \left(\hat{T}_N + \frac{(\omega_a + \omega_b)}{2} \right) \chi_b' - \frac{\mu_{ba}E_0(t)}{2} e^{-i\Delta t} \chi_a'. \tag{6.21b}$$

We now transform back to a frame where the phase in the off-diagonal terms becomes a diagonal entry. Note that since the phase in the off-diagonal terms is now Δ, this transformation does not simply undo the first one, and it leaves the wave function in a (partially) rotating frame. The second transformation is defined by

$$\tilde{\chi}_a(R,t) \equiv \chi_a'(R,t)e^{-i\Delta t/2} \tag{6.22a}$$

$$\tilde{\chi}_b(R,t) \equiv \chi_b'(R,t)e^{+i\Delta t/2}. \tag{6.22b}$$

We again carry out this transformation, moving the additional term from the product rule to the right side and multiplying both sides by $e^{\mp i\Delta t/2}$. With these steps, the TDSE becomes

$$i\frac{\partial \tilde{\chi}_a}{\partial t} = \left(\hat{T}_N + \frac{\omega_a + \omega_b + \Delta}{2}\right)\tilde{\chi}_a - \frac{\mu_{ab}E_0(t)}{2}\tilde{\chi}_b \tag{6.23a}$$

$$i\frac{\partial \tilde{\chi}_b}{\partial t} = \left(\hat{T}_N + \frac{\omega_a + \omega_b - \Delta}{2}\right)\tilde{\chi}_b - \frac{\mu_{ba}E_0(t)}{2}\tilde{\chi}_a. \tag{6.23b}$$

The terms in parentheses can be simplified by noting that $\frac{\omega_a + \omega_b + \Delta}{2} = \omega_a + \frac{\omega_0}{2}$ and $\frac{\omega_a + \omega_b - \Delta}{2} = \omega_b - \frac{\omega_0}{2}$. This means that in the current form of the TDSE, the potential energies are "brought into resonance" by the photon at the midpoint between the two states. We choose to shift this (unimportant) offset so that the upper state is "brought down" to the lower state instead. In our partially rotating frame, this amounts to subtracting off $\omega_0/2$ from each of the diagonal terms. With this shift and dropping the shorthand of $\omega_a = V_a(R)$ and $\omega_b = V_b(R)$, we arrive at the final expression for the two-state TDSE in matrix form:

$$i\frac{\partial}{\partial t}\begin{pmatrix}\tilde{\chi}_a(R,t)\\\tilde{\chi}_b(R,t)\end{pmatrix} = \begin{pmatrix}\hat{T}_N + V_a(R) & -\mu_{ab}E_0(t)/2\\-\mu_{ba}E_0(t)/2 & \hat{T}_N + V_b(R) - \omega_0\end{pmatrix}\begin{pmatrix}\tilde{\chi}_a(R,t)\\\tilde{\chi}_b(R,t)\end{pmatrix}. \tag{6.24}$$

The Hamiltonian in this version of the TDSE is quasi-time-independent, insofar as the only time dependence shows up in the electric field *envelope*, which varies slowly with respect to the carrier frequency of the field (and therefore the energy difference between the states).

6.2.3 An Illustrative Example

Here we present a simple example demonstrating the subtleties of discussing coherence with a superposition of two electronic states (perhaps generated by a laser field). A short laser pulse couples the ground electronic state of a diatomic molecule to an excited electronic state. The pulse transfers a portion of the ground-state wave function to the excited state, creating a vibrational wave packet in the upper PES. At this point, the total wave function is technically no longer separable into a product of vibrational and electronic wave functions, but is rather a sum of products (an entangled state). Specifically, the wave function is a coherent superposition of two electronic states, each with its own vibrational wave packet.[2]

In the case of an isolated atom, the electronic coherence between ground and excited states induced by the laser pulse lasts as long as the spontaneous emission lifetime (typically on the order of nanoseconds). In the molecular case, however, the electronic

---

[2]In the perturbative limit, one assumes negligible change to the ground-state population, and hence the ground-state wave function is typically ignored.

coherence will generally decay rapidly as a result of vibrations: given that the rate at which the phase between two states evolves is determined by their energy separation, the fact that in general two different potential energy surfaces are not parallel leads to a phase evolution between the two electronic states that varies with internuclear separation. Thus, the electronic coherence between ground and excited states tends to "wash out" when averaging over the vibrational coordinate. Although it is common to describe the coherence as being "lost," it is not an actual loss of coherence, but really a dephasing.

6.3 NUMERICAL IMPLEMENTATION

Before considering a perturbative approach to the problem, we address how one can implement time-evolution numerically. This builds directly on Chapter 5, where we numerically propagated the nuclear (vibrational) wave function on a single PES with no field present. The RWA that resulted in Equation 6.24 produced an expression that varies in time with the envelope of the field. This implies that the time steps in a numerical routine only need to be small compared with changes in the envelope (not the carrier), resulting in a code that runs much more quickly.

6.3.1 Single Applied Field

We begin by considering a small time step Δt over which the Hamiltonian is roughly constant. With $\hat{H}$ assumed constant, we can rewrite Equation 6.24 using the usual time-evolution operator $e^{-i\hat{H}\Delta t}$. The result is the wave function at time $t + \Delta t$ in terms of the wave function at time t:

$$
\begin{pmatrix} \tilde{\chi}_a(R,t+\Delta t) \\ \tilde{\chi}_b(R,t+\Delta t) \end{pmatrix} =
$$
$$
\exp \begin{pmatrix} -i(\hat{T}_N + V_a(R))\Delta t & -i(-\mu_{ab}E_0(t))\Delta t/2 \\ -i(-\mu_{ba}E_0(t))\Delta t/2 & -i(\hat{T}_N + V_b(R) - \omega_0)\Delta t \end{pmatrix} \begin{pmatrix} \tilde{\chi}_a(R,t) \\ \tilde{\chi}_b(R,t) \end{pmatrix}. \quad (6.25)
$$

The basic idea is illustrated in Figure 6.1. In general, at time t, there is initial wave function amplitude on both the ground ($\chi_a(R,t)$) and excited ($\chi_b(R,t)$) PESs. Each of these two wave functions evolves into new functions at time $t + \Delta t$ through a combination of two processes: "field-independent" evolution on their respective PESs (through the diagonal $\hat{T}_N$ and $V_i(R)$ terms in Equation 6.25), as well as the field-induced coupling (through the off-diagonal $\mu_{ab}E_0(t)$ terms in Equation 6.25). The field-independent evolution on PES $V_i(R)$ is the same propagation we saw in Chapter 5, while the field-induced coupling is mediated by the near-resonant field represented by the ω_0 arrow. The combination results in the updated wave functions on each state: $\chi_a(R,t + \Delta t)$ and $\chi_b(R,t + \Delta t)$. This process can be continued for many time steps until the desired final time.

The problem with proceeding as we did in Section 5.2 is that the matrix associated with the potential energy portion of the Hamiltonian is no longer diagonal due to the field coupling terms. The time-evolution operator would be much easier to implement if we could rewrite the potential energy operator as a combination of diagonal matrices (that are exponentiated) and nondiagonal matrices (that do not need to be exponentiated). This can be accomplished for a 2×2 matrix with a little bit of algebra. For small time

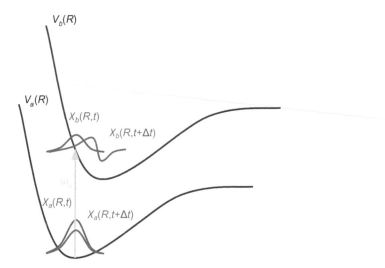

Figure 6.1 (See color insert.) Illustration of the various components of Equation 6.25. $\chi_i(R,t)$ is the initial vibrational wave function on state i, while $\chi_i(R,t + \Delta t)$ is the new wave function after one time step. The ω_0 arrow represents the the field-induced coupling.

steps Δt, we can break up the time-evolution operator according to the usual split-operator approach:

$$e^{-i\hat{H}\Delta t} \approx e^{-i\hat{T}\Delta t} e^{-i\hat{V}\Delta t} = \exp\left(-i \begin{pmatrix} \hat{T}_N & 0 \\ 0 & \hat{T}_N \end{pmatrix} \Delta t\right)$$

$$\times \exp\left(-i \begin{pmatrix} V_a(R) & -\mu_{ab}E_0(t)/2 \\ -\mu_{ba}E_0(t)/2 & V_b(R) - \omega_0 \end{pmatrix} \Delta t\right). \quad (6.26)$$

Now, making use of the results in Section A.6.2, we rewrite the exponentiated, off-diagonal potential matrix in terms of matrix products that can be implemented more readily:

$$\exp\left(-i \begin{pmatrix} V_a(R) & -\mu_{ab}E_0(t)/2 \\ -\mu_{ba}E_0(t)/2 & V_b(R) - \omega_0 \end{pmatrix} \Delta t\right)$$

$$= e^{-i(V_a(R)+V_b(R)-\omega_0)\Delta t/2} \left(\cos\left(\sqrt{D}\Delta t\right) \begin{pmatrix} 1 & 0 \\ 0 & 1 \end{pmatrix}\right. \quad (6.27)$$

$$\left. + \frac{i}{\sqrt{D}} \sin\left(\sqrt{D}\Delta t\right) \begin{pmatrix} V_b(R) - V_a(R) - \omega_0 & \mu_{ab}E_0(t) \\ \mu_{ba}E_0(t) & V_a(R) - V_b(R) + \omega_0 \end{pmatrix}\right),$$

where

$$D = (\mu_{ab}E_0(t))^2 + (V_a(R) - V_b(R) + \omega_0)^2. \quad (6.28)$$

Using this form of the potential matrix with the split-operator approach allows us to calculate the evolution of vibrational wave packets on two electronic states of the molecule coupled by an electric field.

Figure 6.2 shows results in diatomic iodine following this numerical treatment. The molecules are initially prepared in the ground vibrational state of the ground electronic PES (the X state). A pump pulse couples the X and B states of the molecule, launching a vibrational wave packet on the B state. The top panel shows the time dependence of the electric field envelope (gray, right axis) and the population of the B state (black, left axis). The middle panel shows the long-time behavior of the B-state wave packet.

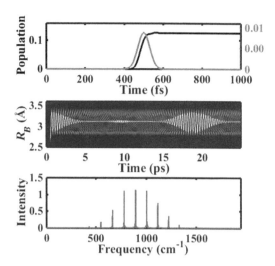

Figure 6.2 (See color insert.) Solution of the TDSE for I_2 interacting with an ultrafast laser pulse resonant between the X and B states of the molecule. The top panel shows the electric field envelope (gray, right axis), along with the population of the B state (black, left axis), versus time. The middle panel shows the time- and space-dependent probability density (gray shading) and expectation value (white curve) on the B state. Note that the time axes for the top and middle panels are different. Finally, the bottom panel shows the spectrum of the B-state wave packet.

The shading represents the probability density, while the white curve plots the expectation value (the first moment). One can clearly see oscillations of the wave packet, dephasing and rephasing of the wave packet, with a revival of the oscillation amplitude of $\langle R(t) \rangle$ near 18 ps. Finally, the bottom panel plots the Fourier transform of the wave packet autocorrelation function, which corresponds to the spectrum of the B-state wave packet. The fact that there are multiple peaks in the spectrum indicates there are multiple vibrational eigenstates in the wave packet. The homework problems at the end of the chapter provide an additional opportunity for numerical implementation.

6.3.2 Multiple Applied Fields

The description of a full pump–probe experiment typically includes a total of three electronic states coupled by two different fields. For the three-state system, there is not an established straightforward analytic expression corresponding to Equation 6.27. Since the diagonal elements of the potential matrix vary with R, while the off-diagonal, field-coupling elements vary with time, we must consider the explicit diagonalization of a 3×3 matrix for every time t and position R. The three-state generalization of the time-evolution operator in Equation 6.25 for a small time step Δt is given by

$$
\begin{pmatrix} \tilde{\chi}_a(R, t+\Delta t) \\ \tilde{\chi}_b(R, t+\Delta t) \\ \tilde{\chi}_c(R, t+\Delta t) \end{pmatrix} =
$$

$$
\exp \begin{pmatrix} -i(\hat{T}_N + V_a(R))\Delta t & -i(-\mu_{ab}E_{01}(t))\Delta t/2 & 0 \\ -i(-\mu_{ba}E_{01}(t))\Delta t/2 & -i(\hat{T}_N + V_b(R) - \omega_{01})\Delta t & -i(-\mu_{bc}E_{02}(t))\Delta t/2 \\ 0 & -i(-\mu_{cb}E_{02}(t))\Delta t/2 & -i(\hat{T}_N + V_c(R) - \omega_{02})\Delta t \end{pmatrix}
$$

$$
\times \begin{pmatrix} \tilde{\chi}_a(R, t) \\ \tilde{\chi}_b(R, t) \\ \tilde{\chi}_c(R, t) \end{pmatrix}, \quad (6.29)
$$

where $E_{01}(t)$ and $E_{02}(t)$ represent the electric field envelopes for the pump and probe pulses, respectively, while ω_{01} and ω_{02} represent the respective carrier frequencies. We have assumed that the pump resonantly couples states a and b, while the probe couples states b and c.

Numerical implementation of Equation 6.29 via the split-operator approach requires factoring the potential and kinetic parts of the time-evolution operator and further diagonalizing the potential matrix. As noted earlier, this diagonalization must be performed for each position on the spatial grid and each time step, because the eigenvalues depend on both the diagonal and off-diagonal terms that vary with space and time. Thus, for each time step, one carries out

$$
\begin{pmatrix} \tilde{\chi}_a(R,t+\Delta t) \\ \tilde{\chi}_b(R,t+\Delta t) \\ \tilde{\chi}_c(R,t+\Delta t) \end{pmatrix} =
$$

$$
\exp \begin{pmatrix} -i\hat{T}_N\Delta t & 0 & 0 \\ 0 & -i\hat{T}_N\Delta t & 0 \\ 0 & 0 & -i\hat{T}_N\Delta t \end{pmatrix}
$$

$$
\times U \exp \begin{pmatrix} -i\lambda_1\Delta t & 0 & 0 \\ 0 & -i\lambda_2\Delta t & 0 \\ 0 & 0 & -i\lambda_3\Delta t \end{pmatrix} U^\dagger
$$

$$
\times \begin{pmatrix} \tilde{\chi}_a(R,t) \\ \tilde{\chi}_b(R,t) \\ \tilde{\chi}_c(R,t) \end{pmatrix} \quad (6.30)
$$

where U is the unitary matrix that diagonalizes the potential matrix, and λ_i is the ith eigenvalue of the diagonalized matrix: $U^\dagger V U = D$, where D is a diagonal matrix with elements λ_i. Reading right to left, $U^\dagger$ transforms the state vector to a basis where the potential matrix can be written in a diagonal form (and its exponential is easily evaluated). Then, U transforms the state vector back to the original basis, where the kinetic portion of the propagator is implemented (typically in momentum space after a Fourier transform of the state vector).

Equation 6.30 can be directly implemented numerically, with U being calculated from the nondiagonal potential matrix that includes the field-induced coupling between states. The process is time consuming compared with the two-state equivalent, due to the fact that one needs to numerically diagonalize the potential matrix at each position and time separately. An additional homework problem in Appendix C makes use of this form of the TDSE.

6.4 PERTURBATIVE DESCRIPTION

As a complement to the numerical approach, it is useful to consider perturbative, analytic solutions describing pump–probe spectroscopy. While these solutions don't preserve the norm of the wave function (and as such are only valid in the limit of negligible ground-state depletion using weak-field pulses), they are nevertheless helpful in providing an analytic framework behind experimental observables. The results also track how the various measurements depend on the strengths and characteristics of the pump and probe fields. We start by deriving first- and second-order perturbative expressions for the wave function after the pump and probe interactions. We then construct the nonlinear polarizations that must be employed to calculate coherent emission of light from a molecular ensemble. In Part III, the resulting expressions will be used to mathematically describe measured quantities in specific implementations of pump–probe spectroscopy. One can think of the general procedure as following a "recipe":

1. Mathematically construct the wave function generated by the pump pulse using time-dependent perturbation theory. In most cases, this will require only first- or second-order perturbation theory (and therefore be first- or second-order in the pump field).

2. Construct a measurement observable after interaction of the probe pulse with the time-dependent wave function generated by the pump. When the probe pulse is resonant and transfers a portion of the wave function from the intermediate to final state, this step is facilitated by calculating the wave function after the probe pulse. When the probe pulse is nonresonant and there is no significant population transfer (e.g. a "parametric process"), it is clearer to calculate the measurement observable directly as part of the interaction with the probe pulse.

As a concrete example, we consider a diatomic molecule with three BO electronic PESs: a, b, and c (in ascending order of energy). States a and b are resonantly coupled by the pump field $E_1(t)$ centered at time t_{pump}, while states b and c are resonantly coupled by the probe field $E_2(t)$ centered at time t_{probe}. We make use of our results from time-dependent perturbation theory in Section A.3. Assuming the initial wave function is entirely in the ground state a, Equation A.34 provides the first-order wave function on the excited b state after interaction with the pump field (we have changed the dummy integration variable from t' to t_1):

$$\chi_b^{(1)}(R,t) = \frac{1}{i} \int_0^t dt_1 e^{-iH_b(t-t_1)} \left(-\mu_{ba}E_1(t_1)\right) e^{-iH_a t_1} \chi_a(R,0). \quad (6.31)$$

Incorporating the subsequent interaction with the probe pulse requires going to second order in perturbation theory. The final wave function on state c then comes from our second-order perturbation theory expression (see Equation A.25), with similar substitutions for H_0 and H_1 in terms of the appropriate matrices. The final result yields

$$\chi_c^{(2)}(R,t) = \frac{1}{i^2} \int_0^t dt_2 \int_0^{t_2} dt_1 e^{-iH_c(t-t_2)} \left(-\mu_{cb}E_2(t_2)\right)$$
$$\times e^{-iH_b(t_2-t_1)} \left(-\mu_{ba}E_1(t_1)\right) e^{-iH_a t_1} \chi_a(R,0). \quad (6.32)$$

Reading right to left, Equation 6.32 says that the initial vibrational wave function $\chi_a(R,0)$ evolves in time on PES a through the (nuclear) time-evolution operator $e^{-iH_a t_1}$ from $t_1 = 0$ to $t_1 = t_{pump}$; if $\chi_a(R,0)$ is a vibrational eigenstate, the dynamics are simply a phase evolution. The pump pulse is localized in time around $t_1 = t_{pump}$ and transfers a portion of the wave function to PES b via a dipole transition. The promoted vibrational wave function then evolves on PES b for a time $\tau = t_{probe} - t_{pump}$, at which point the probe pulse arrives and transfers a portion of this wave function to PES c through a second dipole transition. Finally, this wave function evolves on PES c until the final time t.

It is interesting to consider the limit of very-short duration pump and probe pulses (i.e. delta functions). In some sense this realizes an experiment with perfect time resolution, where the molecular dynamics are not convolved with the finite pulse duration. However, there is some subtlety in doing this, because taking the true delta-function limit implies an infinite spectral bandwidth for the pulses. This violates the RWA and the assumption the field is resonant between potentials over a narrow range of R values. Thus, we note that the ideal pump–probe measurement uses pulses whose durations are much shorter than the vibrational dynamics being measured, but significantly longer than an optical period so that one can define the resonance condition over a narrow range of locations on the PES where the transition between the two electronic states occurs.

With this understanding of the delta-function limit, we write Equation 6.32 using $E_1(t) = E_{01}\delta(t)$ and $E_2(t) = E_{02}\delta(t - \tau)$, where E_{01} and E_{02} are constants describing the strength of the pump and probe fields, respectively:[3]

$$\chi_c^{(2)}(R,t) = \frac{1}{i^2} e^{-iH_c(t-\tau)} \mu_{cb} E_{02} e^{-iH_b \tau} \mu_{ba} E_{01} \chi_a(R,0). \tag{6.33}$$

This equation has a very simple interpretation. A portion of the ground-state wave function (proportional to the pump pulse amplitude, E_{01}) is transferred from state a to state b at $t = 0$, where it evolves for a time τ between pump and probe pulses. A portion of this wave function (proportional to the probe pulse amplitude, E_{02}) is then transferred by the probe to state c and evolves until the final time t.

6.4.1 Incoherent Measurements: Population

Incoherent experiments such as laser-induced fluorescence and ion yield spectroscopy measure the probability of being in the final state c (see Section 4.2). In this case, the measured signal is simply proportional to the total population on state c:

$$\text{Signal} \propto \left\langle \chi_c^{(2)} \middle| \chi_c^{(2)} \right\rangle, \tag{6.34}$$

where the inner product involves a spatial integral over the coordinate R. We can see immediately from Equations 6.32 and 6.33 that the experimental signal is proportional to the integral of the product of the pump and probe fields squared, which corresponds to the product of the pump and probe pulse fluences. This results in a total product of four fields that go into the measured signal. Further development of this result is saved for Chapter 7, where it is presented side by side with the experimental data.

6.4.2 Coherent Measurements: Polarization

For experiments measuring coherent emission of light from a molecular ensemble (e.g. transient absorption), it is useful to calculate the microscopic polarization of the molecules after interaction with both pump and probe pulses. This can then be used to determine the macroscopic polarization and resulting field that propagates according to Maxwell's equations in conjunction with the TDSE. As we saw in Chapter 3, it is the transition dipole that couples a molecule to the radiation field (in agreement with our classical intuition that an oscillating charge radiates). Thus, to calculate the radiation emitted from an ensemble of molecules, one must begin by calculating the transition dipole moment connecting the two states of interest.

---

[3]We note there is also a subtlety with units when using the delta function; the time integral of the product of the dipole moment and the electric field must remain unitless. In addition, we note that the delta function describing the time-of-arrival for the pump pulse and the start of the integral both occur at $t = 0$; if desired, one could think about the integral starting before $t = 0$.

6.4.2.1 Microscopic Polarization

We start by considering the general form of the microscopic polarization for a molecule in a given state Ψ:

$$P(t) \equiv \left\langle \Psi(\mathbf{r}_i, \mathbf{R}_j, t) \,\middle|\, \mu \,\middle|\, \Psi(\mathbf{r}_i, \mathbf{R}_j, t) \right\rangle$$

$$= \left\langle \sum_m \psi_m(\mathbf{r}_i; \mathbf{R}_j) \chi_m(\mathbf{R}_j, t) \,\middle|\, \mu \,\middle|\, \sum_n \psi_n(\mathbf{r}_i; \mathbf{R}_j) \chi_n(\mathbf{R}_j, t) \right\rangle, \tag{6.35}$$

where $\mu = -\sum_k \mathbf{r}_k \cdot \hat{\varepsilon}$ and the braket notation implies integration over all spatial coordinates $\mathbf{r}_i$ and $\mathbf{R}_j$. For a molecule in a coherent superposition of only *two* electronic states, the wave function becomes $\Psi(\mathbf{r}_i, \mathbf{R}_j, t) = \psi_m(\mathbf{r}_i; \mathbf{R}_j) \chi_m(\mathbf{R}_j, t) + \psi_n(\mathbf{r}_i; \mathbf{R}_j) \chi_n(\mathbf{R}_j, t)$, where m and n are now two specific states of the system. In this case we can write the polarization as

$$P(t) = \left\langle \psi_m(\mathbf{r}_i; \mathbf{R}_j) \chi_m(\mathbf{R}_j, t) \,\middle|\, \mu \,\middle|\, \psi_m(\mathbf{r}_i; \mathbf{R}_j) \chi_m(\mathbf{R}_j, t) \right\rangle$$

$$+ \left\langle \psi_n(\mathbf{r}_i; \mathbf{R}_j) \chi_n(\mathbf{R}_j, t) \,\middle|\, \mu \,\middle|\, \psi_n(\mathbf{r}_i; \mathbf{R}_j) \chi_n(\mathbf{R}_j, t) \right\rangle \tag{6.36}$$

$$+ \left\langle \psi_m(\mathbf{r}_i; \mathbf{R}_j) \chi_m(\mathbf{R}_j, t) \,\middle|\, \mu \,\middle|\, \psi_n(\mathbf{r}_i; \mathbf{R}_j) \chi_n(\mathbf{R}_j, t) \right\rangle + c.c.,$$

where $c.c.$ denotes the complex conjugate. Note that μ is independent of nuclear coordinates and comes out of the nuclear integral, implying that the permanent dipole moment contributions from individual states do not oscillate in time and can be ignored. In addition, we simplify the notation for the electronic transition dipole moment between the two states by noting that $\left\langle \psi_m(\mathbf{r}_i; \mathbf{R}_j) \,\middle|\, \mu \,\middle|\, \psi_n(\mathbf{r}_i; \mathbf{R}_j) \right\rangle \equiv \mu_{mn}(\mathbf{R}_j)$. Concentrating on only the first of the two remaining terms, we denote its contribution with subscripts to indicate the two states driving the polarization:

$$P_{mn}(t) = \left\langle \chi_m(\mathbf{R}_j, t) \,\middle|\, \mu_{mn}(\mathbf{R}_j) \,\middle|\, \chi_n(\mathbf{R}_j, t) \right\rangle. \tag{6.37}$$

We will frequently make the Condon approximation, where the R-dependence of the transition dipole moment can be ignored: $\mu_{mn}(\mathbf{R}_j) \to \mu_{mn}$. Note that in this case, μ_{mn} could be pulled out of the remaining integral over R if desired, and the electronic and nuclear contributions to the polarization factorize: $P_{mn}(t) = \mu_{mn} \left\langle \chi_m(\mathbf{R}_j, t) \,\middle|\, \chi_n(\mathbf{R}_j, t) \right\rangle.$[4]

Motivated by this discussion, we begin by calculating the polarization for the case of a diatomic molecule exposed to a short pump pulse $E_1(t)$ resonant between the initial (typically ground) electronic state a and an excited state b. Inserting the perturbative expression for the first-order wave function on state b (see Equation 6.31) into

---

[4]The usual notation retains the μ_{mn} in the integral, and we do so here. In addition to being applicable when the Condon approximation fails, this form of the expression also makes it clearer that the quantity is a polarization.

Equation 6.37, we arrive at a "first-order" polarization between states a and b after interaction with a single pulse:

$$
\begin{aligned}
P_{ab}^{(1)} &= \left\langle \chi_a^{(0)}(R,t) \left| \mu_{ab} \right| \chi_b^{(1)}(R,t) \right\rangle \\
&= \frac{1}{i} \left\langle \chi_a^{(0)}(R,t) \left| \mu_{ab} \int_0^t dt_1 e^{-iH_b(t-t_1)} \left(-\mu_{ba} E_1(t_1)\right) e^{-iH_a t_1} \right| \chi_a^{(0)}(R,0) \right\rangle \\
&= \frac{-1}{i} \left\langle \chi_a^{(0)}(R,0) \left| e^{+iH_a t} \mu_{ab} \int_0^t dt_1 e^{-iH_b(t-t_1)} \mu_{ba} E_1(t_1) e^{-iH_a t_1} \right| \chi_a^{(0)}(R,0) \right\rangle,
\end{aligned}
\tag{6.38}
$$

where in the last line we have included an $e^{+iH_a t}$ so that the wave function on the left is written in terms of χ_a at the initial time (same as on the right).

We next consider the polarization generated by a probe pulse $E_2(t)$ resonant between states b and c, assuming that the pump pulse has already generated a nonstationary, first-order perturbative wave packet on state b (this sequence is directly relevant to many of the spectroscopic approaches we discuss). Since this polarization is typically first-order *in the probe*, we denote it as $P_{\text{probe}}^{(1)}$, and it is assumed to connect states b and c for which the probe is near-resonant. To remind us of the order of the polarization on the total applied field, we choose to keep the order of the wave functions in terms of the total field (both pump and probe). For simplicity in this expression, we take $t_{\text{pump}} = 0$. In this notation, we have

$$
\begin{aligned}
P_{\text{probe}}^{(1)} &= \left\langle \chi_b^{(1)}(R,t) \left| \mu_{bc} \right| \chi_c^{(2)}(R,t) \right\rangle \\
&= \frac{1}{i} \left\langle \chi_b^{(1)}(R,t) \left| \mu_{bc} \int_0^t dt_2 e^{-iH_c(t-t_2)} \left(-\mu_{cb} E_2(t_2)\right) e^{-iH_b t_2} \right| \chi_b^{(1)}(R,0) \right\rangle \\
&= \frac{-1}{i} \left\langle \chi_b^{(1)}(R,t) \left| \mu_{bc} \int_0^t dt_2 e^{-iH_c(t-t_2)} \mu_{cb} E_2(t_2) e^{-iH_b(t_2-t)} \right| \chi_b^{(1)}(R,t) \right\rangle,
\end{aligned}
\tag{6.39}
$$

where in the last line we have inserted the propagator $e^{+iH_b t}$ to write the wave function on the right-hand side at the final time t (same as on the left).[5] Equation 6.39 illustrates how the time-dependent wave function generated by the pump pulse, $\chi_b^{(1)}(R,t)$, modifies the polarization generated by the probe pulse in the sample. This leads to changes in the probe pulse fluence and spectrum as a function of pump–probe delay, and will prove essential for interpreting the coherent nonlinear optical spectroscopies in Chapter 8.

As noted, we have written the polarization in Equation 6.39 as first order, since it is first order in the probe field. However, the full dependence on the total applied field is actually third order (assuming that the wave function on χ_b is first order in the pump field). One can see this explicitly by writing out $\chi_b^{(1)}(R,t)$ as generated by the pump pulse. Note that since χ_b is first-order in perturbation theory (Equation 6.31) and χ_c second order (Equation 6.32), the polarization P_{bc} is therefore *third-order*. Explicitly, the third-order polarization $P_{bc}^{(3)}$ is given by

$$
P_{bc}^{(3)}(t) = P_{\text{probe}}^{(1)} = \left\langle \chi_b^{(1)}(R,t) \left| \mu_{bc} \right| \chi_c^{(2)}(R,t) \right\rangle.
\tag{6.40}
$$

---

[5] We recognize that by taking $t_{\text{pump}} = 0$, there is the difficulty that the integrals in Equation 6.39 only start after the pump pulse has already turned on. This difficulty can be overcome by taking all of the integrals to start slightly before zero - i.e. taking a lower limit of integration to be a few pump-pulse durations less than zero.

Substituting in from above, we have

$$
\begin{aligned}
P_{bc}^{(3)}(t) = \frac{-1^3}{i^3} \Big\langle \chi_a^{(0)}(R,0) \Big| &\int_0^t dt_2 \int_0^{t_2} dt_1 \int_0^t dt_3 \, e^{iH_a t_3} \mu_{ab} E_1(t_3) \\
&\times e^{iH_b(t-t_3)} |\mu_{bc}| e^{-iH_c(t-t_2)} \mu_{cb} E_2(t_2) e^{-iH_b(t_2-t_1)} \\
&\times \mu_{ba} E_1(t_1) e^{-iH_a t_1} \Big| \chi_a^{(0)}(R,0) \Big\rangle .
\end{aligned}
\tag{6.41}
$$

Here t_3 is simply an additional dummy variable required for the integration, as we are combining both a first- and second-order solution. Since the variables t_1 and t_3 both correspond to the time of the pump pulse (they are both for E_1), the functional forms of $E_1(t_1)$ and $E_1(t_3)$ will be identical. The expression in Equation 6.41 describes a process known as "excited-state absorption" and is described in detail in Section 8.2. A similar polarization arising from the coherence between states a and b is known as "ground-state bleach."

6.4.2.2 Macroscopic Polarization

The macroscopic polarization in the medium is related to the microscopic polarization earlier simply by the number of molecules per unit volume, ρ:

$$
P^{\text{macro}}(t) = \rho \left(P_{nm}(t) + c.c. \right),
\tag{6.42}
$$

where P_{nm} and its complex conjugate are added since the macroscopic polarization is a real quantity. As derived in Section A.8.2, the macroscopic polarization in the medium leads to changes in the probe field at the same frequency. From Equation A.77, propagation of the probe pulse envelope $E_0^{\text{probe}}(z,t)$ is given as

$$
\frac{\partial}{\partial z} E_0^{\text{probe}}(z,t) = \frac{2\pi \omega_2 i}{c} P_0^{\text{macro}}(z,t),
\tag{6.43}
$$

where ω_2 is the frequency of the probe field. If we work under the assumption of a low-density, or "optically thin" sample, it is reasonable to assume that the pump and probe pulses do not change significantly with propagation through the sample (changes are less than a few percent and typically much less than 1%). In fact, this is the regime in which one generally wants to carry out experiments, as it allows one to consider all molecules in the ensemble as being prepared and interrogated in the same way. Otherwise, one must take into account the fact that molecules at different locations in the sample see different pump or probe fields. In the limit of small change in the pump and probe fields with propagation, we can integrate Equation 6.43, ignoring the z-dependence of the macroscopic polarization:

$$
\begin{aligned}
E_0^{\text{probe}}(z,t) - E_0^{\text{probe}}(z=0,t) &= \frac{2\pi \omega_2 i}{c} \int_0^z dz' P_0^{\text{macro}}(t) \\
&= \frac{2\pi \omega_2 i}{c} P_0^{\text{macro}}(t) z.
\end{aligned}
\tag{6.44}
$$

We clearly see that the change in the probe field is linear with sample length, density, and polarization from Equation 6.41. Thus, experiments measuring changes in a probe pulse across a thin sample can directly determine the microscopic polarization derived earlier (and provide access to the time-dependent wave functions). Having developed the basic machinery for interpreting time-resolved measurements, in Part III we will turn to discussing specific experimental implementations. First, however, we address an aspect of quantum dynamics that we have thus far avoided: explicit dynamics of the electrons in the system.

6.5 ELECTRON DYNAMICS

6.5.1 Introduction

Thus far, we have focused on vibrational (nuclear) dynamics in molecular systems, avoiding explicit consideration of electron motion. We have been able to neglect the electrons in our treatment of nuclear dynamics by making use of the BO approximation. We assumed that the nuclei moved according to gradients of the PES, which are based on the time-averaged electron distribution. Our earlier discussions thus implicitly averaged over electronic motion to generate the PESs. We are now interested in describing the electron motion explicitly. In this section we focus on single-electron dynamics independent of the nuclei or other electrons in the system (in this sense, the dynamics can be considered "one-dimensional"). In Part IV, we move on to examine coupled electron dynamics, with the coupling being either to the nuclei or to the other electrons.

6.5.2 Atomic Case: Essential States

Similar to our treatment of nuclear dynamics, our goal is to derive the mathematical expressions needed to interpret experimental measurements of electron dynamics in a pump–probe experiment. Although we are eventually interested in molecular systems, the atomic case helps provide a conceptual understanding of the physics in a system where nuclear dynamics are essentially nonexistent. However, like with the two-level system of Section 3.2.3, we note that this treatment can be applied to any system with a discrete spectrum of electronic states for which the eigenstates and transition dipole moments are known or calculable (e.g. a multielectron molecule using *ab initio* electronic structure packages).

In particular, we consider modeling a pump–probe measurement of the time-dependent ionization yield for a one-electron atom such as hydrogen, starting in the ground state. A pump pulse resonantly generates an electronic wave packet by exciting a coherent superposition of electronic eigenstates, and the probe pulse ionizes this wave packet for different pump-probe delays.[6] We perform what is known as an "essential-states calculation," ignoring all states of the atom far from resonance with both the pump and probe fields (this section is a generalization of the one-photon and multiphoton treatments in Sections 3.2.3 and 3.2.4). In the eigenstate picture, the pump pulse prepares an *electronic* wave packet composed of a coherent superposition of electronic eigenstates $\psi_n(\mathbf{r})$:[7]

$$\Psi(\mathbf{r},t) = \sum_{n=0} c_n \psi_n(\mathbf{r}) e^{-i\omega_n t}, \tag{6.45}$$

where c_n is the amplitude coefficient of the nth state, which is determined by both the transition dipole moment from the ground state and the spectral amplitude and phase of the applied field at the resonance frequency. There is a well-defined phase relationship between all of the excited states, as well as between the excited states and the ground state.

In general, the sum in Equation 6.45 extends over all states, including the ground state and states that have no population (where $c_n = 0$). Since the dynamics of interest are

---

[6]Other excitation and probing schemes are of course possible and are discussed further in later chapters.

[7]The convention is to start the sum at $n = 0$, where ψ_0 is the initial ground state.

typically for the laser-excited states, we rewrite Equation 6.45 by separating off the ground state from the excited states:

$$\Psi(\mathbf{r},t) = c_g \psi_g(\mathbf{r})e^{-i\omega_g t} + \sum_{n=1} c_n \psi_n(\mathbf{r})e^{-i\omega_n t}, \tag{6.46}$$

where the sum is assumed to be over only those states resonantly coupled by the pump pulse (the "essential" states). Note that to avoid confusion with our usual notation of ω_0 for the laser frequency, we have chosen to relabel the ground state in the sum of Equation 6.45 using $n = g$ instead of $n = 0$. The basic idea is illustrated in Figure 6.3; only those electronic states $|\psi_n\rangle$ resonantly coupled by the pump pulse (i.e. those within the pump pulse bandwidth) are included in the sum, while off-resonant states such as $|\psi_j\rangle$ are not.

As with the two-level system, we insert Equation 6.46 into the TDSE, project onto a single eigenstate, carry out the RWA, and write the TDSE as a differential equation for the eigenstate coefficients:

$$i\dot{c}_g(t) = -\frac{\mu_{gn}}{2}E_{01}(t)e^{-i(\omega_{ng}-\omega_{01})t}c_n(t) \tag{6.47a}$$

$$i\dot{c}_n(t) = -\frac{\mu_{ng}}{2}E_{01}(t)e^{i(\omega_{ng}-\omega_{01})t}c_g(t), \tag{6.47b}$$

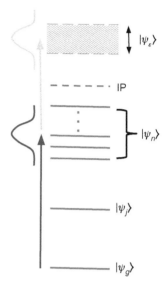

Figure 6.3 Essential states for the calculation of an atomic wave packet.

where $E_{01}(t)$ is the time-dependent envelope of the pump field and ω_{01} its frequency. In the limit of perturbative excitation (Section A.3), one can neglect ground-state depletion and assume that $\dot{c}_g(t) = 0$. By fixing $c_g(t) = 1$, we can write down a simple expression for the excited-state coefficients:

$$c_n(t) = -\frac{\mu_{ng}}{2i}\int_{-\infty}^{t} e^{i(\omega_{ng}-\omega_{01})t_1}E_{01}(t_1)dt_1. \tag{6.48}$$

In the limit of long times after the pump pulse is gone, we can let the upper limit go to $t \to \infty$, and the expression for $c_n(t)$ becomes

$$\begin{aligned}c_n(t \to \infty) &= -\frac{\mu_{ng}}{2i}\int_{-\infty}^{\infty}e^{i\Delta_n t_1}E_{01}(t_1)dt_1\\&= -\frac{\mu_{ng}}{2i}\sqrt{2\pi}E_{01}(\Delta_n),\end{aligned} \tag{6.49}$$

where $\Delta_n \equiv \omega_{ng} - \omega_{01}$ is the detuning of state n, and the second line comes from the fact that the integral in the first line is a Fourier transform of the time-domain field envelope. We see that the contribution of each excited state to the wave packet is proportional to the Fourier component of the excitation pulse envelope at the detuning frequency. As illustrated in Figure 6.4, this is equivalent to the Fourier component of the full field (including the carrier frequency) at the resonant frequency between the ground state and each state n in the wave packet.

We next consider the evolution of the electronic wave packet between pump and probe pulses. Since the wave packet is already represented as a coherent superposition of energy eigenstates, evolution between pulses is straightforward: a simple phase advance for each eigenstate proportional to the time delay. Assuming the pump pulse arrives at $t = 0$ and the probe a time delay τ later, the excited-state electronic wave packet just before the probe pulse is

$$\Psi(\mathbf{r},\tau) = \sum_n c_n \psi_n(\mathbf{r})e^{-i\omega_n \tau} = \sum_n -\frac{\mu_{ng}}{2i}\sqrt{2\pi}E_{01}(\Delta_n)\psi_n(\mathbf{r})e^{-i\omega_n \tau}, \tag{6.50}$$

Figure 6.4 Laser spectrum with transition frequencies between the ground state and excited states in the electronic wave packet generated by the pump pulse. The central frequency of the pulse, ω_{01}, is assumed to be resonant with the $g \rightarrow n$ transition.

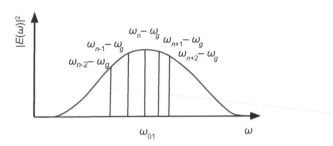

where we have dropped the ground-state term from the sum since it is assumed constant in the perturbative limit.

At a delay τ the probe pulse arrives and ionizes the atom by creating an electronic wave packet in the continuum. The continuum portion of the electron wave function after interaction with the probe pulse can be written in analogy to Equation 6.45 as

$$\Psi_e(\mathbf{r},t) = \int_0^\infty c(\varepsilon)\psi_\varepsilon(\mathbf{r})e^{-i\varepsilon t}d\varepsilon, \tag{6.51}$$

where the coefficient c is now a function of the continuous variable ε (instead of the discrete index n), and Ψ_e is a function of a single-electron coordinate $\mathbf{r}$ (in three dimensions). As earlier, our work is in finding the coefficients $c(\varepsilon)$. Like with the pump, integration of the TDSE in the perturbative limit leads to an expression for the excited-state coefficients after interaction with the probe pulse. Equation 6.48 can be generalized to include the initial states being those populated by the pump, evolution for a time τ between pulses, and the final states being in the continuum. This results in a time-dependent expression for the continuum-state coefficients, $c(\varepsilon,t)$:

$$\begin{aligned}
c(\varepsilon,t) &= -\frac{1}{2i}\sum_n \mu_{\varepsilon n}c_n e^{-i\omega_n\tau}\int_0^t e^{i(\omega_{\varepsilon n}-\omega_{02})t_2}E_{02}(t_2)dt_2 \\
&= \left(\frac{1}{2i}\right)^2 \sum_n \mu_{\varepsilon n}\mu_{ng}\sqrt{2\pi}E_{01}(\Delta_n)e^{-i\omega_n\tau}\int_0^t e^{i(\omega_{\varepsilon n}-\omega_{02})t_2} \\
&\quad \times E_{02}(t_2)dt_2,
\end{aligned} \tag{6.52}$$

where ω_{02} is the frequency of the probe laser, $\omega_{\varepsilon n} \equiv \omega_\varepsilon - \omega_n = \varepsilon - \omega_n$, and the lower limit of integration ($t = 0$) is assumed to begin after the pump pulse is gone. Note that $c(\varepsilon,t)$ implicitly depends on the pump–probe delay τ through phase evolution of the electronic eigenstates during the time between the pump and probe pulses. Taking the long time limit of Equation 6.52 ($t \rightarrow \infty$ after the probe pulse is gone), extending the lower limit to long times before the probe (but well after the pump), and using the Fourier transform relation yields the following:

$$c(\varepsilon) = \left(\frac{1}{2i}\right)^2 \sum_n \mu_{\varepsilon n}\mu_{ng}\sqrt{2\pi}E_{01}(\Delta_n)e^{-i\omega_n\tau}\sqrt{2\pi}E_{02}(\omega_{\varepsilon n}-\omega_{02}). \tag{6.53}$$

Equation 6.53 expresses the coefficients $c(\varepsilon,t)$ in terms of the Fourier components (frequencies) of the probe pulse. The measurement of a photoelectron yield at energy ε is proportional to $|c(\varepsilon,t)|^2$:

$$|c(\varepsilon,t)|^2 \propto \sum_{n,m} \mu_{\varepsilon n}\mu_{ng}\mu_{\varepsilon m}\mu_{mg}$$

$$\times E_{01}(\Delta_n)E_{01}^*(\Delta_m)E_{02}(\omega_{\varepsilon n}-\omega_{02})E_{02}^*(\omega_{\varepsilon m}-\omega_{02}) \qquad (6.54)$$

$$\times e^{-i(\omega_n-\omega_m)\tau}.$$

Equation 6.54 shows how the photoelectron yield as a function of pump-probe delay, τ, is modulated at frequencies corresponding to energy *differences* between states in the wave packet. We will return to this when modeling experimental observables for electron dynamics in Section 7.4.4.

6.6 HOMEWORK PROBLEMS

1. Numerically, launch a wave packet from the X-state of Na_2 to the A-state via a weak, ultrafast laser pulse on resonance for the $v = 0$ to $v' = 0$ transition. Consider a molecular ensemble for which only the ground vibrational state of the X-state is thermally populated. Calculate the spectrum of vibrational states excited for pulse durations of 10, 100, and 1000 fs (FWHM). Use the Morse potential $V(R) = D_e(1 - e^{-a(R-R_e)})^2$, with approximate parameters for the X and A states of

   X-state: $D_e = 0.027$ a.u. $a = 0.42$ a.u. $R_e = 5.82$ a.u.

   A-state: $D_e = 0.036$ a.u. $a = 0.29$ a.u. $R_e = 6.86$ a.u.

   Since you are not provided with the strength of the laser field, don't worry about the absolute amount of population transferred to the A state, just the shape of the wave function. You can find the R-dependent dipole moment for this transition in Table I of *Journal of Chemical Physics* **124**, 084308 (2006). Compare a calculation where you make the Condon approximation with one where you use the full, R-dependent dipole moment.

2. Rewrite Equation 6.31, expanding $\chi_a(R,0)$ in terms of the vibrational eigenstates of the *excited* electronic state, $\chi_{bm}(R)$ (i.e. $\chi_a(R,0) = \sum_m c_m \chi_{bm}(R)$, where m labels the vibrational eigenstate number). For simplicity, start with the ground vibrational eigenstate in electronic state a: $\chi_{a0}(R)$. Assuming a Gaussian envelope for the applied field with duration τ_p, show that the final wave function on the excited state after the end of the pulse can be written as

$$\chi_b^{(1)}(R,t) = \frac{-1}{i}\mu_{ba}\sum_m c_m \chi_{bm}(R)e^{-iE_{bm}t}$$

$$\times \int_0^t dt_1 e^{-(t_1^2/\tau_p^2)}e^{i(E_{bm}-E_{a0}-\omega_0)t_1}, \qquad (6.55)$$

   where E_{bm} is the energy of the mth vibrational eigenstate on PES b, E_{a0} the energy of the ground vibration eigenstate on PES a, and ω_0 the frequency of the applied field. Make use of the RWA and the fact that the Hamiltonian acting on an eigenstate yields an eigenenergy. Discuss your result, and use the answer to interpret Problem 1.

3. Write a program to calculate the evolution of the vibrational wave function on the B state of I_2 after excitation by a pump pulse. Start with the molecule in

the third vibrational state of the X-state potential. A pump pulse with a central frequency of 0.0735 a.u. and a pulse duration of 40 fs (FWHM) promotes population to the B state. Assume the B state is 0.071 a.u. higher in energy than the X state. Use the Morse potential $V(R) = D_e(1 - e^{-a(R-R_e)})^2$, with parameters for the X and B states of

X-state: $D_e = 0.0567$ a.u. $a = 0.9872$ a.u. $R_e = 5.04$ a.u.

B-state: $D_e = 0.020$ a.u. $a = 0.97$ a.u. $R_e = 5.72$ a.u.

Make two-dimensional plots of the wave function amplitude squared on both the B and X states as a function of R and t after excitation by the pump. Hint: Don't save the wave function at every time and space point required for the calculation since that will use a lot of memory. For the two-dimensional plot, sample your wave function coarsely using time steps of 10 fs and roughly 100 samples in R.

4. Consider creating a Rydberg wave packet in atomic sodium by shining a weak, broadband laser pulse resonant from the $3s$ ground state to a series of p states centered about $n = 20$. Assuming the laser pulse has a duration of 20 fs (FWHM), calculate the Rydberg wave packet amplitude as a function of time and the expectation value $\langle r(t) \rangle$. Hint: use the time-independent Schrödinger equation with quantum-defect energies to numerically calculate the eigenstates that comprise the wave packet. Based on the laser spectrum, you can determine the relative amplitude of each state in a perturbative excitation. Then add these states together with the phase evolution for each state dictated by its energy.

PART III

Measurements of One Dimensional Dynamics

CHAPTER 7

Incoherent Measurements in 1D

7.1 INTRODUCTION

Constructing experiments capable of performing time-resolved measurements of quantum dynamics is a formidable task. In Part III, we consider approaches for following one-dimensional dynamics by building on the prototypical techniques introduced in Chapter 4. The goal is to develop an understanding of the various experimental implementations, the types of information accessible with each, and the mathematical descriptions that allow for interpretation of the measurements. We begin this chapter by examining two incoherent approaches for measuring vibrational dynamics in one dimension: laser-induced fluorescence (LIF) and time-resolved ionization spectroscopy (TRIS). We then finish the chapter with a discussion of how one measures *electron* dynamics in atomic and molecular systems, focusing on examples that can be thought of as one dimensional.

As discussed in Chapter 4, in all techniques the measurement is initiated by a pump pulse that creates a nonstationary wave function on an excited potential energy surface (PES). The molecule then typically interacts with a probe pulse, after which the measurement is made. The observable can be any one of a number of quantities, including fluorescence from an excited state, changes in absorption of the probe pulse, photoelectron or ion yield, or scattered electrons. The signal depends on the pump–probe delay, providing information about the underlying molecular dynamics.

7.2 LASER-INDUCED FLUORESCENCE

7.2.1 Introduction

We first encountered the technique of LIF in Section 4.2.1. LIF is a background-free, incoherent detection technique that can be readily applied even in samples with low molecular density (e.g. gas cell or molecular beam). Figure 7.1 provides a detailed view of the process. A pump pulse launches a vibrational wave packet on an excited-state PES, promoting a portion of the well-localized, ground-state wave function on PES *a* to the excited PES *b* through a vertical Franck–Condon transition (see Section 3.3). If the molecule is cold (i.e. $kT \ll \hbar\omega_{vib}$), the initial wave function is simply the ground vibrational eigenstate of PES *a*. If the molecule is not cold, the initial wave function will be an incoherent sum of vibrational eigenstates.

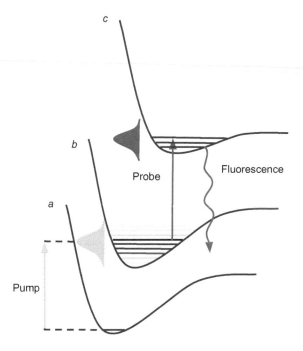

Figure 7.1 Illustration of pump–probe excitation in LIF. The pump pulse launches a vibrational wave packet on an excited PES, where the vibrational eigenstates of PES b are populated in proportion to their Franck–Condon factors and the spectral intensity of the pump pulse at the resonance frequency. After some time delay, the probe pulse arrives and promotes a portion of the oscillating wave function to PES c, from which it subsequently fluoresces. The fluorescence serves as the measured observable.

The wave function in the intermediate state has a central energy determined by the photon energy of the pump pulse (schematically shown by the length of the pump-pulse arrow in Figure 7.1), and is composed of eigenstates within the pump spectral bandwidth around this central energy. We assume that the bandwidth of the pump pulse is sufficient to populate a number of vibrational eigenstates on PES b. Since both the minimum of the excited PES and its curvature differ from the ground PES, the wave function will evolve and disperse in time after excitation by the pump (as seen in the numerical results of Chapter 5). This wave packet evolves on the intermediate PES until the probe pulse arrives.

At any particular time delay, the probe transfers a portion of the wave function to the excited fluorescent state c, with the amount depending critically on the overlap between the wave packet on the intermediate PES and the shape of the eigenstates on the upper PES at the energy of the probe photon (technically, all the eigenstates within the bandwidth of the probe pulse). For significant transfer to occur, there must be vibrational eigenstates on PES c with good Franck–Condon overlap with the wave function on PES b at the time the probe arrives. For example, in Figure 7.1, the central frequency of the probe pulse is tuned to be resonant with PES c near the outer turning point of surface b. Away from the outer turning point, absorption of the probe is much less likely due to low Franck–Condon overlap of the vibrational wave function on surface b with the vibrational eigenstates of surface c at the resonance frequency.[1] Therefore, the portion of the wave function on PES b near the outer turning point when the probe pulse arrives has a relatively high probability of being excited to PES c,

---

[1]This really can be interpreted in a number of ways. As described in the text, one can think of changes in the overlap between the time-dependent wave function on PES b and the vibrational eigenstates on PES c. If one wants to think about eigenstates on both PESs b and c, it is the constructive/destructive interferences between the various pairwise Franck–Condon factors of all the different eigenstates that lead to the time-dependent excitation probability. Finally, one can think about the wave function on PES b being off-resonance with PES c when it is away from the outer turning point.

while the wave function amplitude away from the outer turning point at the time of probe arrival has a much smaller probability of excitation. The fact that the likelihood of excitation to the fluorescent state depends on the location of the wave packet on the intermediate-state potential is the key to "clocking" vibrational dynamics with LIF.

7.2.2 Experimental Setup

The top panel of Figure 4.1 in Chapter 4 shows the typical experimental arrangement. The pump and probe beams cross in the sample with a small angle such that they have significant focal volume overlap.[2] Population that reaches the final electronic PES will spontaneously decay and emit fluorescent light into all 4π steradians of solid angle. By placing a fast (small f-number) lens near the sample, one can collect on the order of 10% of the emission. The collected fluorescence is directed through a chromatic filter and onto a photodetector, such as a photomultiplier tube or avalanche photodiode. The filter serves to block scattered light from the pump and probe pulses so that the signal measured during the short time window of the experiment is primarily due to fluorescence. The time-integrated fluorescence yield is proportional to the molecular population in the fluorescent state and serves as the signal for the experiment. The process is repeated at different delay times, and the fluorescence yield as a function of pump–probe delay is the data from which one extracts dynamics occurring on the intermediate PES.

7.2.3 Data and Basic Interpretation

An early experiment showing the promise of LIF and other time-resolved techniques measured dynamics on the B state of diatomic iodine I_2 [16]. The left panel of Figure 7.2 plots the fluorescence yield as a function of pump–probe delay. There are two primary features in the data as noted by the different timescales. The first is the fast oscillation at 300 fs, reflecting the largely coherent oscillation of the wave packet on the B state. The LIF signal is maximum if the probe pulse arrives when the wave packet is at large R, where the probe pulse most efficiently transfers the wave packet to the final state (c), from which one measures LIF. As the wave function oscillates back in toward smaller R again, the signal decreases, reaching a minimum when the wave function is at the inner turning point. This oscillation continues with a period of

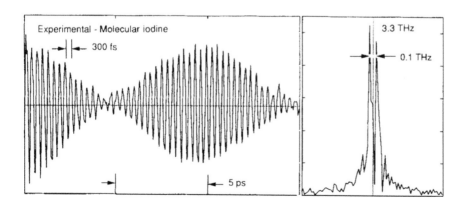

Figure 7.2 Experimental data for LIF measuring excited-state vibrational dynamics in I_2. Left panel: fluorescent signal (zero at bottom) as a function of pump–probe time delay (zero at left). Timescales note both the fast oscillation and the slower revival time. Right panel: Fourier transform of the data where the vertical scale shows magnitude. Both the central frequency and spacing are noted. Figure adapted from Ref. [16].

[2]The polarization of the probe beam is typically rotated using a half-wave plate such that it makes an angle of 54.7° with respect to the pump polarization. This is the so-called "magic angle" and is used to suppress rotational effects in the data (see Section B.1).

about 300 fs, providing a measurement of the round-trip time for a wave packet of this energy in the B-state potential.[3]

The second primary feature in the data is the decay and revival of the oscillation amplitude that occurs on a timescale of 10 ps. The initial decay in the oscillation amplitude is due to a spreading (dephasing) of the wave function on the B state. As the wave packet becomes more spread out, there is less temporal variation of wave function amplitude within the Franck–Condon window where the probe pulse couples to the fluorescent state. This agrees with our results from one-dimensional wave function propagation in Chapter 5; for a completely delocalized (or dephased) wave function, there is no significant change of the probability density in time that would lead to oscillation in excitation to the fluorescent state. The depth of modulation reaches a minimum around 5 ps, at which point it begins to increase again. By 10 ps after time zero, the oscillations are nearly fully restored, indicating the wave function is once again well localized (the wave function has "rephased"). This rephasing of the wave function is known as a wave packet revival, since localization of the wave function revives after a temporary absence.

The data shown in the left panel of Figure 7.2 reminds one of a beat pattern between two closely spaced frequencies, with a carrier oscillation at the mean frequency and a modulation in the oscillation amplitude at the difference frequency. This is illustrated in the right panel of Figure 7.2, which shows the magnitude of the Fourier transform of the data (the "spectrum"). The spectrum shows two primary peaks centered at 3.3 THz and spaced by 0.1 THz. These correspond to periods of $1/3.3 = 0.30$ ps and $1/0.1 = 10$ ps, as expected from examining the temporal data.

In fact, the frequency-domain picture of the dynamics comes right out of the time-domain data. The 3.3 THz average frequency is equivalent to 110 cm^{-1} (wavenumbers). This vibrational frequency agrees with complementary spectroscopic measurements of the vibrational eigenstates on the B-PES expected to be populated by the photon energy of the pump pulse. The frequency splitting of 0.1 THz is due to the anharmonicity of the B-state potential. A superposition of vibrational eigenstates will show oscillations at all *difference* frequencies between the eigenstates. In a harmonic potential, the difference frequencies are all identical, and the wave packet oscillates with no dephasing. In this case, the LIF data shown are well described by a vibrational wave packet containing two difference frequencies, implying the population of at least three vibrational eigenstates. One of the homework exercises in Appendix C asks the reader to simulate these experimental results using the numerical approach developed in Chapter 6.

It is worth noting the similarity to our discussion in Section 5.5.2.1, where we saw that the Fourier transform of the wave packet autocorrelation function produced a spectrum showing the frequencies of the eigenstates contained in the wave packet (the "wave packet spectrum"). Figure 7.2 plots the Fourier transform of the LIF pump–probe signal. This spectrum shows the frequencies of oscillation in the signal, which correspond to frequency *differences* between eigenstates.

7.2.4 Mathematical Description

In the perturbative treatment of pump–probe spectroscopy (Section 6.4.1), we found that the fluorescence signal was proportional to the population on PES c: $\langle \chi_c^{(2)} | \chi_c^{(2)} \rangle$.

---

[3]We note that with the data presented in Figure 7.2 without a clear zero along the time axis, it is difficult to determine the phase of the oscillations.

The vibrational wave function $\chi_c^{(2)}$ is second-order (bilinear) in the pump and probe fields and is given by Equation 6.33 (assuming the short-pulse limit where the wave function does not have time to evolve during the pulse duration):

$$\chi_c^{(2)}(R,t) = \frac{1}{i^2} e^{-iH_c(t-\tau)} \mu_{cb} E_{02} e^{-iH_b\tau} \mu_{ba} E_{01} \chi_a(R,0). \tag{7.1}$$

The population on PES c is therefore a function of pump–probe delay τ due to the implicit τ dependence in $\chi_c^{(2)}$:

$$\begin{aligned}
\text{Pop}_c(\tau) = \left\langle \chi_c^{(2)} \,\middle|\, \chi_c^{(2)} \right\rangle &= \frac{1}{i^4} \left\langle \chi_a(R,0) E_{01} \mu_{ab} e^{+iH_b\tau} E_{02} \mu_{bc} e^{+iH_c(t-\tau)} \right| \\
&\quad \left| e^{-iH_c(t-\tau)} \mu_{cb} E_{02} e^{-iH_b\tau} \mu_{ba} E_{01} \chi_a(R,0) \right\rangle \\
&= \left\langle \chi_a(R,0) E_{01} \mu_{ab} e^{+iH_b\tau} E_{02} \mu_{bc} \,\middle|\, \mu_{cb} E_{02} e^{-iH_b\tau} \mu_{ba} E_{01} \chi_a(R,0) \right\rangle.
\end{aligned} \tag{7.2}$$

In the perturbative limit, the population involves a product of four fields and is linear in both the pump and probe *intensities*. Note that the phase acquired from any propagation on state c cancels - the c state population is independent of dynamics on this state. However, the remaining time-evolution factors do not cancel, as one cannot move the dipole and field operators relative to the propagation on each of the electronic states. This comes from the fact that the expressions above are shorthand for the full matrix representation of both the state vector and the field and propagation operators (see Section A.4). Due to the matrix multiplication involved, the ordering of these operators matters.

7.2.5 LIF with Nonadiabatic Coupling

In our discussion thus far, we considered tracking vibrational wave packet dynamics on an isolated, bound PES. In this section, we consider using LIF to measure dynamics in a one-dimensional system where there is coupling between two (adiabatic) PESs [17]. Figure 7.3 shows the relevant potentials for the diatomic molecule NaI. The pump pulse launches a wave packet on the inner wall of the intermediate state potential ($E_2(r)$), which then propagates toward larger internuclear separation. Before the outer turning point of the potential, there is an avoided crossing with the ground electronic state ($E_1(r)$). The kinetic coupling (see Equation 2.87) between the two states allows a portion of the wave packet to cross back over to the ground state. The remaining vibrational wave packet on $E_2(r)$ continues to oscillate, with a portion of it switching to the ground state every time the wave packet passes the avoided crossing. The remaining wave packet on $E_2(r)$ is monitored by the probe pulse that excites the wave packet to a final state $E_3(r)$ (from which the molecule fluoresces).

These dynamics are captured in the plot of LIF intensity as a function of pump–probe delay shown in Figure 7.4. At a pump–probe delay of τ_0, the LIF signal increases dramatically as the wave packet launched by the pump pulse first passes through the Franck-Condon (FC) region for the probe. The signal then decreases as the wave packet moves out of the FC region toward the outer turning point of the potential. A second local maximum in the LIF signal comes at a pump–probe delay of about 750 fs, corresponding to the wave packet returning to the FC region on its way back to smaller internuclear distances. At a pump–probe delay of one vibrational period (τ_{osc}), the LIF signal is at a local minimum since the wave packet is once again at the inner turning point. Note that the two local maxima within each τ_{osc} correspond to the

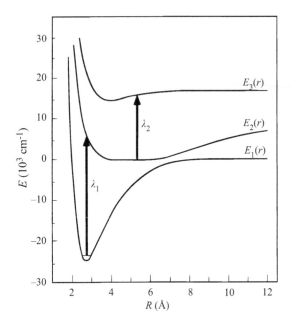

Figure 7.3 Relevant PES and excitation for LIF in NaI. Figure from Ref. [17].

Figure 7.4 LIF signal in NaI showing both wave packet oscillation in the intermediate PES and coupling back down to the ground PES. There are two peaks within each oscillation period since the Franck-Condon point of the probe is located near the center of the well. The signal decreases with each oscillation due to probability density returning to the ground state during each transit of the level crossing. Figure from Ref. [17].

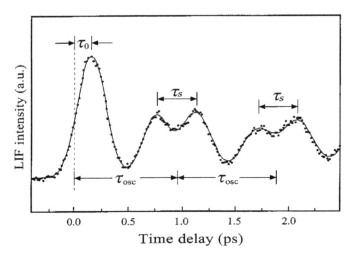

wave packet passing through the FC region twice per oscillation; the FC region for the probe is near the center of the potential well and is traversed by the wave packet twice as it oscillates back and forth. The LIF signal does not go all the way to zero at the first τ_{osc} (the time when the wave packet has returned to the inner turning point). This is mainly due to the fact that the FC region of the probe is shifted toward the inner turning point and (to some extent) that the wave packet is spreading as it propagates in the anharmonic potential.

Importantly, the LIF signal decreases with every vibrational period τ_{osc}. This is because a portion of the wave function proceeds to the ground state via nonadiabatic coupling at the point of the avoided crossing shown in Figure 7.3. This nonadiabatic (or kinetic) coupling is due to terms in the molecular Hamiltonian that are neglected in the Born–Oppenheimer approximation (as discussed surrounding Equation 2.80 in

Section 2.5.2). In the case of NaI, the population that returns to the ground PES disso-ciates (NaI $\to$ Na + I) due to the excess vibrational energy arising from the electronic–nuclear coupling; the electronic potential energy is converted to nuclear kinetic energy during the crossing.

7.2.6 Advantages and Disadvantages: Why Choose LIF?

In LIF, the molecules excited to the upper PES fluoresce independently of each other. Because of this, incoherent techniques such as LIF can be applied to relatively small ensembles of molecules. In fact, one of the major advantages of LIF is that it does not require high sample densities and can therefore be carried out in a diverse array of environments, from gas phase to living cells. In addition, LIF is typically background-free, since the fluorescent wavelength usually differs from both the pump and probe colors and can be spectrally and spatially filtered. Measuring small signals relative to zero is much easier than measuring those small changes on top of a larger background, where fluctuations in the background can mask the signal.

A disadvantage of LIF and other fluorescent techniques is that the light comprising the signal is (incoherently) emitted into 4π solid angle. This precludes using directionality for discriminating between different types of signals. As we will see in our discussion of coherent optical techniques, the direction of the emitted radiation can be an impor-tant tool for differentiating between signals that convey alternative information about the underlying dynamics.

Another limitation common to all optical techniques is that they do not directly mea-sure atomic positions or spacings. It is therefore necessary to have prior knowledge of the system in order to design the experiment and interpret the measurement. For example, one must have a general idea of the energies of the ground, intermediate, and final PESs at the location of the ground state minimum (Franck–Condon point), as well as their general variation with R. Furthermore, one should ideally have reason-able estimates of the transition dipole moments to predict the effectiveness of coupling different PESs with the laser pulses. This is all in contrast to more "direct" measure-ments such as ultrafast electron diffraction (see Section 9.1), where one does not need as detailed information about the PESs to connect the measurements to time-dependent changes in the molecular geometry. A discussion of different examples of LIF can be found in reference [18].

7.2.7 Fluorescence Gating

Closely related to LIF is the technique of fluorescence gating (FG), where the time dependence of the wave packet in the excited state is monitored by time resolving the fluorescence emission itself. As before, a pump pulse launches a wave packet on an excited electronic PES. However, in contrast with LIF, the probe pulse does not inter-act with the molecule directly, but instead is used to "gate" the fluorescence from the excited PES through a nonlinear interaction separate from the molecular sample (see Figure 7.5). Conceptually, wave packet evolution on the excited-state surface leads to temporal modulations in both the fluorescence frequency and intensity due to the vary-ing energy separation and transition dipole moment between the excited and ground surfaces. Simply time-integrating the total fluorescence yield after the pump pulse masks information about the excited-state dynamics; however, temporally and spec-trally resolving the fluorescence yield can provide a picture of the molecular dynamics similar to LIF. FG is simpler than LIF in that it only involves two electronic PESs and

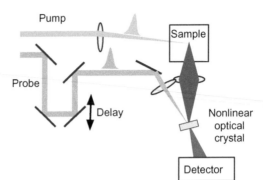

Figure 7.5 Schematic diagram of the experimental setup for FG. The time-delayed probe pulse "gates" the fluorescence light though nonlinear mixing in a crystal.

is thus easier to apply in complex systems where characterization of the states involved is difficult.

Gating of the fluorescent light is typically accomplished by an optical cross-correlation in a nonlinear medium (two of the most prominent techniques being fluorescence upconversion and optical Kerr gating). The cross-correlation signal from the fluorescence is only observed during temporal overlap with the short duration of the probe pulse. By scanning the time delay between the pump and probe pulses, one maps out the temporal (and potentially spectral) profile of the fluorescent signal. FG is typically used with larger molecular systems to study dynamics that are inherently multidimensional, such as internal conversion via conical intersections. We therefore leave a detailed discussion of typical FG data to Part V (see Section 12.3.2).

As noted in Chapter 1, if it were possible to both time and frequency resolve the fluorescence emission directly with a fast detector, there would be no need for a probe pulse to measure excited-state dynamics. In fact, when the excited state lifetime is long enough (typically hundreds of picoseconds or more), the fluorescent signal can be resolved directly with a fast photodetector, allowing one to skip the cross-correlation process (for example, see Section 12.4.1). However, for changes on the subpicosecond timescale, gating the fluorescence with a probe pulse is required to obtain the desired temporal resolution.

7.3 TIME-RESOLVED IONIZATION SPECTROSCOPIES

7.3.1 Introduction

In LIF, the final state of the molecular system is neutral (or at least the original charge state), and the probability of being in that state is measured via fluorescence. Here, we consider experiments where the final molecular state is ionic, and the probability of excitation to the final state is measured by collecting the electron or ion yield after ionization by the probe pulse. These techniques are generally referred to as time-resolved ionization spectroscopies, and they use time-of-flight and/or momentum-resolved detection of charged particles to obtain dynamical information about the molecular system. We focus on two different techniques that collect either positively charged ions or negatively charged electrons.

Figure 7.6 illustrates a typical TRIS experiment. As with LIF, a pump pulse launches a vibrational wave packet on an excited PES. A time-delayed probe pulse promotes a portion of this wave packet to an ionic state (it "ionizes the molecule"). Since the ionization probability and/or properties of the detected particles vary with the location and spread of the wave packet on the intermediate state, the ion or electron signal as a function of pump–probe delay provides information regarding dynamics on PES b. The primary difference as compared to LIF is that the final state is in the electronic continuum, leading to a molecular cation and an emitted electron. If the final ionic state is dissociative, there will be multiple fragments.

The ability to measure the kinetic energy and vector momentum of the charged particles produced during ionization allows time-resolved ionization spectroscopies to extract significant information about excited-state dynamics. In addition, since the final state is in the electronic continuum, selection rules for ionization are somewhat relaxed as compared to bound-to-bound transitions. In particular, transition dipole matrix elements to continuum states do not vanish due to symmetry conditions, so essentially any neutral molecular state can be ionized (i.e. there are no "dark states" in photoionization).

This has an important consequence regarding the probe pulse. Unlike LIF, with ionization spectroscopies, one is not limited to a probe pulse wavelength that matches the energy difference between the intermediate and final state. In principle, the only requirement on the probe frequency (and hence energy) is that it be greater than the ionization potential along the path the wave packet explores. While it may therefore seem attractive to choose a probe photon energy well above the ionization potential, there are a few considerations that make it unfavorable to go too high. One is that ionization of ground state molecules can create a large background on top of which a small excited state ionization signal can be difficult to extract. Another is that one

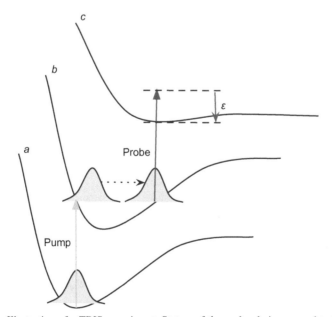

Figure 7.6   Illustration of a TRIS experiment. State c of the molecule is assumed to be ionic, and the kinetic energy of the ejected electron is depicted by ε.

can have ionization to multiple cationic states which is difficult to sort out. Finally, the cross section for ionization tends to decrease quickly as the photon energy moves above the ionization potential (or "threshold"). Figure 7.7 shows the ionization cross section of H_2 as a function of photon energy. The rapid turn-on at threshold is followed by a gradual decay in cross section with increasing photon energy. The energy scale for the decay is typically ~ 10 eV, which is largely driven by electronic wave function overlap between the initial bound state and the continuum. This is very similar to Franck–Condon overlap between vibrational wave functions; although the ionization cross section involves a matrix element with the **r**-operator, the decay in cross section at high photon energy is similarly driven by a wave function overlap. So, while one is relatively free to choose the probe frequency in an ionization experiment, going too far above threshold significantly reduces the efficiency, can involve multiple ionic states, and lead to a large background in the signal.

Looking at the probe transition in Figure 7.6, it is clear that, above threshold, there is extra energy available to the system after absorption of the probe. A natural question concerns how much of this excess energy goes into the photoelectron as compared to the molecular cation. Given momentum conservation and the large mass ratio between the ion and electron, translational energy acquired by the molecular cation during ionization can generally be neglected. Instead, any energy left in the cation is vibrational and is determined by the Franck–Condon factors for the vibrational eigenstates comprising the wave packet. This tends to limit population of highly excited vibrational states in the cation, and thus most of the energy is imparted to the outgoing electron. In other words, since there is essentially no vibrational momentum transferred directly by photoexcitation in a vertical transition, there cannot be significant energy deposited into vibrations. Any additional energy above threshold during ionization by the probe is taken away by the outgoing electron (depicted by ε in Figure 7.6). In many electron-based ionization spectroscopies, it is the electron yield as a function of time and energy that serves as the experimental signal.

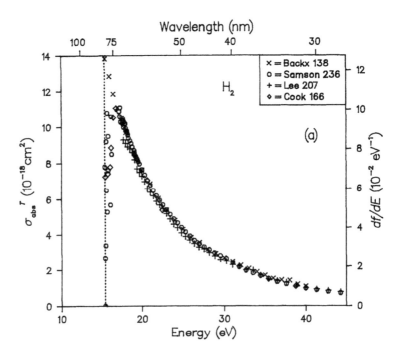

Figure 7.7 Ionization cross section for H_2 as a function of photon energy. Figure adapted from Ref. [19].

7.3.2 **Experimental Setup**

The basic experimental set-up is similar to LIF and is shown in Figure 4.2. The need to ionize the molecules and collect the resulting charged particles adds a few additional requirements to the experimental procedure as compared to LIF. We briefly address these issues here.

7.3.2.1 Sample Preparation

Ionization-based spectroscopies are usually performed under high vacuum (10^{-6} Torr or below) in order to collect and measure the electrons and ions produced by the pump and probe pulses. The sample is typically a molecular beam, and ideally the molecules in the sample are vibrationally and rotationally cold. This can be achieved using a supersonic molecular beam, expanding the sample from a region of high pressure into vacuum via a small nozzle (the high pressure is often from an additional "carrier" gas such as helium) [20]. As the molecules expand into vacuum, they lose rotational and vibrational energy via collisions with carrier gas atoms.

While a rotationally and vibrationally cold sample is ideal because the initial state of the molecule is well determined, creating such a sample is not always necessary. For example, when studying dynamics on timescales faster than $1/k_B T$, the initial distribution is of less importance.[4] Also, if the intermediate-state PES is relatively steep near the Franck–Condon region, the intermediate-state dynamics are not particularly sensitive to the initial temperature of the molecules. A cold molecular sample is most important in cases where the intermediate PES is relatively flat in the Franck–Condon region, since thermal motion in the ground state will influence wave packet evolution on the intermediate state after transfer by the pump pulse. In addition, it is important to begin with a cold molecular ensemble when making high-resolution measurements. This ensures the distribution of initial molecular kinetic energies does not smear out the energy distribution of the measured ions or electrons.

7.3.2.2 Charged Particle Detection

In order to measure the generated electrons and/or ions, the pump and probe pulses typically intersect the molecular sample in an apparatus designed to collect and detect charged particles. Examples include a time-of-flight mass spectrometer (TOFMS), a velocity map imaging (VMI) device, or a magnetic bottle spectrometer. Figure 7.8 shows the basic elements of a VMI apparatus. The pulses intersect a molecular beam between a series of metal plates with circular holes. By carefully choosing the voltages on the plates, one can map the initial transverse velocity of the charged particles to position on a spatially resolved detector.[5] Collection of the charged particles can be performed with high efficiency while retaining information about their initial momentum and energy.

Detection of single electrons and ions is possible using electron multipliers such as microchannel plates (MCP). Similar to a photomultiplier tube for the case of photons, a single electron or ion incident on the front surface of an MCP results in a shower of approximately 10^6 electrons at the back. The electron distribution at the back of the MCP is localized to a transverse region a few microns wide, allowing the electrons

---

[4]Note that in atomic units, energy and frequency have the same units, and thus the units of $1/k_B T$ are time.

[5]Technically it is the square root of the mass times the transverse velocity which is mapped to position. This means that electrons and ions with the same energy are mapped to the same distance away from the center of the detector.

Figure 7.8 VMI of electrons and ions. The pump and probe pulses intersect a molecular beam inside a series of metal plates that are used to map the initial transverse velocity of the charged particles to position on a spatially resolved microchannel plate (MCP) detector at the end of the apparatus. A phosphor screen after the MCP can be used to convert the electron distribution into a photon distribution, which is then detected with a camera. Figure courtesy of Peter Sandor.

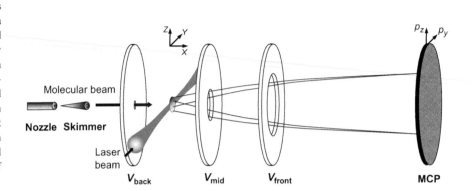

to be measured with spatial as well as temporal resolution. This provide both time-of-flight information and the (transverse) initial velocity of the particle. Typically, an oscilloscope is used to measure the flight time and a phosphor screen with camera to record the location.

7.3.2.3 Pulses For Ionization

As a final experimental note, it can be challenging to generate and deliver a short pulse with sufficient photon energy to ionize the molecule directly from the intermediate state. An alternative is to use a multiphoton process for ionization, where more than one photon is absorbed from the probe pulse. Probing with multiphoton ionization has both advantages and disadvantages as compared to its single-photon counterpart. It is generally easier to generate and deliver probe pulses for multiphoton ionization since the frequencies are in the visible (or near-IR) portion of the spectrum. So although the pulses require extra intensity to drive the multiphoton process, they can be directly produced with standard laser oscillators and amplifiers. In addition, these pulses propagate through air and optical materials without significant absorption and only minimal dispersion. In contrast, the probe pulses required for single-photon ionization are typically in the deep or vacuum ultraviolet region of the spectrum and suffer significant absorption and dispersion when propagating through air or materials. These pulses must be generated via nonlinear optical approaches since standard ultrafast lasers do not operate directly in these frequency ranges.

On the other hand, the mathematical formalism required to describe experiments using single-photon ionization is more straightforward than for multiphoton experiments, where the interaction Hamiltonian is more complicated. Thus, there is a trade-off between experimental implementation (easier for multiphoton) and a theoretical description (easier for single photon). The process of multiphoton absorption is discussed in Section 3.2.4, where we use adiabatic elimination to derive an expression for the two-photon absorption strength.

7.3.3 Measuring Ions: Data and Basic Interpretation

The simplest charged-particle measurement records the total ion yield as a function of pump–probe delay. We therefore begin with a discussion of this experiment before moving on to techniques that characterize additional information.

7.3.3.1 Bound Potential Energy Surface

To facilitate comparison with LIF, we again consider wave packet dynamics in the B state of I_2 [22, 21]. Figure 7.9 shows the relevant PESs and photon energies. As with LIF, a pump pulse launches a vibrational wave packet that evolves on the B-state. This wave packet is probed via ionization to the ground state of the molecular cation (X^+) after two-photon absorption from the probe pulse.

The measured signal in Figure 7.10 consists of the total I_2^+ yield as a function of pump–probe delay. Similar to Figure 7.2, the data show oscillations of the wave packet over several vibrational periods. The amplitude of the modulations decreases to zero as the wave packet dephases, before once again reviving when the eigenstates in the wave packet rephase. At this point in time, the wave packet has relocalized, once again showing a modulation in the ion yield as the wave packet moves in and out of the region on the B state where ionization is most effective. The fact that there are multiple vibrational eigenstates contributing to the wave packet is reflected in the Fourier transform of the time-dependent ion yield (inset). The multiple peaks correspond to several frequency differences that beat against each other to produce the dephasing and revival of the modulated ion yield. The modulations in the ion yield of Figure 7.10 show a maximum near zero time delay (and integer numbers of vibrational periods later), when the wave packet is localized near the inner turning point. This is consistent with the picture of Figure 7.9, where the most efficient ionization of the wave packet occurs near the inner turning point.

7.3.3.2 Dissociative Potential Energy Surface

Ionization measurements can also be used to follow dynamics on dissociative PESs (i.e. vibrationally unbound). Figure 7.11 shows pump–probe measurements of the polyatomic molecule CH_2I_2 excited to a dissociative PES by a 4.8 eV pump pulse. Excitation at 4.8 eV is known to cause C–I fission along a dissociative state that has unpaired electrons in both a bonding σ orbital and an antibonding σ^* orbital (as opposed to the ground PES that has a doubly occupied σ orbital and an unoccupied σ^* orbital). While the dynamics in this system are, in principle, multidimensional, the projection of the reaction coordinate onto local modes (i.e. bond lengths or angles) is dominated by the C–I distance. So while there is some motion along other degrees of

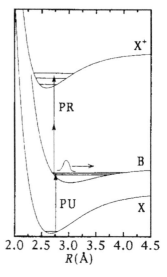

Figure 7.9 Relevant PESs for ionization spectroscopy in I_2, along with arrows depicting the pump ("PU") and probe ("PR") processes. Note the probe induces multiphoton ionization. Figure from Ref. [21].

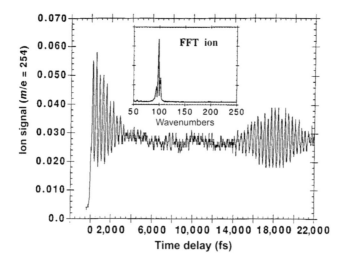

Figure 7.10 TOFMS data for ionization spectroscopy in I_2, showing oscillations and revival in the B state. Compare to LIF measurement of similar dynamics in Figure 7.2. Note that the rephasing timescale depends on the specific superposition of eigenstates excited by the pump pulse. Figure adapted from Ref. [22].

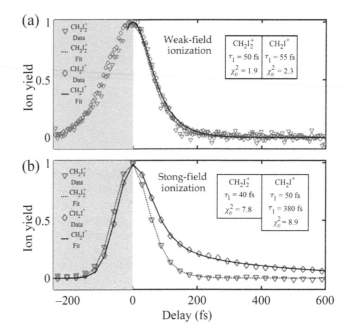

Figure 7.11 Pump–probe ionization measurements of dissociation in CH_2I_2. Data for both weak-field and strong-field ionization are shown. Measurements courtesy of Spencer Horton.

freedom, it is possible to construct a reasonable picture of the dynamics from consideration of the one-dimensional C–I stretching (asymmetric C–I stretching coordinate).[6]

Figure 7.11 shows measurements of both single-photon and multiphoton ionization by probe pulses at 8 and 1.6 eV, respectively. The two measurements produce qualitatively similar results, the ion yields show decay on similar timescales as the molecule dissociates. The signal decays as the electronic potential energy is converted into nuclear kinetic energy and the probe cannot ionize the separating molecular fragments. However, there are quantitative differences that reflect the two probing mechanisms. For example, as the wave packet moves, it can transit locations where the multiphoton ionization cross section is enhanced by intermediate resonances. This means that for multiphoton probing, the coupling between the excited state and the ionization continuum depends sensitively on laser intensity and the location of the wave packet on the excited-state potential. This can significantly influence the signal in a way that is not directly connected to the excited-state dynamics we wish to measure. With single-photon ionization measurements, the ionization cross section tends not to vary significantly as the wave packet moves along the vibrational coordinate. Thus, while probing with multiphoton ionization is a sensitive and versatile approach for following excited-state dynamics, it is less straightforward to interpret than the single-photon measurements, which can be modeled more easily.

7.3.4 Measuring Electrons: Data and Basic Interpretation

We now consider experiments that detect the negatively charged photoelectrons produced during ionization. Figure 7.12 illustrates the basic idea of the technique known

---

[6]We note that trajectory surface hopping calculations reveal some internal conversion between absorption of the pump light and dissociation of the molecule. These dynamics are very rapid (taking place in ∼10 fs), and therefore not reflected in the measurements shown. This allows us to include the results in our discussion of measurements highlighting one-dimensional dynamics.

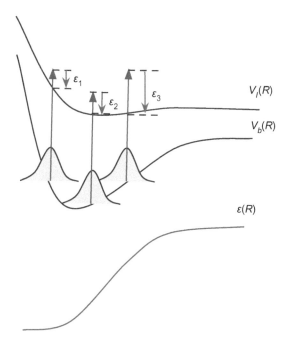

Figure 7.12 Illustration of how TRPES can be used to follow the evolution of a wave packet on an excited-state PES (the pump process from the ground PES has been omitted). The top portion of the figure shows the PES for both the excited neutral state b and ionic state c. At each R, the photoelectron energy ε_i depends on the difference between the two potentials. The bottom portion illustrates how the photoelectron energy might depend on R.

as time-resolved photoelectron spectroscopy (TRPES).[7] From the intermediate state b, the molecule absorbs an amount of energy corresponding to the photon energy of the probe pulse, leaving an excess energy equal to the probe photon energy minus the difference in energy between the states c and b. The excess energy, depicted by ε_i in Figure 7.12, is carried away as kinetic energy by the outgoing photoelectron. The kinetic energies of the electrons are measured as a function of pump–probe delay, typically using a magnetic-bottle time-of-flight detector or VMI apparatus (Figure 7.8).

Assuming the ionic PES c is not parallel to the intermediate PES b on which the wave packet evolves, the photoelectron energy remaining after ionization correlates with the position R on potential b. In fact, if the difference between the intermediate and ionic states is monotonic with R, the position of the wave packet on the intermediate state has a one-to-one relationship to the photoelectron energy. This implies a significant advantage of measuring photoelectrons when compared with ions: the photoelectron spectrum has the ability to follow wave packet motion throughout its entire range, rather than capturing it within only a particular Franck–Condon window. Note that one must use a probe photon of sufficient energy such that the wave packet can be ionized from anywhere on the intermediate-state potential.

We examine a TRPES measurement carried out in Na_2 [23, 24]. Figure 7.13 shows the potentials of the states involved, as well as the photon energies of the pump and probe. One complication in this particular system is that the $2^1\Pi_g$-PES is accessible via a two-photon process from the ground state at the same wavelength that the A-PES is one-photon resonant. Therefore, the pump pulse at 620 nm actually creates time-dependent wave packets on both PESs. A probe pulse (also at 620 nm) arrives after some time delay and promotes portions of the wave packets from both neutral

---

[7]For simplicity, we have omitted the ground PES and pump process from the diagram. As with all techniques, we assume a time-dependent wave packet on the intermediate state that has been generated by a pump pulse.

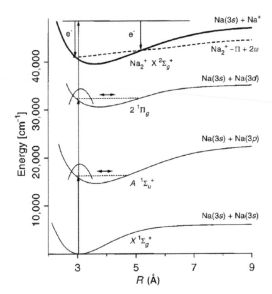

Figure 7.13 Relevant PESs and photon energies for TRPES in Na$_2$. Figure from Ref. [23].

PESs to the ground electronic state in the ionic continuum. This involves a direct, one-photon ionization process from the $2^1\Pi_g$ state, as compared to a two-photon, resonant-enhanced process from the A state. By analyzing the energy differences as a function of internuclear separation R between the three participating states, the signal from the $2^1\Pi_g$ state can be isolated from that of the A state, allowing an independent analysis of wave packet dynamics on this state.[8]

Figure 7.14 shows the intensity of the ionized electron signal as a function of pump–probe delay and internuclear distance, where the internuclear distance has been mapped from the photoelectron kinetic energy. It is worth highlighting the fact that, in this simple case, the data is essentially a direct measure of the time-dependent nuclear probability density on the excited $2^1\Pi_g$ state. One can see the probability oscillating back and forth in the excited-state potential. Based on the slopes of the PESs, the spectrometer resolution, and the bandwidth of the laser used, the measurement has an effective spatial resolution of approximately 0.5 Å. Simulations of the photoelectron spectra based on wave packet calculations in Na$_2$ show good agreement with the measurement.

7.3.5 Measuring Electrons: Mathematical Description

In order to connect the measured photoelectron spectrum to the wave function on PES b, we begin with Equation 6.32 that describes the second-order wave function on PES c after both pump and probe pulses:

$$\chi_c^{(2)}(R,t) = \frac{1}{i^2} \int_0^t dt_2 \int_0^{t_2} dt_1 e^{-iH_c(t-t_2)} \left(-\mu_{cb}E_2(t_2)\right)$$
$$\times e^{-iH_b(t_2-t_1)} \left(-\mu_{ba}E_1(t_1)\right) e^{-iH_a t_1} \chi_a(R,0). \tag{7.3}$$

---

[8] We note that this particular experimental implementation required significant knowledge of the system in order to obtain and interpret the measurements. In this sense, the primary purpose of the experimental data is to verify the theoretical model.

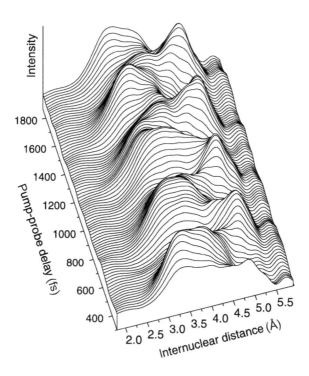

Figure 7.14 Measured photoelectron spectra as a function of pump–probe delay mapped onto internuclear coordinate. Figure from Ref. [24].

We would like to rewrite this expression in terms of the wave function of interest: $\chi_b^{(1)}(R,t_2)$, the time-dependent wave function on PES b generated by the pump pulse (see Equation 6.31):

$$\chi_b^{(1)}(R,t_2) = \frac{1}{i} \int_0^{t_2} dt_1 e^{-iH_b(t_2-t_1)} \left(-\mu_{ba}E_1(t_1)\right) e^{-iH_a t_1} \chi_a(R,0). \tag{7.4}$$

Using this, we can write the final, second-order wave function on state c in terms of $\chi_b^{(1)}$ as

$$\chi_c^{(2)}(R,t) = \frac{1}{i} \int_0^t dt_2 e^{-iH_c(t-t_2)} \left(-\mu_{cb}E_2(t_2)\right) \chi_b^{(1)}(R,t_2). \tag{7.5}$$

For simplicity, we assume that the pump pulse is localized in time around $t_1 = 0$, so that the initial wave function on state b is $\chi_b^{(1)}(R,0)$. With this, Equation 7.5 becomes

$$\chi_c^{(2)}(R,t) = \frac{1}{i} \int_0^t dt_2 e^{-iH_c(t-t_2)} \left(-\mu_{cb}E_2(t_2)\right) e^{-iH_b t_2} \chi_b^{(1)}(R,0). \tag{7.6}$$

Since the final state is ionic, we write $H_c = H_I + \varepsilon = T + V_I(R) + \varepsilon$, where T is the nuclear kinetic energy operator, H_I is the Hamiltonian for the ionic state, and ε is the electron kinetic energy. In addition, we write H_b as $T + V_b(R)$ and the probe field as $E_2(t_2) = \frac{1}{2}E_{02}(t_2)(e^{+i\omega_2 t} + e^{-i\omega_2 t})$, where $E_{02}(t_2)$ is the amplitude envelope of the probe pulse. This allows us to rewrite Equation 7.6, expressing the vibrational wave function on the ionic state associated with a photoelectron energy ε as

$$\chi_{c,\varepsilon}^{(2)}(R,t) = \frac{-1}{2i} e^{-i(H_I+\varepsilon)t} \int_0^t dt_2 e^{i(T+V_I(R)+\varepsilon)t_2} \mu_{cb} E_{02}(t_2)$$
$$\times \left(e^{+i\omega_2 t} + e^{-i\omega_2 t}\right) e^{-i(T+V_b(R))t_2} \chi_b^{(1)}(R,0). \tag{7.7}$$

In order to simplify this expression, we assume the case of an idealized experiment, in which one can neglect evolution of the wave function during the probe pulse. This allows us to neglect the commutators between T and V for both H_I and H_b inside the integral (see Section 5.2), and therefore cancel the kinetic energy operators. We again make the rotating-wave approximation, where we keep only the on-resonance $e^{-i\omega_2 t}$ term from the applied field (leaving in the exponential the terms that make the RWA clear):

$$\chi^{(2)}_{c,\varepsilon}(R,t) = \frac{-1}{2i} e^{-i(H_I+\varepsilon)t}$$
$$\times \int_0^t dt_2 e^{i(V_I(R)-V_b(R)-\omega_2+\varepsilon)t_2} \mu_{cb} E_{02}(t_2) \chi^{(1)}_b(R,0). \tag{7.8}$$

Note that while our idealized experiment uses a probe pulse short enough that the wave function does not evolve during the pulse, it is important that the probe pulse has a finite duration (and hence finite spectral bandwidth) to preserve selectivity in the final state and obtain useful information from the photoelectron spectrum. A true delta-function probe pulse would produce an infinite number of photoelectron energies, removing any information about wave packet location.

We condense notation in Equation 7.8 by defining the function F that includes all the time-dependent factors inside the integral (along with a leading constant):

$$F(R,\varepsilon,t,\tau) \equiv \frac{-1}{2i} \int_0^t dt_2 e^{i(V_I(R)-V_b(R)-\omega_2+\varepsilon)t_2} \mu_{cb} E_{02}(t_2), \tag{7.9}$$

where F depends on the internuclear coordinate R, the electron energy ε, the pump–probe delay τ, and the (final) time t. Note that the dependence on the pump–probe delay τ is buried in the temporal position of the probe pulse envelope $E_{02}(t_2)$, which is only nonzero for time values around $t_2 = \tau$ (e.g. Gaussian centered at time $t_2 = \tau$). With this definition, Equation 7.8 can be written as

$$\chi^{(2)}_{c,\varepsilon}(R,t) = e^{-i(H_I+\varepsilon)t} F(R,\varepsilon,t,\tau) \chi^{(1)}_b(R,0). \tag{7.10}$$

The measured photoelectron spectrum for a particular pump–probe delay τ is given by the probability of finding photoelectrons with a kinetic energy ε. This probability is determined by the norm (or population) in the ionic state, which is an inner product (integral over R) of the wave function on the intermediate state and the function $F(R,\varepsilon,t,\tau)$:

$$P(\varepsilon,t,\tau) = \left\langle \chi^{(2)}_{c,\varepsilon}(R,t) \middle| \chi^{(2)}_{c,\varepsilon}(R,t) \right\rangle$$
$$= \langle \chi^{(1)}_b(R,0) F(R,\varepsilon,t,\tau) e^{+i(H_I+\varepsilon)t} | e^{-i(H_I+\varepsilon)t} F(R,\varepsilon,t,\tau)$$
$$\times \chi^{(1)}_b(R,0) \rangle \tag{7.11}$$
$$= \left\langle \chi^{(1)}_b(R,0) F(R,\varepsilon,t,\tau) \middle| F(R,\varepsilon,t,\tau) \chi^{(1)}_b(R,0) \right\rangle.$$

We note that since the probe pulse has zero amplitude at long times (i.e. $t \gg \tau$), we can take the limit of t being large in Equation 7.9, and thereby omit the dependence on t in the expression for F, focusing instead on the τ dependence:

$$P(\varepsilon,\tau) = \left\langle \chi^{(1)}_b(R,0) F(R,\varepsilon,\tau) \middle| F(R,\varepsilon,\tau) \chi^{(1)}_b(R,0) \right\rangle. \tag{7.12}$$

From Equation 7.9, we see that $F(R, \varepsilon, \tau)$ (and therefore $P(\varepsilon, \tau)$) will be large for ε values such that $V_I(R) - V_b(R) - \omega_2 + \varepsilon \sim 0$; as expected, the energy of the photoelectron produced, ε, roughly follows the difference between the potentials where the wave packet is located:

$$\varepsilon = \omega_2 - \left(V_I(R) - V_b(R) \right). \tag{7.13}$$

If the R-dependent difference between the two potentials $V_I(R)$ and $V_b(R)$ is single-valued, one can use the time-dependent photoelectron spectrum to map the evolution of the wave packet on the intermediate state surface. As the wave packet evolves on $V_b(R)$, the probe pulse produces a photoelectron energy that varies with time delay, as the continuum state energy that is "resonant" also varies with R (this is illustrated in Figure 7.12 and used to plot the data in Figure 7.14). We note that in many situations, including most polyatomic molecules, the interpretation of TRPES data is more complicated than this particular case, since there are many factors that influence the photoelectron yield and energy. More involved TRPES experiments are discussed in Section 12.3.

7.3.6 Advantages and Disadvantages: Why Choose TRPES?

Similar to LIF, TRPES provides a background-free measurement with high detection efficiency, allowing TRPES to be used with extremely small sample sizes and provide single-molecule sensitivity. As discussed previously, two additional advantages come from the final state being in the electronic continuum. First, the resonance condition is automatically met and there are no "dark states," as there is always a continuum state available with nonzero transition dipole moment within a few electron volts of the ionization potential. Furthermore, by recording the photoelectron spectrum rather than just the total ion or electron yield, TRPES allows one to follow the evolution of a wave packet throughout its entire range of travel, rather than just observing it during its passage through an observation window. Some of the difficulties or limitations of TRPES are that it must be carried out in vacuum, it requires probe pulses capable of ionizing the molecule (possibly using multiphoton ionization), and it still necessitates prior knowledge of the electronic structure to interpret the measurements. For reviews of ionization spectroscopies, see [25, 26, 27, 28].

7.4 MEASURING ELECTRON DYNAMICS

7.4.1 Introduction

We now move on to consider measurements that explicitly track electron dynamics. We focus on cases that can be considered "one-dimensional," in that they involve a single electron without coupling to other electrons or the nuclei. We start with an atomic example, building off the discussion in Section 6.5.2. As we will see, in such a simple system, the time-domain approach does not add any information beyond what could be obtained by a frequency-domain measurement of the spectrum if one knows the eigenfunctions. Given an initial wave function defined by the amplitudes and phases of the eigenfunctions, one can calculate the subsequent dynamics by adding the eigenfunctions together, each with a time-dependent phase that advances according to the eigenenergy. Nevertheless, in addition to illustrating the correspondence between the eigenstate and time-domain interpretations, it is worth developing the mathematical framework and basic principles with the atomic case before moving on to more complex systems.

7.4.2 Experimental Setup

The first experiments we discuss measure electron dynamics in an alkali atom where one only need to consider the valence electron (it is an effective one-electron system). The experimental set-up is essentially TRIS, as discussed in Section 7.3. Here, a pump pulse excites an *electronic* wave packet, and a time-delayed probe arrives and ionizes the atom, producing a photoelectron and photoion. The photoions produced by the probe are collected, and the total ion signal is plotted as a function of pump–probe delay.

7.4.3 Data and Basic Interpretation

Figure 7.15 shows TRIS measurements in atomic potassium, where panel (a) plots the experimental data and panel (b) plots a simulation [29]. One sees a modulation in the ion yield as a function of pump–probe delay, where the timescale of modulation over the first portion of the plot corresponds to the period of wave packet oscillation. As the delay time increases, the depth of modulation decreases and one sees a doubling of the frequency, reflecting a double-peaked structure of the wave function as it begins to dephase.

The results can be interpreted in terms of the dynamics presented in Section 6.5.2. Excitation by the pump pulse creates a radially localized wave packet whose properties are determined by the central frequency and bandwidth of the pump pulse. The radial expectation value subsequently oscillates in time as the wave packet evolves. As

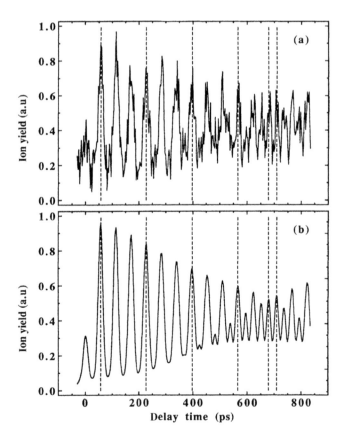

Figure 7.15 Pump–probe ionization signal from an electronic wave packet in atomic potassium. Panel (a): experimental data. Panel (b): calculations. Figure from Ref. [29].

explained later, the probability for ionization by the probe pulse is maximized when the wave packet is localized near the core (and minimized when it is farthest from the core). Thus, radial wave packet motion is reflected in the signal modulations.

We note that the timescales in this experiment are long compared to even the nuclear dynamics discussed earlier. This is because the pump frequency was chosen so that the electronic states comprising the wave packet would evolve on timescales slow enough that the motion could be resolved with the relatively long-duration pulses available at the time.

7.4.4 Mathematical Description

We outlined a mathematical treatment of electron dynamics in Section 6.5.2. In particular, we found for the electronic wave packet in the continuum after excitation by the pump and probe pulses:

$$\Psi_e(\mathbf{r},t) = \int_0^\infty c(\varepsilon)\psi_\varepsilon(\mathbf{r})e^{-i\varepsilon t}d\varepsilon, \tag{7.14}$$

where ε is the photoelectron energy. The coefficients $c(\varepsilon)$ are given by Equation 6.53:

$$c(\varepsilon) = \left(\frac{1}{2i}\right)^2 \sum_n \mu_{\varepsilon n}\mu_{ng}\sqrt{2\pi}E_{01}(\Delta_n)e^{-i\omega_n\tau}\sqrt{2\pi}E_{02}(\omega_{\varepsilon n} - \omega_{02}), \tag{7.15}$$

where the μ_{ij} are the dipole moments connecting the different electronic states, $E_{01}(\Delta_n)$ is the Fourier component of the pump pulse envelope at the detuning frequency, τ is the pump–probe delay, and $E_{02}(\omega_{\varepsilon n} - \omega_{02})$ is the Fourier component of the probe pulse envelope at the frequency difference $\omega_{\varepsilon n} - \omega_{02}$.[9]

While the experimental observable is the total ion yield, we are interested in describing electron dynamics and our corresponding mathematical expressions are for electron wave functions. Fortunately, we can directly connect the electron wave function to the observable of the total ion yield: for every ion produced, there is a corresponding electron in the continuum. Integrating the electron probability density in the continuum over all space is an equivalent measure to counting the ion yield. We therefore write the experimentally measured ion yield as follows:

$$\text{Yield} = \int_{\mathbf{r}} |\Psi_e(\mathbf{r},t\to\infty)|^2\,d\mathbf{r} = \int_{\mathbf{r}} \left|\int_0^\infty c(\varepsilon)\psi_\varepsilon(\mathbf{r})e^{-i\varepsilon t}d\varepsilon\right|^2 d\mathbf{r}, \tag{7.16}$$

where we have let $t \to \infty$ since we want the ion yield after the probe pulse is gone. Expanding the modulus-squared in Equation 7.16 and making use of orthonormality for the continuum wave functions, we can carry out the integration over space, giving the yield as

$$\text{Yield} = \int_\varepsilon\int_{\varepsilon'} d\varepsilon d\varepsilon'\,\delta_{\varepsilon\varepsilon'}c(\varepsilon)c^*(\varepsilon')e^{-i\varepsilon t}e^{i\varepsilon' t} = \int_\varepsilon d\varepsilon|c(\varepsilon)|^2. \tag{7.17}$$

---

[9]We note that this expression only describes the state of the electron that was removed from the atom or molecule, but does not consider the state of the remaining atomic or molecular cation. The atom/molecule can be left in different cationic states, and in general, the system needs to be described in terms of an entangled state of the cation and free electron (i.e. a sum over product states for the cation and free electron).

The modulus squared in Equation 7.17 was derived in Chapter 6 (cf. Equation 6.54):

$$|c(\varepsilon,t)|^2 \propto \sum_{n,m} \mu_{\varepsilon n} \mu_{ng} \mu_{\varepsilon m} \mu_{mg} \tag{7.18}$$

$$\times E_{01}(\Delta_n) E_{01}^*(\Delta_m) E_{02}(\omega_{\varepsilon n} - \omega_{02}) E_{02}^*(\omega_{\varepsilon m} - \omega_{02}) e^{-i(\omega_n - \omega_m)\tau}.$$

The cross-terms in Equation 7.18 lead to modulation in the ion yield with delay time τ at frequencies corresponding to energy *differences* between the bound electronic states (which survive the integral over all continuum electron energies because the delay dependence is separate from the electron energy in the continuum). Similar to what we found for vibrational wave packets in molecules, the eigenstate energy differences determine the period of wave packet oscillation.

Physically, the modulation as a function of pump–probe delay in the ionization signal can be understood from either a simplified classical or quantum picture. Removal of the electron during ionization must satisfy both energy and momentum conservation. Classically, it is challenging to impart the required change in momentum of the electron given the small linear momentum of a visible photon ($p = h\nu/c$). However, near the atomic core, the electron can exchange momentum more readily with the nucleus, enhancing the probability for ionization.

Quantum mechanically, the ionization probability depends on the relative phases of constituent eigenstates in the wave packet. Assuming the signs of the various transition dipole moments between the initial bound eigenstates and the continuum do not change across the spectrum of initial states, the ion signal will be maximized when the phase differences between the bound states are zero (e.g. all positive contributions from the different bound eigenstates).[10] These phase differences are zero exactly when the wave packet is localized near the core, and contributions to the ion yield from the various bound states add up in-phase at the inner turning point of the wave packet. We note that the ion signal measurement is not sensitive to the phase relationship between the ground and excited states, but only to the phase differences between the various excited states.

7.4.5 Attosecond Electron Dynamics

The electronic wave packet measurements in Figure 7.15 followed dynamics on picosecond timescales due to the small energy differences between highly excited (Rydberg) states of the atom. One can perform similar measurements on more tightly bound states that evolve on much faster timescales (attoseconds) using shorter pump and probe pulses. An example of attosecond electron dynamics in N_2 is illustrated in Figure 7.16 [30].

The pump is an attosecond extreme ultraviolet (XUV) pulse produced via high-harmonic generation (see Section B.4) that excites a superposition of vibrationally bound electronic states in both the neutral and cation. Panel B plots the relevant PESs, along with the photon energy and spectral intensity of each harmonic order in the attosecond pulse. A subsequent attosecond probe pulse promotes a fraction of the

---

[10]It could be that the probe photon energy corresponds to a Cooper minimum, where the transition dipole moment as a function of photon energy goes through zero due to a change in the sign of the dipole moment [31]. In this case, the ionization yield would be maximized not when the electron is near the core, but at a different phase of its motion (and our simple classical intuition breaks down).

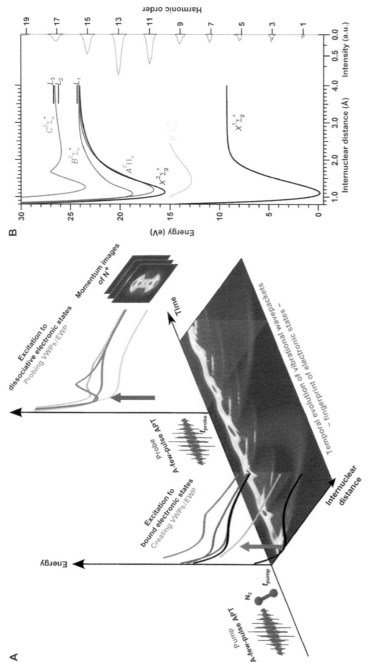

Figure 7.16 Pump–probe measurement of attosecond electronic wave packets in N_2. Figure from Ref. [30].

probability amplitude to dissociative states in the cation that produce the ionic fragment N^+.[11] The fragment ion yield is measured as a function of pump–probe delay in a
TRIS arrangement. To isolate contributions from well-defined final states, the authors
use VMI to resolve the kinetic energy of the dissociated N^+ fragments, selecting fragment ions with a kinetic energy release of about 0.2 eV.

Panel a of Figure 7.17 plots the yield from these fragment ions as a function of pump–
probe delay, while panel b plots the Fourier transform of the time-domain data. Note
the conceptual similarity to the experiment on atomic potassium in Figure 7.15. Both
looked at electron wave packets described in terms of a coherent superposition of
eigenstates prepared by linear interactions with the pump and probe pulses. In addition, both experiments used ion yields as the observable, with the modulations in the

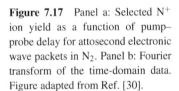

Figure 7.17 Panel a: Selected N^+ ion yield as a function of pump–probe delay for attosecond electronic wave packets in N_2. Panel b: Fourier transform of the time-domain data. Figure adapted from Ref. [30].

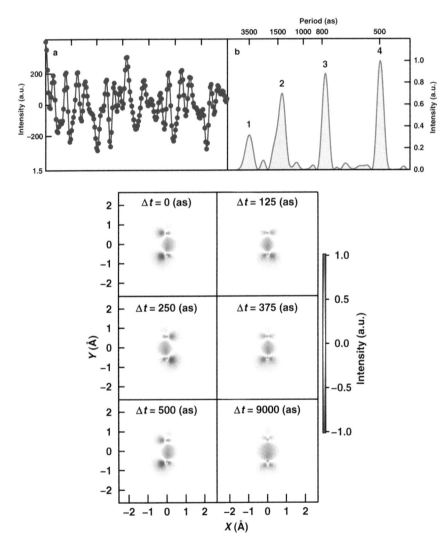

Figure 7.18 Calculated differential electron density for different pump–probe time delays. Figure adapted from Ref. [30].

[11] We note that both the pump and probe pulses are actually attosecond pulse trains since the pulses are
generated via high order harmonic generation with multicycle drive pulses which produce very short trains
of attosecond pulses separated by half an optical cycle of the drive pulse.

ion yield occurring at frequencies that correspond to energy *differences* between electronic states in the wave packet. In particular, the four peaks in panel b of Figure 7.17 can be assigned to energy differences between four separate pairs of electronic states. Peak 1 corresponds to the beating between the $X^2\Sigma_g^+$ and $A^2\Pi_u$ states (ground and first-excited states in the cation), while peak 4 matches the difference between the $A^2\Pi_u$ and $C^2\Sigma_u^+$ states (third excited state in the cation). There is some ambiguity for peaks 2 and 3, as they match energy differences between multiple pairs of states in the neutral and ion. However, using known transition probabilities the authors assign peak 2 to $b'^1\Sigma_u^+$ (high-lying neutral state) and $A^2\Pi_u^+$. Similarly, peak 3 is assigned to $b'^1\Sigma_u^+$ and $B^2\Sigma_u^+$ (second excited state in the cation).

We note that the measurements do not provide direct spatial information about the electron density or wave function, but rather give timescales on which the wave function changes. As with the LIF and TRPES experiments earlier, one needs additional information about the system to fully interpret the measurements. In particular, to produce a picture of how the wave function actually evolves, one must know the form of individual electronic eigenstates for the atom or molecule. One can then generate a picture of the time-dependent wave function by adding the eigenfunctions (with appropriate time-dependent phases), where the amplitudes of the individual eigenfunctions were determined by the measurement. Figure 7.18 shows the time-dependent electron density for a superposition of two different electronic states of N_2^+. For reviews of attosecond physics and other strong-field phenomena, see [32, 33, 34, 35, 36].

CHAPTER 8

Coherent Optical Measurements in 1D

8.1 INTRODUCTION

In the previous chapter, we considered measurements of excited-state vibrational dynamics in molecules using the incoherent detection techniques of laser-induced fluorescence (LIF) and time-resolved photoelectron spectroscopy. We discussed typical experimental setups and the resulting data, as well as interpretations of the measurements. Analytic expressions from models provide a connection with the physical picture, while numerical integration of the equations lets one reproduce the experimental data and "watch" the wave functions evolve in real time.

In this chapter, we consider measuring similar dynamics using coherently scattered light via nonlinear optical spectroscopy. While these techniques typically measure the fluence (or spectrally resolved fluence) of the probe beam, there are many variations that involve slightly different resonance conditions and beam geometries. The techniques go by different names (e.g. transient absorption [TA] or coherent anti-Stokes Raman scattering [CARS]), some of which detect scattered light in directions other than the incident probe. Nevertheless, they all measure coherently scattered light from the sample as a function of pump–probe delay. The data can be modeled by calculating the delay-dependent polarization in the sample after interaction with the pulses. The basic perturbative approach for determining the polarization was outlined in Section 6.4.2; in the following sections, we build on this description while considering a few of the primary techniques.

8.2 TRANSIENT ABSORPTION

8.2.1 Experimental Setup

While the formal description and model of TA has some subtleties, the experimental implementation is perhaps one of the simplest. The measurement can be carried out in condensed or gas-phase samples, and in the simplest configuration, one measures the transmitted fluence of the probe pulse as a function of pump–probe delay. More common is to spectrally resolve the transmitted probe light.

The experimental apparatus is shown in Figure 4.3 of Chapter 4. The pump and probe pulses cross in the sample and are typically linearly polarized, with the angle between their polarization vectors often set at the magic angle of 54.7° to eliminate rotational contributions to the signal (see Section B.1). The experiment measures the spectrally resolved, transmitted probe light, both with and without the pump pulse present. By comparing the two, one obtains the change in absorption of the probe (ΔA) as a function of both wavelength and pump–probe delay.

Despite the simple-looking arrangement, a perturbative description of the measured probe spectrum is surprisingly subtle and features a somewhat counterintuitive dependence on pump and probe pulses. This is due to the fact that absorption of the probe is a coherent optical effect. For any coherent, nonlinear optical measurement, one must calculate the nonlinear polarization of the medium and how this polarization subsequently interacts with the propagating probe pulse.

8.2.2 Data and Basic Interpretation

Before delving into the details regarding how the molecular polarization gives rise to the TA signal, we first present results for two different experimental configurations: excited-state absorption (ESA) and ground-state bleach (GSB). As mentioned in Chapter 4, scattering of the probe occurs due to a generated polarization between either the intermediate state and ground state, or between the intermediate state and a third, higher-lying excited state. In general, these both measure dynamics on the intermediate potential energy surface (PES) b. The former is known as GSB due to the decreased probe absorption (or "bleach") from the action of the pump, whereas the latter is known as ESA since one probes the absorption of an excited state populated by the pump.

8.2.2.1 Excited-State Absorption

Figure 8.1 illustrates the basic idea behind ESA. The pump pulse launches a vibrational wave packet on an excited state b. The probe pulse generates a polarization P_{bc} between states b and c that depends on the amount of wave function present within the Franck–Condon window (i.e. the region where the probe pulse is resonant between states b and c) when the probe arrives. As described in Section 6.4.2, this polarization leads to absorption of the probe pulse as it propagates through the sample. The probe

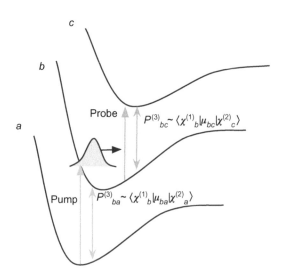

Figure 8.1   Illustration of a pump–probe experiment with a third-order polarization that leads to ESA (via $P_{bc}^{(3)}$) and GSB (via $P_{ba}^{(3)}$).

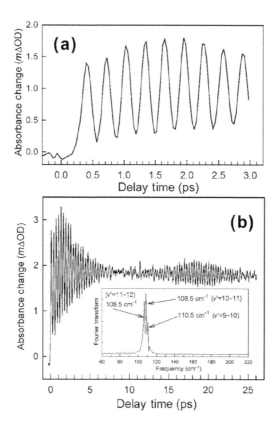

Figure 8.2 ESA data in I_2 showing the change in absorption on both short (a) and long (b) timescales. The change in absorbance is measured in mΔOD, or the change in optical density in units of milli-OD (OD = $10\log(I_{out}/I_{in})$). The inset on the bottom shows eigenstate frequency differences obtained via Fourier transform of the time-domain data. Figure from Ref. [37].

absorption as a function of pump–probe delay, therefore, translates into a measurement of the wave packet density within the Franck–Condon window at each delay.

For comparison with other approaches, we once again examine vibrational dynamics on the B-state of gas-phase I_2 [37]. A pump pulse at 613 nm promotes a portion of the initial ground-state wave function to the excited B state. This wave function subsequently evolves until a time-delayed probe pulse at 400 nm creates a polarization with a highly excited state. Figure 8.2 shows the total (not spectrally resolved) change in absorption of the probe due to the pump pulse. A general increase in the absorption is visible in panel (a) after the pump pulse arrives at time zero and excites the molecule to the B state. The ESA signal is modulated at the oscillation frequency of the B-state wave packet (period of roughly 310 fs). Panel (b) shows the data over long times, where the depth of modulation in the ESA signal goes to zero around 8–9 ps as the B-state wave packet becomes completely delocalized. The modulations return (to some degree) as the wave packet rephases around 19 ps. The inset of panel (b) plots a Fourier transform of the time-domain data. As with the other techniques, the peaks in the transform correspond to frequency differences between pairs of vibrational eigenstates states that comprise the wave packet. In particular, this TA data can be compared to that obtained with LIF (Figure 7.2) and TRIS (Figure 7.10).

8.2.2.2 Ground-State Bleach

Similar information can be obtained by probing the *b* to *a* transition with the probe pulse (also shown in Figure 8.1). In this case, the probe generates a polarization between states *b* and *a*, which again leads to a change in probe absorption that depends on pump–probe delay. Figure 8.3 shows GSB data from gas-phase I_2, where a 529-nm pump pulse generates a time-dependent wave packet on the B state at time zero [38]. The probe pulse, also at 529 nm, is similarly resonant between the X and B states (see panel (b) of Figure 8.3). The change in probe absorption (with and without the pump) is shown as a function of pump–probe delay in panel (a). Note that for GSB, the overall change in absorption is negative: there is less absorption of the probe when the pump is present. Similar to ESA, the GSB signal shows modulations as the wave packet on the B-state potential moves in and out of resonance with the X state. The long-time behavior in panel (a) shows multiple dephasings and revivals of the wave packet.[1]

These particular measurements were performed in an optical cavity. Cavities can be used to enhance both the pump and probe processes. For the pump, it leads to an increase in the excitation fraction, while for the probe, it improves the sensitivity to the excited-state wave packet. The enhancement factors for both processes are given by $\mathcal{F}/\pi$, where $\mathcal{F}$ is the finesse of the cavity. While cavity-enhanced spectroscopy enables very high sensitivity, it comes with a number of technical challenges, including dispersion management of the broad bandwidths inside the cavity.

Given that both pulses are resonant between states a and b, it is natural to wonder how one can distinguish between competing processes. Specifically, in addition to the third-order process shown in Figure 8.1, there could also be absorption of the probe due to a *first-order* polarization $P_{ab}^{(1)}$ generated by two other means: (1) by the pump pulse alone, and (2) the probe-generated polarization independent of the pump. In general both of these processes occur, and it is really a question of how the measurement discriminates between these different cases in the signal.

In the first instance, the pump pulse generates a microscopic polarization in the medium. The phase relationship between various molecules in the ensemble is dictated by the pump pulse, and over a distance of one wavelength (along its k-vector), the phase of the polarization for different molecules cycles through 2π. Thus, when a second pulse propagates through the ensemble with a different frequency or k-vector, averaging over the relative phases of different molecules in the ensemble leads to zero net change in the absorption. Even if the pump and probe pulses had the same frequency and $k-$vector (ensuring a well-defined phase relationship over the molecular ensemble), they would require a stable phase relationship from one shot of the laser to the next for the absorption/emission not to average to zero over multiple laser shots. In either case, there is no change in absorption of the probe due to the polarization generated by the pump pulse alone. Finally, if the pump and probe pulses were the same frequency and had a stable phase relationship between them, then they could essentially be considered as a single combined field, and it becomes similar to case (2), for which any polarization and absorption of the probe independent of pump–probe delay simply contributes a nonzero background to the signal and does not contain any information about the dynamics. We will find that it is the third-order polarization that leads to modulations in the TA signal.

---

[1]The reason the parallel and perpendicular polarization signals differ at early times is due to geometric alignment during the pump interaction: the excitation probability is proportional to the square of the cosine of the angle between the pump polarization axis and the transition dipole moment. The molecules continue to rotate with different angular speeds due to the finite temperature of the sample, and after some time, the molecular ensemble is randomly aligned. For parallel pump and probe polarizations, this leads to a slow decay of the initial signal, while for perpendicular polarizations, the signal slowly increases. At long times, this rotational dephasing leads to equal signals for the two polarizations.

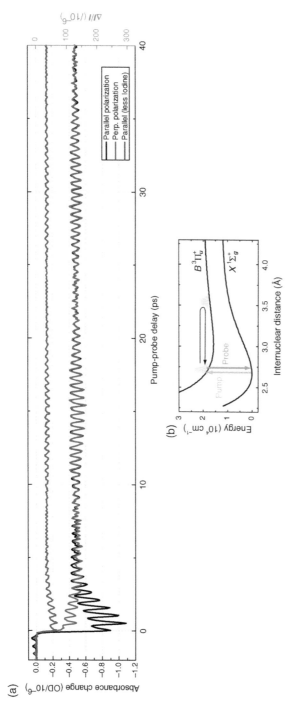

Figure 8.3 **(See color insert.)** GSB data in I_2 using cavity-enhanced spectroscopy. Panel (a): long-time GSB signal for different polarization configurations. Panel (b): illustration of wave packet dynamics and pump/probe photons for the relevant PESs. Figure adapted from Ref. [38].

8.2.3 Mathematical Description

We are now ready to mathematically connect the induced polarization to the signal measured in the laboratory. In particular, TA measures changes in probe absorption due to the interaction of the propagating probe pulse and the induced polarization in the medium oscillating at the same frequency. We separately consider mathematical descriptions for both ESA and GSB, building on the treatment from Section 6.4.2.

8.2.3.1 Excited-State Absorption

We begin with the polarization P_{bc} induced by the probe pulse that leads to ESA. This comes from Equation 6.39 describing the first-order polarization for the probe: $P_{probe}^{(1)}(t)$.[2] Assuming that the pump pulse arrives at $t = 0$, the initial wave function on the excited state is $\chi_b^{(1)}(R,0)$. When keeping the propagators on the b state, Equation 6.39 can be written as

$$P_{probe}^{(1)}(t) = \left\langle \chi_b^{(1)}(R,t) \middle| \mu_{bc} \middle| \chi_c^{(2)}(R,t) \right\rangle$$
$$= \frac{-1}{i} \left\langle \chi_b^{(1)}(R,0) \middle| e^{+iH_bt} \mu_{bc} \int_0^t dt_2 e^{-iH_c(t-t_2)} \left(\mu_{cb}E_2(t_2) \right) \right.$$
$$\left. \times e^{-iH_bt_2} \middle| \chi_b^{(1)}(R,0) \right\rangle, \tag{8.1}$$

where $E_2(t_2)$ is the probe field. Once again, this expression can be simplified for interpretation. Inside the integral (for time t_2), we write out the Hamiltonian operators in terms of kinetic and potential contributions: $H_b = T + V_b$ and $H_c = T + V_c$. We also consider the limit of a short probe pulse such that one can neglect evolution of the wave function during the pulse duration. Substituting the Hamiltonians into Equation 8.1, separating the field envelope from the carrier, and neglecting the commutator of T and V inside the integral (valid for short time durations), we have

$$P_{probe}^{(1)}(t) = \frac{-1}{i} \left\langle \chi_b^{(1)}(R,0) \middle| e^{+iH_bt} \mu_{bc} \int_0^t dt_2 e^{-iH_ct} e^{+i(T+V_c)t_2} \right.$$
$$\times \mu_{cb}E_{02}(t_2)\frac{1}{2} \left(e^{+i\omega_2t_2} + e^{-i\omega_2t_2} \right) e^{-i(T+V_b)t_2} \middle| \chi_b^{(1)}(R,0) \right\rangle$$
$$= \frac{-1}{i} \left\langle \chi_b^{(1)}(R,0) \middle| e^{+iH_bt} \mu_{bc} e^{-iH_ct} \int_0^t dt_2 e^{+i(V_c(R))t_2} \right.$$
$$\times \mu_{cb}E_{02}(t_2)\frac{1}{2} \left(e^{+i\omega_2t_2} + e^{-i\omega_2t_2} \right) e^{-i(V_b(R))t_2} \middle| \chi_b^{(1)}(R,0) \right\rangle \tag{8.2}$$
$$= \frac{-1}{2i} \left\langle \chi_b^{(1)}(R,0) \middle| e^{+iH_bt} \mu_{bc} e^{-iH_ct} \int_0^t dt_2 e^{+i(V_c(R)-V_b(R)-\omega_2)t_2} \right.$$
$$\times \mu_{cb}E_{02}(t_2) \middle| \chi_b^{(1)}(R,0) \right\rangle,$$

where in the second step, we canceled the kinetic energy operators in the propagators inside the integral, and in the last step, we made the rotating wave approximation: $V_c(R) - V_b(R) - \omega_2 \ll V_c(R) - V_b(R) + \omega_2$. The remaining exponential inside the integral in the last line makes it clear that for a finite-duration probe pulse, the polarization is significant only if the probe frequency is resonant between the two potentials for some R (essentially a statement of conservation of energy).

---

[2] Note that as earlier, this polarization will actually be third order in the *total* applied field.

To simplify Equation 8.2 further, we define the function F to encompass the resonance condition factor (as in Equation 7.9):

$$F(R,t,\tau) \equiv \frac{-1}{2i} \int_0^t dt_2 e^{+i(V_c(R)-V_b(R)-\omega_2)t_2} \mu_{cb} E_{02}(t_2), \qquad (8.3)$$

where the dependence on probe delay time τ is buried in the probe-pulse envelope $E_{02}(t_2)$, which is only nonzero for time values around $t_2 = \tau$. Rewriting Equation 8.2 with F, we have

$$P_{\text{probe}}^{(1)}(t) = \left\langle \chi_b^{(1)}(R,0) \left| e^{+iH_bt} \mu_{bc} e^{-iH_ct} F(R,t,\tau) \right| \chi_b^{(1)}(R,0) \right\rangle. \qquad (8.4)$$

The function $F(R,t,\tau)$ is peaked near R values where $(V_c(R) - V_b(R) - \omega_2)$ is roughly zero. Since $F(R,t,\tau)$ multiplies the wave function inside the spatial integral, it picks out the portion of the wave function for R values where this resonance condition is met. Thus, the polarization is large for time delays τ when the wave packet is at the resonance location between the two PESs when the probe pulse arrives.

In the preceding equations, we wrote $P_{bc}(t)$ in terms of a first-order polarization for the probe. For reference, we provide here the full, third-order polarization between states b and c in terms of the original wave function on state a (see Equation 6.41):

$$P_{bc}^{(3)}(t) = \frac{-1^3}{i^3} \left\langle \chi_a^{(0)}(R,0) \left| \int_0^t dt_2 \int_0^{t_2} dt_1 \int_0^t dt_3 e^{iH_a t_3} \left(\mu_{ab} E_1(t_3) \right) \right.\right.$$
$$\times e^{iH_b(t-t_3)} |\mu_{bc}| e^{-iH_c(t-t_2)} \left(\mu_{cb} E_2(t_2) \right) e^{-iH_b(t_2-t_1)}$$
$$\times \left. \left(\mu_{ba} E_1(t_1) \right) e^{-iH_a t_1} \left| \chi_a^{(0)}(R,0) \right. \right\rangle. \qquad (8.5)$$

We note the dependence on the pump and probe fields in this expression: $P_{bc}^{(3)}$ depends linearly on the probe field (E_2) and quadratically on the pump field (E_1).

Having determined the induced polarization in the medium, we are now interested in how the probe pulse changes due to the coherent interaction between this polarization and the propagating probe field. Both the phase and amplitude of the polarization are modulated by the evolution of the wave packets, which in turn modify the propagating probe field. Recalling Equations 6.42–6.44, we have:

$$E_0^{\text{probe}}(z,t) - E_0^{\text{probe}}(z=0,t) = \frac{2\pi\omega_2 i}{c} P_0^{\text{macro}}(t)z, \qquad (8.6)$$

where $P_0^{\text{macro}}(t)$ is the envelope of the macroscopic polarization (see Section A.8.2).

A TA experiment typically measures one of the three quantities: the probe fluence, $\int |E_{\text{probe}}(t)|^2 dt$; the time-integrated, spectrally resolved probe intensity, $|E_{\text{probe}}(\omega)|^2$; or the time-integrated intensity of the probe field plus a reference field (i.e. a heterodyne measurement), $\int |E_{\text{probe}}(t) + E_{\text{ref}}(t)|^2 dt$. If one simply measures the probe fluence as a function of pump–probe delay (like the data of Figures 8.2 and 8.3), the measured signal, S_{ESA}, is given by[3]:

---

[3] As mentioned earlier, one typically looks at the *change* in this signal with and without the pump pulse.

$$S_{ESA} = \int dt \left| E_0^{probe}(z,t) \right|^2$$

$$= \int dt \left| E_0^{probe}(z=0,t) + \frac{2\pi\omega_2 i}{c} P_0^{macro}(t)z \right|^2$$

$$= \int dt \left| E_0^{probe}(z=0,t) \right|^2 - \int dt \frac{4\pi\omega_2 z}{c} E_0^{probe}(z=0,t)$$

$$\times Im\{P_0^{macro}(t)\} + \int dt \left| \frac{2\pi\omega_2}{c} P_0^{macro}(t)z \right|^2. \tag{8.7}$$

Our assumption of an optically thin sample implies that the probe undergoes very little change as it propagates through the sample, and so the last term in Equation 8.7 is much smaller than the first two and can be neglected. Thus, the measured TA signal can be written as:

$$S_{ESA} \approx \int dt \left| E_0^{probe}(z=0,t) \right|^2 - \int dt \frac{4\pi\omega_2 z}{c} E_0^{probe}(z=0,t)$$

$$\times Im\{P_0^{macro}(t)\}. \tag{8.8}$$

The first term leads to a large, delay-independent offset that is modulated by the smaller, second term that varies with pump–probe delay.

We can draw a number of conclusions from our results. Since changes in the probe fluence are small, one needs to make sensitive measurements, and fluctuations in the probe fluence should be avoided as much as possible since they can easily overwhelm any pump-induced changes. Furthermore, the polarization is proportional to the second power of the pump field amplitude (the intensity), but only the first power of the probe field amplitude (square root of the intensity). Thus the signal, which is proportional to the product of the polarization and probe field, is linear in both the pump and probe fluences. Therefore, the *fractional* change in the probe fluence, $S_{ESA}/(\int dt |E_0^{probe}(z,t)|^2)$, is independent of probe fluence. In other words, increasing the pump-pulse energy can improve the signal, but increasing the probe energy will not. Finally, since the change in fluence is proportional to the product of the probe field times the polarization (which itself is proportional to three field amplitudes), the measured signal involves four fields and a third-order nonlinearity. Nonlinear optical processes of this type are known as "four-wave mixing." There are many possible variations of the general technique, one of which is CARS (see Section 8.3).

Although we considered a single, local-in-time interaction with the pump pulse during ESA, the interaction is, in fact, second order in the pump field. As we shall see, in CARS this second-order interaction can naturally be considered an up-down process, returning wave function population to the ground state. In the case of TA, however, this second-order interaction with the pump pulse only goes to a single state.[4] Of course, there are many responses of the molecular sample to the total applied field that enter at different orders in perturbation theory (second, third, fourth, etc.), and what dominates a given experimental measurement is not necessarily the first nonvanishing term. In many cases, coherent buildup (e.g. phase matching) of a given signal means

---

[4]One way to interpret the second-order dependence is that the interaction with the pump must "leave enough wave function behind" in state b after interaction with the probe to create the coherent polarization with the wave function on state c. In other words, in the perturbative picture, the fact that there is negligible ground-state depletion implies that you cannot "use" the excited-state wave function for an additional process.

it can overwhelm the others. In particular, in TA spectroscopy, the coherent signal detected after interaction with a sufficient path length of molecules dominates over any contributions that are lower order in the pump and probe fields.

8.2.3.2 Ground-State Bleach

In this section, we mathematically describe a GSB measurement following a similar treatment as for ESA. The primary difference between GSB and ESA arises due the frequency of the probe pulse. In ESA, the probe resonantly couples states b and c, and the probe frequency is typically well separated from that of the pump coupling states a and b. In a GSB measurement, the frequency of the probe is resonant with the a to b transition as well, implying that the pump and probe frequencies are similar (if not identical). The result is that in GSB, the second-order wave function of interest after interaction with the pump and probe fields is back on the ground-state surface.

We begin by writing down the second-order wave function describing the molecule after interaction with both pump and probe pulses. A modification of Equation 6.32 for this case reads:

$$\chi_a^{(2)}(R,t) = \frac{1}{i^2} \int_0^t dt_2 \int_0^{t_2} dt_1 e^{-iH_a(t-t_2)} \left(-\mu_{ab}E_2(t_2)\right)$$
$$\times e^{-iH_b(t_2-t_1)} \left(-\mu_{ba}E_1(t_1)\right) e^{-iH_a t_1} \chi_a(R,0). \tag{8.9}$$

Once again writing this in terms of $\chi_b^{(1)}$, or the wave function produced by the pump pulse (see Equation 6.31), we have

$$\chi_a^{(2)}(R,t) = \frac{1}{i} \int_0^t dt_2 e^{-iH_a(t-t_2)} \left(-\mu_{ab}E_2(t_2)\right) \chi_b^{(1)}(R,t_2)$$
$$= \frac{1}{i} \int_0^t dt_2 e^{-iH_a(t-t_2)} \left(-\mu_{ab}E_2(t_2)\right) e^{-iH_b t_2} \chi_b^{(1)}(R,0), \tag{8.10}$$

where in the last line, we have again used the propagator on state b to write the expression in terms of the wave function immediately after the pump pulse arrives at time zero. Substituting this into the corresponding version of Equation 8.1 for a polarization between states a and b yields

$$P_{\text{probe}}^{(1)} = \left\langle \chi_b^{(1)}(R,t) \left| \mu_{ba} \right| \chi_a^{(2)}(R,t) \right\rangle$$
$$= \frac{-1}{i} \left\langle \chi_b^{(1)}(R,0) \left| e^{+iH_b t} \mu_{ba} \int_0^t dt_2 e^{-iH_a(t-t_2)} \left(\mu_{ab}E_2(t_2)\right) \right. \right.$$
$$\times e^{-iH_b t_2} \left| \chi_b^{(1)}(R,0) \right\rangle. \tag{8.11}$$

We now follow the same steps as with ESA, writing out the Hamiltonian operators to visualize the resonance condition in the rotating-wave approximation. We define an analogous expression for F to encompass the resonance condition factor for the case of GSB:

$$F(R,t,\tau) \equiv \frac{-1}{2i} \int_0^t dt_2 e^{-i(V_b(R)-V_a(R)-\omega_2)t_2} \mu_{ab}E_{02}(t_2). \tag{8.12}$$

Rewriting Equation 8.11 with F, we have

$$P_{\text{probe}}^{(1)} = \left\langle \chi_b^{(1)}(R,0) \left| e^{+iH_b t} \mu_{ba} e^{-iH_a t} F(R,t,\tau) \right| \chi_b^{(1)}(R,0) \right\rangle. \tag{8.13}$$

For GSB, the function $F(R,t,\tau)$ deviates from zero for R values where $(V_b(R) - V_a(R) - \omega_2)$ is roughly zero, indicating that the polarization is large whenever the probe pulse is resonant between the lowest two PESs. For these values (and in the Condon approximation when μ_{ba} does not depend on R), the expression in the bracket can be thought of as a correlation function of two wave packets: one evolving on state b and the other on state a. As with ESA, Equation 8.13 is linear in the probe field but quadratic in the pump. If one measures the probe-pulse fluence at each delay τ, the signal is given by (see Equations 8.7 and 8.8):

$$
\begin{aligned}
S_{\text{GSB}} &= \int dt \left| E_0^{\text{probe}}(z,t) \right|^2 \\
&= \int dt \left| E_0^{\text{probe}}(z=0,t) + \frac{2\pi\omega_2 i}{c} P_0^{\text{macro}}(t)z \right|^2 \\
&\approx \int dt \left| E_0^{\text{probe}}(z=0,t) \right|^2 - \int dt \frac{4\pi\omega_2 z}{c} E_0^{\text{probe}}(z=0,t) \\
&\qquad \times \operatorname{Im}\{P_0^{\text{macro}}(t)\},
\end{aligned}
\tag{8.14}
$$

where $P_0^{\text{macro}}(t)$ is the envelope of the macroscopic polarization corresponding to the microscopic polarization of Equation 8.13.

In summary, both ESA and GSB measure the change in transmission of the probe due to the presence of the pump pulse. In the case of ESA, the action of the pump pulse leads to an increased absorption of the probe, as the pump creates a wave packet on PES b that the probe can excite to PES c. In GSB, the action of the pump leads to less absorption of the probe between states a and b due to the decreased ground-state population remaining after the pump pulse. Note that the difference in sign between GSB and ESA is contained in the indices of the polarizations: $P_{ba}^{(3)}(t)$ for GSB versus $P_{bc}^{(3)}(t)$ for ESA. The sign of $\operatorname{Im}\{P_0^{\text{macro}}(t)\}$ depends on the ordering of the indices, and in the case of GSB, the higher energy state is the first index, whereas the opposite is true for ESA. This difference corresponds to the fact that the polarizations lead to either increased probe intensity (GSB) or decreased probe intensity (ESA) due to the action of the pump pulse.

8.2.4 TA with Core-Level Excitation

Although not explicitly discussed, the TA experiments described in the previous sections assumed the probe field interacted primarily with electrons in valence orbitals. TA experiments can also be carried out with higher energy photons that probe core-to-valence transitions, as the absorption frequency of these transitions can vary significantly along the reaction coordinate. In addition, exciting core-to-valence transitions can provide atom-specific information, since the core electrons are localized on a particular atom (rather than being delocalized over the molecule as valence electrons tend to be). Since core orbitals are much more tightly bound, core-to-valence transitions typically involve photons with energies 50–500 eV, necessitating the use of extreme ultraviolet (XUV) radiation.

Figure 8.4 illustrates the basic idea of a TA experiment using a core-to-valence transition during the probe step of the experiment. The particular experiment considered here used a pump pulse to excite a vibrational wave packet on an ionic PES by removing the valence electron from the highest occupied molecular orbital (HOMO). An XUV probe then excites a core electron to the lowest unoccupied molecular orbital

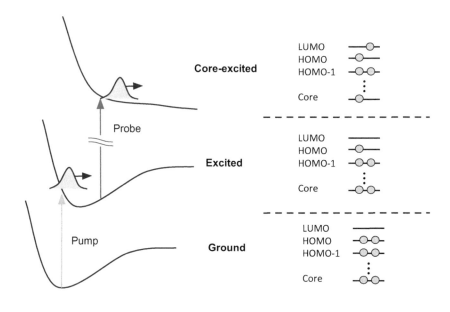

Figure 8.4 Diagram showing TA with core-level excitation from both an electronic-state picture (left) and a molecular orbital picture (right). In this particular case, the pump excitation ionizes the molecule (as indicated in the orbital picture).

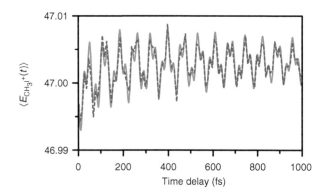

Figure 8.5 The first moment of the absorption signal as a function of pump–probe delay in CH_3I. The probe pulse initiates wave packet motion in the molecular cation through strong field ionization and the XUV probe drives a core-to-valence transition centered near 47 eV. Figure adapted from Ref. [39].

(LUMO). The probe process can be atom specific based on known core-to-valence transition energies in different elements (for example, in the data of Figure 8.5, the core excitation is known to come from a $4d$ electron of iodine).

Figure 8.5 shows measurements of wave packet motion in CH_3I probed by XUV TA [39]. An intense, near-infrared laser pulse is used to ionize the molecule, producing a vibrational wave packet on the ground electronic state of the molecular cation. The minimum energy of this PES is displaced from the ground-state equilibrium geometry, and the subsequent wave packet motion is probed by XUV absorption. This probe is sensitive to wave packet motion due to small shifts in the core-to-valence absorption frequency with displacement along vibrational coordinates. Figure 8.5 plots the first moment of the absorption signal (photon energy) for a core-to-valence transition that takes a $4d$ core iodine electron to an unoccupied valence orbital. The data show modulations at two frequencies: one corresponds to motion along the C–I stretching mode, while the other to motion of the CH_3 umbrella mode. While TA with core level excitation is experimentally more challenging than TA with valence excitation, it relies on the same basic principles, and can be described with the same formalism.

8.2.5 Advantages and Disadvantages: Why Choose TA?

Coherent optical spectroscopies typically offer relatively large signals, as they involve a coherent buildup of the signal over the sample. This means that photon or electron/ion counting are generally not required. In addition, these techniques are well suited to condensed-phase samples since they do not require vacuum (as photoelectron spectroscopy does). As we explore in detail in the next section, a primary advantage of coherent optical spectroscopies is that they allow for selection of particular nonlinear contributions to the signal using phase matching.

On the other hand, coherent optical spectroscopies are experimentally more complex than a technique such as LIF. In addition, they are not well suited for studying systems where a limited number of molecules contribute to scattering, since the coherent signal must build up in the sample. Therefore, dilute, gas-phase systems are typically incompatible with TA experiments unless the apparatus is sensitive enough to detect extremely small changes in absorption (for example, the data from Figure 8.3). Finally, standard TA measurements are not background free, and one is typically trying to measure small changes in the probe signal on top of a larger background.[5] For reviews of TA, see [40, 41, 42].

Given the widespread use of spectrally-resolved transient absorption experiments, it is natural to ask whether one could accomplish the same outcome by simply adding together the results of multiple experiments using narrow-band (long) pulses. After all, if one coherently adds together the wave packets formed by multiple, narrow-band pump pulses with slightly different frequencies, the resulting wave packet is the same as that produced by a short pulse whose spectral bandwidth covers the full range of the various long-duration pulses.[6] However, time-domain measurements for the two cases using coherent, nonlinear spectroscopy will generally not be the same. This stems from the fact that the typical pump–probe measurement is nonlinearly dependent on the total applied field, and so any measured quantity summed over a series of long-pulse measurements is not the same as for a single, short pulse (or pulse pair).

For example, in a pump–probe experiment such as laser-induced fluorescence, the probability for being in the final state as a function of pump–probe delay for both short-pump and short-probe pulses cannot be expressed as a sum of the probabilities for measurements carried out with long pump and probe pulses. In particular, for pulses whose duration is longer than the vibrational period on the excited state, the delay-dependent yields for long pulses cannot add constructively or destructively to yield the modulations that one observes with a pulse whose duration is shorter than the vibrational period on the excited state. This reinforces the idea that time-resolved spectroscopies that yield information beyond what is available from frequency-domain measurements (the limit of very long pulses with infinitesimal bandwidths) are inherently nonlinear spectroscopies, where the measured quantity is always a nonlinear function of the total applied field.

---

[5]As we discuss in the next section, other coherent optical techniques can achieve background-free conditions through phase matching.

[6]We assume each pulse interacts with the molecule weakly in a manner that is well described by first-order perturbation theory.

8.3 COHERENT ANTI-STOKES RAMAN SCATTERING

8.3.1 Introduction

In all the molecular examples discussed so far, we considered techniques for measuring dynamics on an *excited* electronic PES (state *b*). Here we consider a coherent optical experiment designed to probe vibrational dynamics on the *ground* electronic state of a molecule (state *a*). Energetically, it is impossible for optical radiation to directly couple vibrational states in a single electronic state, as these transitions appear in the mid-infrared region of the electromagnetic spectrum (wavelengths ~ 3–$10\,\mu$m). An attractive alternative to direct infrared excitation is to use Raman scattering to generate a vibrational wave packet on the ground electronic state (see Section 3.3.2 for a discussion of infrared and Raman interactions). In this section, we discuss a technique known as CARS for probing ground-state molecular dynamics.

Like standard TA described earlier, CARS is a third-order nonlinear spectroscopy involving a coherent interaction of applied fields and fields emitted by the induced molecular polarization. Unlike TA, CARS is a fully phase-matched process, and particular implementations can result in background-free signals. Our discussion of CARS also provides an opportunity to consider how the basic pump–probe framework we have developed can be extended to experimental situations that, at first glance, appear significantly more complicated.

The interaction with the pump pulse in CARS is a two-photon Raman process that involves absorption of one photon (known as the "pump") and emission of a second photon (the "Stokes") at a different frequency. Both frequencies can be present if the pump pulse has sufficient bandwidth. [7] This interaction leaves the molecule in the ground electronic state but with some vibrational energy/excitation due to the difference in frequencies of the pump and Stokes photons. In many CARS experiments the pump and Stokes photons are provided by separate pulses incident on the sample from different directions and at potentially different times. Nevertheless, it can be useful to consider the net effect of the pump and Stokes interactions in terms of a single interaction which occurs during the Stokes pulse if the two are separated in time. Thus, from a microscopic perspective, the CARS process is still conveniently described in terms of two separate interactions: one with a pump (that includes the "pump" and "Stokes" photons) and one with a probe. As with TA, we consider the experiment from both microscopic and macroscopic perspectives. We shall see that, for CARS, relating to the macroscopic viewpoint is even more important given the phase-matching considerations necessary to describe the experiment.

8.3.2 Experimental Setup

While there are a number of different experimental arrangements for a CARS measurement, we choose to focus on the case where the pump and probe frequencies are near resonant with an excited electronic state in the molecule, and the beams are configured in what is known as the box-CARS geometry. Figure 8.6 shows the basic idea of such a resonant, box-CARS experiment. The top portion is a plot of the energy-level diagram for the process. Molecules first undergo a stimulated, two-photon Raman process driven by the pump and Stokes photons. From the perspective of a given molecule, the fact that the pump and Stokes photons potentially come from two different laser pulses

---

[7]This is known as "impulsive" Raman scattering.

Figure 8.6 (See color insert.) Illustration of energy and momentum conservation in a box-CARS experiment. The top panel shows the PESs involved, along with the pump and probe interactions (including the Stokes and anti-Stokes fields). The bottom-left depicts the experimental beam geometry: three beams are focused into the sample, and coherently scattered probe light at the anti-Stokes frequency serves as the signal. The bottom-right illustration shows a two-dimensional representation of the phase-matching condition, in which the wave vectors for the beams combine so that $\Delta\mathbf{k} = 0$.

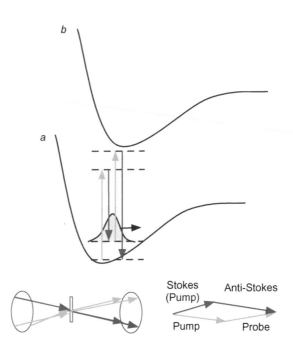

is unimportant, as the molecule simply experiences the composite field of the two pulses. We consider this combined interaction to be the pump process in our pump–probe framework. The Raman scattering transfers a portion of the initial nuclear wave function into higher-lying vibrational state(s) on PES a, resulting in a time-dependent wave packet in the ground electronic state. We note that choosing a pump-photon frequency near the transition to an excited PES as depicted in Figure 8.6 enhances the process and is known as a "resonant Raman" interaction. This is not strictly necessary, and many femtosecond CARS experiments use a pump frequency well below the first excited electronic state.

After some time delay, the probe pulse arrives (although not required, the probe is typically the same frequency as the pump). Similar to TA, the measurement involves a coherent detection of light from the sample, and our description will once again rely on the induced polarization in the medium. However, unlike TA, it is not the transmitted probe light that is measured. While a number of different interactions are possible, the one of interest in a CARS experiment involves the molecule absorbing a probe photon and emitting a higher-energy photon (the "anti-Stokes") that returns the molecule back to the ground electronic state. On the microscopic level, the anti-Stokes emission is initially a spontaneous process, as there is no light present in the incident fields at the anti-Stokes frequency. Subsequent molecules further downstream in the macroscopic sample see incident radiation at the anti-Stokes frequency, and the process becomes stimulated. On the macroscopic level, it is this coherently stimulated light at the anti-Stokes frequency that serves as the experimental signal.

The beam arrangement in a box-CARS experiment is shown in the lower-left portion of Figure 8.6. Three different laser pulses are focused into the sample through a single lens (the name box-CARS comes from the fact that three pulses sit at three corners of a square or box). Two of these pulses have central frequencies at the pump and Stokes colors and together comprise the pump portion of the process. The third pulse acts as the probe. The coherently stimulated anti-Stokes light that serves as the signal is emitted along a separate direction from the three incident beams and arrives at the fourth corner of the box after the sample. The fact that the anti-Stokes signal comes out in a different direction is due to phase matching and is dictated entirely by macroscopic considerations. Specifically, coherent production of light at the anti-Stokes frequency is enhanced when the light generated downstream is emitted in phase with the existing radiation. This occurs when a condition known as phase matching is achieved.

In phase matching, the total wavevector mismatch of the process is zero (or at least minimized):

$$\Delta \mathbf{k} = \mathbf{k}_{pump} - \mathbf{k}_{Stokes} + \mathbf{k}_{probe} - \mathbf{k}_{anti\text{-}Stokes} = 0, \qquad (8.15)$$

where the wavevectors of the "emitted" fields (Stokes and anti-Stokes) combine with a negative sign. There will only be significant buildup of light at the anti-Stokes frequency in the direction where this condition is met.[8] This is shown geometrically in the lower-right portion of Figure 8.6, where the four wavevectors are added vectorially. This geometry provides a major advantage of box-CARS: it is an essentially background-free measurement, since the coherently emitted anti-Stokes beam can be spatially selected before detection. In the macroscopic picture, one can think about the interference of the crossing pump and Stokes pulses setting up a complicated transient grating in the molecular sample; variations in the index of refraction due to the modulated excitation leads to scattering of the probe beam into the anti-Stokes signal with a specific wavevector.

In a typical box-CARS experiment, the pump and Stokes pulses arrive at the same time. The delay of the probe pulse is then scanned while the intensity of the anti-Stokes signal is monitored. The change in the anti-Stokes signal as a function of pump–probe delay serves as the signal and yields information about molecular dynamics on the ground electronic state.

8.3.3 Data and Basic Interpretation

Figure 8.7 shows resonant box-CARS data from gas-phase I_2, where the pump wavelength of 525 nm is resonant with the excited B state [43]. The Stokes pulse (wavelength 545 nm) arrives coincident with the pump and together they create a vibrational wave packet on the ground X state through a two-photon Raman transition. The probe pulse at 525 nm scatters from the sample, and the integrated CARS signal is plotted as a function of delay between the pump and probe pulses. A modulation with a period of 160 fs is observed in the data, corresponding to the round-trip time of the wave packet induced by the pump interaction on the *ground* electronic state. The slower decay in the signal is due to vibrational and rotational dephasing.[9] In the eigenstate picture, the 209 cm^{-1} modulation frequency (160 fs period) corresponds to the frequency difference between the vibrational eigenstates comprising the X-state wave packet ($v = 3, 4$).

---

[8]Phase matching is only present in TA in that the probe field interacts coherently (in phase) with the emitted light due to P_{bc} (ESA) or P_{ba} (GSB). So although TA is a coherent process (similar to stimulated emission in a laser cavity), the pump and probe are not phase matched in TA, and so the total wavevector mismatch is not zero as in box-CARS.

[9]The rotational dephasing is due to the fact that initially, molecules are excited in proportion to the $\cos^2$ of the angle between their transition dipole moment and the pump and Stokes polarization vectors. However, the molecules are still rotating with different angular velocities given the thermal distribution. With time they will incoherently come out of alignment with the probe pulse (i.e. their transition dipole moments will be randomly aligned relative to the probe polarization vector); this is known as rotational dephasing.

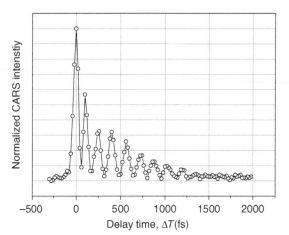

Figure 8.7 Box-CARS data for gas-phase I_2. Periodicity of 160 fs represents vibrational dynamics on the ground electronic PES. Figure adapted from Ref. [43].

8.3.4 Mathematical Description

As always, we begin by calculating the wave function after the pump interaction. Unlike our previous examples, in CARS, the wave function after the pump is formed by a two-photon, Raman interaction with both pump and Stokes beams (and is therefore second order in perturbation theory). In fact, the wave function after the pump and Stokes interaction is the same as for the case of GSB in Section 8.2.3 with the probe pulse replaced by the Stokes (see Equation 8.9):

$$\chi_a^{(2)}(R,t) = \frac{1}{i^2} \int_0^t dt_2 \int_0^{t_2} dt_1 \, e^{-iH_a(t-t_2)} \left(-\mu_{ab}E_{St}(t_2)\right)$$
$$\times e^{-iH_b(t_2-t_1)} \left(-\mu_{ba}E_{pu}(t_1)\right) e^{-iH_a t_1} \chi_a(R,0), \quad (8.16)$$

where $E_{pu}(t_1)$ is the pump field, $E_{St}(t_2)$ is the Stokes field, and the superscript (2) on χ specifies that the combined pump and Stokes process is a second-order interaction. If the pump and Stokes fields occur at the same time (as in the experiment of Figure 8.7), there is no actual propagation time on the excited state: $t_2 - t_1 = 0$ in the b-state propagator. For this derivation, we leave the expression general so that the pump and Stokes could arrive at different times, although as noted earlier, we can still consider their combined effect as one "interaction" where the pump-probe delay is determined by the time between the Stokes and probe pulses.

The next step is to calculate the interaction of the probe pulse with the second-order wave function of Equation 8.16. The probe process is first order, and in resonant CARS takes the wave function $\chi_a^{(2)}(R,t)$ back up to the excited PES b. We therefore use Equation 6.31 for the general, first-order wave function, where the initial wave function is the one *after* interaction with the pump and Stokes (Equation 8.16). Substituting this, we find that the wave function after interaction with the probe is third order in the total applied field:

$$\chi_b^{(3)}(R,t) = \frac{1}{i^3} \int_0^t dt_3 \int_0^{t_3} dt_2 \int_0^{t_2} dt_1 \, e^{-iH_b(t-t_3)} \left(-\mu_{ba}E_{pr}(t_3)\right)$$
$$\times e^{-iH_a(t_3-t_2)} \left(-\mu_{ab}E_{St}(t_2)\right) \quad (8.17)$$
$$\times e^{-iH_b(t_2-t_1)} \left(-\mu_{ba}E_{pu}(t_1)\right) e^{-iH_a t_1} \chi_a(R,0).$$

It is the wave function in Equation 8.17 that gives rise to the nonlinear polarization connecting states a and b (which subsequently generates the anti-Stokes signal we measure in the lab):

$$
\begin{aligned}
P_{ab}^{(3)}(t) &= \left\langle \chi_a^{(0)}(R,t) \middle| \mu_{ab} \middle| \chi_b^{(3)}(R,t) \right\rangle \\
&= \frac{-1^3}{i^3} \left\langle \chi_a^{(0)}(R,t) \middle| \mu_{ab} \int_0^t dt_3 \int_0^{t_3} dt_2 \int_0^{t_2} dt_1 \right. \\
&\quad \times e^{-iH_b(t-t_3)} \left(\mu_{ba} E_{\mathrm{pr}}(t_3) \right) e^{-iH_a(t_3-t_2)} \left(\mu_{ab} E_{\mathrm{St}}(t_2) \right) \\
&\quad \times e^{-iH_b(t_2-t_1)} \left(\mu_{ba} E_{\mathrm{pu}}(t_1) \right) e^{-iH_a t_1} \left| \chi_a(R,0) \right\rangle .
\end{aligned}
\tag{8.18}
$$

While it is still a third-order polarization that generates the signal in CARS, it comes from an inner product of zeroth-order and third-order wave functions (as opposed to first- and second-order wave functions in TA). It is worth noting that if the pump and Stokes excitations are both considered part of the single pump interaction, the expression we arrive at is again second order in the pump and first order in the probe. This allows us to treat all time-resolved experiments as pump–probe measurements that simply depend in different ways on the pump and probe fields. In the case of CARS, the second-order Raman interaction for the pump is necessitated by the fact that one wishes to measure dynamics on the ground electronic state.

As we did earlier for ESA, one can write the full, third-order nonlinear polarization in terms of a first-order polarization for the probe:

$$
P_{\mathrm{probe}}^{(1)}(t) = \left\langle \chi_a^{(0)}(R,0) \middle| e^{+iH_a t} \mu_{ab} e^{-iH_b t} F(R,t,\tau) \middle| \chi_a^{(2)}(R,0) \right\rangle,
\tag{8.19}
$$

where the resonance condition factor is given by

$$
F(R,t,\tau) \equiv \frac{-1}{2i} \int_0^t dt_3 e^{+i(V_b(R)-V_a(R)-\omega_{\mathrm{pr}})t_3} \mu_{ba} E_{0\mathrm{pr}}(t_3),
\tag{8.20}
$$

where $E_{0\mathrm{pr}}$ is the probe pulse envelope and ω_{pr} its frequency. Note that for simplicity, we have assumed that the pump and Stokes fields both arrive at the same time ($t = 0$) in Equation 8.19. This is not necessary, and introducing a time delay between them "pumps" a wave packet to the excited state and lets it evolve for some time before "dumping" it back to the ground electronic state with the Stokes pulse. In that case, the zero of time would correspond to when the Stokes field arrives and the evolution on the ground state begins. Although it is useful to think of the pump and Stokes fields as a combined "pump pulse" that initiates wave packet dynamics on the ground state, in the box-CARS geometry, they are required to be separate pulses with different k-vectors to produce the phase-matched CARS signal on the fourth corner of the box. In some implementations, the pump and Stokes fields are identical copies of a single, short pulse with a spectral bandwidth broader than the vibrational level spacing; in this way, both pump and Stokes frequencies are present in two pulses.

As earlier, the $F(R,t,\tau)$ term is large whenever the argument of the exponential is close to zero. This occurs when the wave packet is at a position (at the time of arrival of the probe pulse) where the probe frequency is resonant between the two potentials. This is similar to what we saw in ESA and GSB, and it relates to energy conservation. The difference now is that in box-CARS, one can enforce momentum conservation by looking at a particular k-vector direction that satisfies the phase-matching condition discussed in Figure 8.6. This allows for a background-free measurement of a signal field that tracks wave packet dynamics. The momentum component of the signal that is phase matched in box-CARS contains light shifted in frequency from the probe pulse by the energy separation between vibrational levels of the ground electronic state (i.e. by a Stokes shift to the high-frequency, or anti-Stokes side of the probe spectrum). This is why the approach is known as CARS.

We note that CARS measurements can be carried out for probe pulses that are both shorter and longer than a vibrational period, and Equations 8.19 and 8.20 can be interpreted to understand the effect on the probe pulse in both cases. As the wave packet generated by the pump and Stokes fields moves back and forth on the ground-state potential, it modulates the polarization as it comes in and out of the resonance location between the two potentials. In the limit where the probe pulse duration is longer than the vibrational period (narrow-band), this modulation occurs multiple times during the probe pulse. This implies that the wave packet dynamics modulate the phase of the probe field in time, thereby creating new frequencies (Stokes and anti-Stoke sidebands). If the probe pulse duration is comparable to a vibrational period, there are already frequency components at the Stokes and anti-Stokes in the pulse spectrum. In this case, motion of the wave packet causes the resonance factor to weight different portions of the probe spectrum, shifting it slightly to higher or lower frequency. When the probe pulse duration is much shorter than a vibrational period (extremely broadband), the intensity of the probe is simply modulated as a function of delay, while the spectrum of the probe is not significantly altered.

As noted, our treatment of CARS assumes that both pump and probe pulses are close to resonance between electronic states a and b. CARS can also be performed when the pump and probe pulses are far from resonance. In this case, the signal is weaker, and mathematically one cannot make the rotating-wave approximation as we did earlier to arrive at Equation 8.20. Rather, one needs to perform an adiabatic elimination of off-resonant states as we did in Section 3.2.4 (we also follow this approach with transient reflectivity in Section 8.4). The resulting signal and interpretation are nevertheless similar, as the signal is modulated by wave packet motion on the ground electronic state.

Before moving on, we note that CARS is only one term of the general third-order nonlinear polarization (TA provided others). Calculating the *general* result provides one with all possible interactions and outcomes at third order, which, while complete, is more than what one desires for any given experiment or process. Our approach was to identify a particular experiment (or dynamics) and describe it in terms of analytic expressions. A complementary approach is to calculate all nonlinear polarizations at a given order and identify a measurement or dynamic associated with each one [44].

8.3.5 Advantages and Disadvantages: Why Choose CARS?

One of the primary advantages of CARS is that it is well suited for studying vibrational dynamics on the ground electronic state, as the combined pump and Stokes interactions initiate a vibrational wave packet on the ground electronic state of the molecule via impulsive Raman scattering. In addition, phase matching can be used both to focus on particular ground-state dynamics and to provide a background-free signal (e.g. in a box-CARS geometry). Finally, CARS is easily applicable to a wide range of gas and condensed-phase samples, provided one has a sufficient number of sample molecules within the interaction region. For reviews of femtosecond CARS techniques, see [45, 46].

8.4 TRANSIENT REFLECTIVITY

8.4.1 Introduction

We finish this chapter by considering a coherent optical approach that is particularly well suited for measuring vibrational dynamics in solids (i.e. "phonons"). The technique of "transient reflectivity" (as opposed to TA) measures changes in sample reflectivity of a probe pulse as a function of pump–probe delay. It is useful to frame the process in solids as an extension of our treatment for isolated molecular systems. As with the examples already discussed, the phonons can be excited by direct absorption of the pump pulse to an electronically excited state of the sample, or through a two-photon Raman transition back to the ground electronic state. In a single molecule, one makes a transition from the ground PES to an excited PES whose minimum is displaced from the ground-state equilibrium geometry. Assuming this occurs in a time shorter than the vibrational period, a time-dependent vibrational wave packet is generated on the excited PES. In a solid-state system, the vibrations are driven by the fact that the electronically-excited state has a different equilibrium configuration of atoms. Therefore, electronic excitation of the solid leads to vibrational motion since the sample finds itself displaced from equilibrium. If this displacement is generated in a time shorter than the phonon period, absorption of the pump results in a coherent oscillation that can be detected by the probe. In particular, consider a transition from the valence band to the conduction band (similar to a HOMO–LUMO transition in a molecule). If the minimum in the potential energy of the conduction band is displaced from the minimum in the valence band, the excitation launches a vibrational wave packet (a coherent phonon). This is illustrated in Figure 8.8, which shows the energy levels for diamond as a function of lattice constant (i.e. interatomic spacing).

8.4.2 Experimental Setup

We consider the case of an off-resonance probe pulse whose reflection from the surface is modulated by coherent phonons generated by the pump. Figure 8.9 shows the

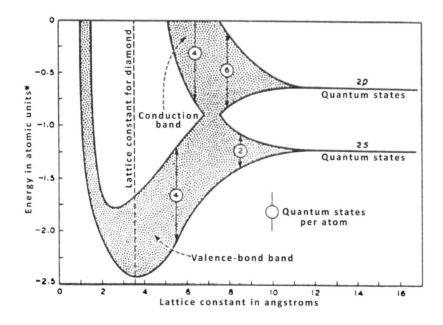

Figure 8.8 Energy bands in diamond. Figure from [47].

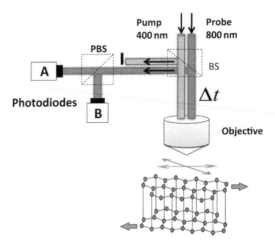

Figure 8.9 Experimental transient reflectivity setup for measuring coherent phonons in few-layer graphene. Figure from Ref. [48].

experimental setup for measuring such phonons in a sample of graphene a few layers thick [48]. The probe pulse is polarized at 45° with respect to the pump, and the two pulses are combined collinearly and incident on the sample with an adjustable delay. The reflected portion of the probe pulse is sent through a polarizing beam splitter and measured using a balanced photodetector. This arrangement provides two important benefits. The first is that the modulation of probe reflectivity by the pump-induced, shear-mode phonon in graphene has the opposite sign for light polarized parallel to the pump as compared to perpendicular. So the components of the probe field parallel and perpendicular to the pump are modulated π out of phase with each other. Detecting these together enhances the effective signal-to-noise.

The other advantage of balanced detection is that it helps mitigate the fact that the overall reflection of the probe is strongly affected by electronic excitation of the sample by the pump (independent of any phonon dynamics). Fluctuations in the pump pulse energy drive reflectivity changes of the probe unrelated to phonon dynamics that can overwhelm the small changes in reflectivity resulting from phonon oscillation. However, any changes in reflectivity unrelated to phonon motion are independent of the probe polarization, and so the balanced detection arrangement subtracts these off. In addition, measurements are performed using lock-in detection with the pump beam modulated at 20 KHz, allowing for more sensitive detection of reflectivity changes.

8.4.3 Data and Basic Interpretation

Figure 8.10 shows measurements (panel a) and illustrations (panel b) of shear-mode phonon oscillations in few-layer graphene samples of different thicknesses. For the measurements shown in panel a, the vertical axis plots the difference in reflection of the two polarization components of the probe pulse as a function of pump–probe delay for samples with different numbers of layers. All samples display an oscillation in the signal illustrating coherent phonon motion driven by the pump pulse. While the damping time of the signal is independent of the number of layers, the oscillation frequency increases monotonically with sample thickness in a way that can be modeled using series of coupled harmonic oscillators.

8.4.4 Mathematical Description: Off-Resonance Probe

Similar to other coherent optical techniques, we use the induced polarization in the sample to provide a mathematical description of transient reflectivity. In this case,

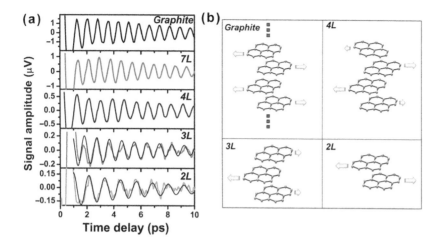

(a)

Signal amplitude (μV)

Graphite

7L

4L

3L

2L

Time delay (ps)

Figure 8.10 Measurements (panel a) and illustrations (panel b) of shear-mode phonons in few-layer graphene driven by ~400-nm pump pulses and measured via transient reflection of ~800-nm probe pulses. Figure from Ref. [48].

however, we model the probe frequency as being far off-resonance with any transitions in the molecular sample, as transient reflectivity measurements are typically performed with the probe light primarily reflected or transmitted (as opposed to absorbed). Our treatment follows that of TA in Section 8.2, where we calculate a polarization that is modulated by wave packet dynamics on an excited electronic state. In particular, we begin with the first-order polarization for the probe, $P^{(1)}_{\text{probe}}(t)$, where the initial wave function is $\chi_b^{(1)}(R,0)$ on the excited electronic state after the pump pulse arrival at $t = 0$.[10] This time, however, the probe is not resonant with any transition in the molecule, and therefore transfers little, if any, wave function out of the b state. Nevertheless, it is the polarization, or "coherence," between two states that leads to the observed signal. For simplicity, we assume the $b \rightarrow a$ transition is the closest to resonance and can be used to model the signal (in reality, there are contributions from all off-resonant states). The first-order expression for the polarization from the probe pulse is therefore given by (see Equation 8.11 from GSB)

$$
\begin{aligned}
P^{(1)}_{\text{probe}} = \frac{-1}{i} \Big\langle \chi_b^{(1)}(R,0) \Big| e^{+iH_b t} \mu_{ba} \int_0^t dt_2 e^{-iH_a(t-t_2)} \left(\mu_{ab} E_2(t_2) \right) \\
\times\ e^{-iH_b t_2} \Big| \chi_b^{(1)}(R,0) \Big\rangle .
\end{aligned}
\tag{8.21}
$$

As earlier, this expression can be simplified for interpretation if we write out the Hamiltonian operators for time t_2 in terms of kinetic and potential terms ($H_a = T + V_a$ and $H_b = T + V_b$), as well as consider the limit of a short probe pulse such that evolution of the wave function during the pulse duration can be neglected. With these assumptions, separating the field envelope from the carrier, and neglecting the commutator of T and V inside the integral (valid for short time durations), Equation 8.21 can be written as

---

[10]For consistency, we continue to use R as the one-dimensional coordinate even through the phonon displacement is more complicated than the separation of two atoms in a diatomic molecule.

$$
\begin{aligned}
P^{(1)}_{\text{probe}} =& \frac{-1}{i} \left\langle \chi^{(1)}_b(R,0) \middle| e^{+iH_b t} \mu_{ba} \int_0^t dt_2 e^{-iH_a t} e^{+i(T+V_a)t_2} \right. \\
& \times \mu_{ab} E_{02}(t_2) \frac{1}{2} \left(e^{+i\omega_2 t_2} + e^{-i\omega_2 t_2} \right) e^{-i(T+V_b)t_2} \left| \chi^{(1)}_b(R,0) \right\rangle \\
=& \frac{-1}{i} \left\langle \chi^{(1)}_b(R,0) \middle| e^{+iH_b t} \mu_{ba} \int_0^t dt_2 e^{-iH_a t} e^{+i(V_a(R))t_2} \right. \\
& \times \mu_{ab} E_{02}(t_2) \frac{1}{2} \left(e^{+i\omega_2 t_2} + e^{-i\omega_2 t_2} \right) e^{-i(V_b(R))t_2} \left| \chi^{(1)}_b(R,0) \right\rangle \\
=& \frac{-1}{2i} \left\langle \chi^{(1)}_b(R,0) \middle| e^{+iH_b t} \mu_{ba} e^{-iH_a t} \int_0^t dt_2 \left(e^{-i(V_b(R)-V_a(R)+\omega_2)t_2} \right. \right. \\
& \left. + e^{-i(V_b(R)-V_a(R)-\omega_2)t_2} \right) \mu_{ab} E_{02}(t_2) \left| \chi^{(1)}_b(R,0) \right\rangle ,
\end{aligned}
\tag{8.22}
$$

where in the second step we canceled the kinetic energy operators in the propagators inside the integral. In contrast to our earlier treatment, we cannot make the rotating-wave approximation, since the probe frequency ω_2 is not resonant with the energy difference between states a and b. Fortunately, we can actually do the integral in the case when the probe is far off-resonant. We can perform the integral by parts, and in the limit that $V_b(R) - V_a(R) \pm \omega_2 \gg \frac{1}{E_{02}} \frac{\partial E_{02}(t)}{\partial t}$, we can drop the vdu term, only keeping the uv term. This leads to

$$
\begin{aligned}
\int_0^t dt_2 & e^{-i(V_b(R)-V_a(R)\pm\omega_2)t_2} E_{02}(t_2) \\
& \approx \frac{e^{-i(V_b(R)-V_a(R)\pm\omega_2)t} E_{02}(t) - E_{02}(0)}{-i(V_b(R)-V_a(R)\pm\omega_2)}.
\end{aligned}
\tag{8.23}
$$

If we assume that $E_{02}(t=0)=0$ (i.e. probe pulse is off when the pump arrives), we can drop the last term on the right side. Putting this into our expression for the polarization yields

$$
\begin{aligned}
P^{(1)}_{\text{probe}} =& \frac{-1}{2} \left\langle \chi^{(1)}_b(R,0) \middle| e^{+iH_b t} \mu_{ba} e^{-iH_a t} \mu_{ab} \right. \\
& \times \left(\frac{e^{-i(V_b(R)-V_a(R)+\omega_2)t}}{(V_b(R)-V_a(R)+\omega_2)} + \frac{e^{-i(V_b(R)-V_a(R)-\omega_2)t}}{(V_b(R)-V_a(R)-\omega_2)} \right) \\
& \times E_{02}(t) \left| \chi^{(1)}_b(R,0) \right\rangle.
\end{aligned}
\tag{8.24}
$$

For interpretation, it is helpful to write out $H_a = T + V_a(R)$, cancel the V_a terms in the exponent, and combine $T + V_b(R)$ to make H_b (justified by the fact that the polarization is only nonzero when the probe pulse is on, and the probe is only on very briefly):

$$
\begin{aligned}
P^{(1)}_{\text{probe}} =& \frac{-1}{2} \left\langle \chi^{(1)}_b(R,0) \middle| e^{+iH_b t} \mu_{ba} \mu_{ab} \right. \\
& \times \left(\frac{e^{-i\omega_2 t}}{(V_b(R)-V_a(R)+\omega_2)} + \frac{e^{+i\omega_2 t}}{(V_b(R)-V_a(R)-\omega_2)} \right) \\
& \times E_{02}(t) e^{-iH_b t} \left| \chi^{(1)}_b(R,0) \right\rangle.
\end{aligned}
\tag{8.25}
$$

Finally, recognizing that $e^{-iH_b t} \chi_b^{(1)}(R,0) = \chi_b^{(1)}(R,t)$, we write

$$
\begin{aligned}
P_{\text{probe}}^{(1)} = \frac{-1}{2} \Big\langle \chi_b^{(1)}(R,t) \Big| \mu_{ba}\mu_{ab} \\
\times \left(\frac{e^{-i\omega_2 t}}{(V_b(R) - V_a(R) + \omega_2)} + \frac{e^{+i\omega_2 t}}{(V_b(R) - V_a(R) - \omega_2)} \right) \\
\times E_{02}(t) \Big| \chi_b^{(1)}(R,t) \Big\rangle .
\end{aligned}
\tag{8.26}
$$

As noted, in the case of an off-resonant probe, the polarization is only nonzero during the duration of the probe pulse. This is essentially dictated by energy conservation: since the resonance condition is not satisfied, the probe does not transfer any population and cannot create a coherence that exists after the pulse is gone. In addition, the polarization is modulated by the expectation value of the energy denominator. As the wave packet in the excited state moves in R (i.e. the phonon propagates), the magnitude of this expectation value changes with time, leading to a measured change in the polarization. Finally, note that this expression is very similar to what we found for the Raman scattering polarizability and two-photon excitation (Sections 3.3.2 and 3.2.4), both of which make use of adiabatic elimination.

8.4.5 Advantages and Disadvantages: Why Choose Transient Reflectivity?

Transient reflectivity is well suited for studying surfaces, particularly engineered samples such as quantum dots with features that are difficult to interrogate with nonoptical approaches. While it is quite sensitive to small changes in sample properties (e.g. lattice spacing in a crystal), transient reflectivity is not a background-free measurement, and is thus best implemented using a high-repetition-rate, stable probe beam.

CHAPTER 9

Coherent Diffractive Measurements in 1D

9.1 INTRODUCTION

While the experiments discussed in Chapters 7 and 8 measured different outcomes (fluorescence, photoelectrons, or probe absorption), they all utilized a pulse of light for the probe. In this chapter, we broaden our perspective to examine measurements that use charged particles instead of light as the probe. Specifically, we consider probing quantum dynamics using either electron or X-ray diffraction from a molecular sample that has been excited by a pump pulse. This technique was introduced in Section 4.3.2, and the basic principle takes advantage of the much smaller de Broglie wavelength of energetic electrons or X-rays (angstroms) as compared to the wavelength of optical light (hundreds of nanometers). Optical spectroscopic measurements must infer molecular structure by comparing experimental data to a molecular model. In contrast, diffraction measurements using wavelengths smaller than the interatomic spacings in a molecule can, at least in principle, directly determine molecular structure.

Another significant difference between spectroscopic and diffractive measurements relates to the type of information obtained. Diffraction experiments ideally provide atomic positions as a function of time (e.g. $[x(t), y(t), z(t)]$ for each atom in the molecule), whereas spectroscopic measurements tend to yield energies as a function of time (e.g. $E_n(t)$). In some cases, it is more useful to have direct access to the atomic positions; for example, in a ring-opening or isomerization reaction, how the molecular structure changes in time is especially relevant. In other cases, it may be more useful to have direct information about the time-dependent energies, such as a solid whose band structure changes after excitation. It is not that these two pictures contain completely separate information, but rather that different forms of the information may be more useful. In addition, it is typically possible to perform calculations that switch between the two pictures. If one knows the positions of all atoms as a function of time, one can calculate the energies of the states as well. Conversely, one can use energetic information to infer structure. However, both kinds of calculations involve significant effort and lead to additional uncertainties.

A final difference between spectroscopic and diffractive measurements involves the relationship between the spectral bandwidth and the time resolution. In spectroscopic measurements, the absolute spectral and temporal resolutions are what matter (i.e. the central frequency does not affect the resolution). In diffractive measurements the

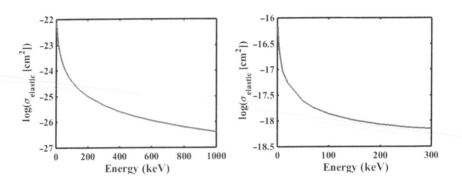

Figure 9.1 Elastic scattering cross sections for photons (left) and electrons (right) incident on sodium atoms. The logarithm of the cross section (in cm^2) per atom is plotted as a function of photon/electron energy. Data from the National Institute of Standards and Technology [49].

fractional bandwidth of the probe also matters, as this is what limits the spatial resolution. Thus, by working at sufficiently high electron (or X-ray) energy, one can achieve improved spatial resolution without sacrificing time resolution.

In comparing electrons to X-rays, electrons suffer from Coulomb repulsion that tends to stretch an electron pulse in time. X-ray photons do not repel each other, and are thus easier to bunch together in a short pulse. However, a major drawback of X-rays is that their scattering cross sections are much lower than for electrons, so many more photons are required in the probe pulse. This can be seen in Figure 9.1, which plots the elastic scattering cross section from sodium atoms as a function of energy for both photons (left) and electrons (right). While the cross section falls off with increasing energy for both, it is several orders of magnitude larger for electrons than for photons. This is because the charged electrons interact directly with both the protons and electrons in an atom, and the scattering pattern is primarily due to the nuclear charge density. In contrast, the acceleration of the protons due to the applied field of the photon is negligible, and X-rays essentially interact only with the electrons in the atom.

In this chapter, we consider time-resolved diffractive measurements from one-dimensional systems, beginning with experiments using electrons and followed by similar experiments with X-rays. Interpretation of the data in diffractive measurements is not as straightforward as with optical probes. In particular, diffraction data are typically processed before plotting pump–probe results, and we therefore delay our interpretation of the data until after developing a mathematical description of the scattering and the resulting experimental signal.

9.2 MATHEMATICAL DESCRIPTION

9.2.1 Introduction

As usual, dynamics are initiated by a pump pulse that launches a wave packet on an excited electronic state (as described by Equation 6.31). The observable in most diffraction experiments is the position dependent intensity of the two-dimensional diffraction pattern after the electron (or X-ray) probe beam passes through the molecular sample at a particular pump–probe delay. Our goal is to develop a mathematical description that allows us to calculate the expected diffraction pattern given a specific geometric configuration of the molecule. We initially apply our results to electron diffraction, but note that the expressions we derive relating measured diffraction intensities to molecular structure are equally valid for both electrons and X-rays. As mentioned, the main difference between X-ray and electron diffraction is that X-rays are scattered exclusively from the electrons, whereas electron scattering contains contributions from the nuclear charge density. This means that the atomic angle-dependent scattering amplitudes ($f(\theta)$) introduced in Equation 9.2 will be different for X-rays and electrons, but it does not fundamentally change the structural information about the molecule contained in the diffraction pattern.

The primary quantity of interest is the scattering vector $\mathbf{s} \equiv \mathbf{k} - \mathbf{k_0}$, or the momentum change of the scattered electron after interaction with the molecular sample (Figure 9.2). Here $\mathbf{k_0}$ is the wavevector of the incident electrons, which can be represented by a plane wave of the form $e^{ik_o z}$ for electrons traveling in the z-direction. Similarly, $\mathbf{k}$ is the wavevector for the scattered electrons. Geometrically, the magnitude of $\mathbf{s}$ can be written as

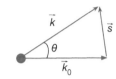

Figure 9.2 Illustration of the k-vectors for the incident ($\vec{k}_0$) and scattered ($\vec{k}$) electrons (or X-rays), as well as the scattering vector, ($\vec{s}$).

$$\begin{aligned} s &= |\mathbf{k} - \mathbf{k_0}| \\ &= 2k_o \sin(\theta/2) \\ &= \frac{4\pi}{\lambda_o} \sin(\theta/2), \end{aligned} \qquad (9.1)$$

where θ is the angle between $\mathbf{k_0}$ and $\mathbf{k}$ (the scattering angle).

We consider atomic scattering in the independent atomic model, where it is assumed that the scattering is dominated by spherical potential atoms at specific locations within the molecule. In this model, one essentially neglects the scattering from the bonding electrons, and incident electrons scatter primarily off the charge centers corresponding to the positions of the atomic nuclei and their core electrons.[1] The scattered electrons are modeled as outgoing spherical waves from the individual atoms of the form:

$$\psi(R, \theta) = f(\theta) \frac{e^{ikR}}{R}, \qquad (9.2)$$

where $f(\theta)$ is the angular-dependent scattering amplitude and R the distance from the atom.[2] Equation 9.2 describes the scattering from an individual atomic center. In general, one needs to *coherently* add the scattering amplitudes from all atoms in the sample, with the detected scattering intensity being the square of this coherent sum. For a sample with N total atoms, Equation 9.2 leads to the follow expression (written in terms of the scattering vector, $\mathbf{s} = \mathbf{k} - \mathbf{k_0}$)

$$I_{\text{sample}}(\theta) \propto \left| \sum_{n=1}^{N} f_n(\theta) \frac{e^{i\mathbf{s}\cdot\mathbf{r}_n}}{R} \right|^2, \qquad (9.3)$$

where $\mathbf{r}_n$ is the vector connecting the nth atom to the detector, and R the average distance from the atoms in the sample to the detector. We have assumed that the distance R to the detector is much greater than the spatial extent of the sample, so that the $1/R$ amplitude dependence is the same for all atoms.

9.2.2 Coherent vs. Incoherent Contributions

While Equation 9.3 is formally true for any case, its usefulness very much depends on the type of molecular sample and the properties of the electron or X-ray probe pulse.

---

[1] We note that in the case of X-ray scattering, the electrons must be considered explicitly.

[2] The scattering amplitude $f(\theta)$ is related to the differential scattering cross section, $D(\theta)$, via $D(\theta) = |f(\theta)|^2$.

For the case of a spatially uniform sample and a spatially coherent probe beam, the full, coherent addition of Equation 9.3 leads to the well-known Bragg peaks in crystal diffraction. However, it is often the case experimentally that one or both of these spatial requirements is not met. For example, in gas-phase samples, the different molecules are randomly positioned throughout the sample (not spatially uniform). In addition, in electron diffraction, the transverse spatial coherence of the incident scattering wave is often small compared with the intermolecular spacing and sample size. When either (or both) of these situations occurs, Equation 9.3 becomes impractical, and it is more useful to break it up into coherent and incoherent contributions.

We begin by focusing on the case of a gas-phase molecular sample. Because of the random location of molecules within the sample, the contribution of each molecule to the total diffraction signal adds with a random overall phase (in other words, one will never satisfy the Bragg condition). Therefore, the contributions from all the individual molecules in the gas add incoherently, implying they add as the sum of squares as opposed to the square of a sum. Nevertheless, contributions from individual atoms in a single molecule do add coherently, and there is information regarding molecular structure embedded in the diffraction signal. This is analogous to powder diffraction, in which each small microcrystal is equivalent to a molecule in a gas-phase sample.

Figure 9.3 illustrates the basic idea. The sample on the left consists of randomly aligned, gas-phase diatomic molecules, while the sample on the right is a perfectly ordered crystal. The small and large shaded areas show two different possible transverse coherence widths of the probe beam: a small coherence beam as is typical in ultrafast electron diffraction (UED), and a large-coherence beam as is typical for X-rays. When the transverse coherence width is small compared with the intermolecular spacing, the signal from different molecules adds incoherently and contributes a uniform background (the coherent addition is only over atoms within one molecule). In contrast, when the transverse coherence width is large compared with the intermolecular spacing, whether or not one coherently adds contributions from all atoms depends on the spatial uniformity of the sample. Ordered samples such as crystals necessitate the full coherent sum, while unordered ones such as gas-phase molecules still only require coherent addition of atoms within a particular molecule.

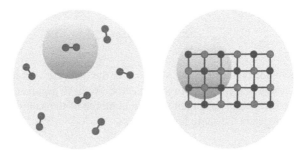

Figure 9.3   **(See color insert.)** Illustration of diffraction from gas-phase molecules (left panel) and a crystal lattice (right panel). The small and large shaded areas indicate the typical transverse coherence widths of electron and X-ray beams, respectively. Note that in the case of the lattice, the atoms or molecules are arranged in a regular array such that the diffracted X-rays or electrons from each molecule or unit cell add coherently. This is in contrast with the gas-phase sample, for which the molecules are randomly arranged and oriented, leading to an incoherent addition of the signals from each molecule, independent of the coherence width of the beam.

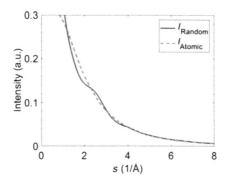

Figure 9.4 UED simulation of gas phase $C_2F_4I_2$. The figure shows the calculated atomic diffraction intensity (I_{Atomic}), as well as the simulated molecular diffraction intensity with random orientations (I_{Random}). Modulations in the molecular diffraction intensity arise from the interference in the scattering from pairs of atoms. Figure adapted from [50].

As an example, consider the case of randomly-aligned, gas-phase $C_2F_4I_2$ molecules. Figure 9.4 shows the simulated diffraction intensity as a function of momentum transfer, s, for both a randomly-aligned molecular sample (I_{Random}) and only the atomic scattering component (I_{Atomic}). I_{Atomic} is the signal from a randomly distributed collection of atoms that comprise the molecules (but with no molecular structure). Note that the difference between the randomly aligned sample and the atomic contribution is very subtle. Due to the random alignment of the molecules, the coherent interference between scattering amplitudes from different atomic centers in a given molecule is a small perturbation on the atomic scattering background; it is this small oscillation in I_{Random} that is extracted in the analysis. In the following section, we discuss the important role of molecular alignment in diffractive imaging experiments.

9.2.3 Role of Molecular Alignment

As seen in Figure 9.4, while the diffraction signal for a randomly aligned (or unaligned), gas-phase sample contains information about molecular structure, it is typically weak and marginally different from the atomic scattering signal. In addition, the signal from an unaligned sample only provides the one-dimensional radial (or pair) distribution function known as $f(r)$ (see Figure 9.6 and Equation 9.20). This function must be compared with calculations to determine specific molecular structures; for complex molecules, this can be a formidable task. Aligning the molecules in the sample has the potential to significantly improve the experiment. While the effects of alignment are primarily important for polyatomic molecules (see Chapter 14), the general idea we introduce here holds true for one-dimensional systems.

In practice, partial field-free alignment of molecular samples can be accomplished using intense, off-resonant laser fields (see Section B.1.1). However, it is a nontrivial task with a number of hurdles. For significant improvement in the signal, one must achieve a degree of alignment roughly corresponding to $\langle \cos^2(\theta) \rangle \geq 0.6$. Achieving such a high degree of alignment requires a very cold molecular sample ($T \sim 10$ K), thereby limiting the sample density for diffraction since molecular cooling is typically accomplished via a supersonic expansion. Furthermore, one typically wants the alignment to occur in a field-free environment (without the alignment laser on) so that the molecule is not affected by the presence of the alignment field. This "impulsive alignment" is temporary and lasts for only a small fraction of a rotational period, making pump–probe experiments a challenge. We note that even without active alignment, the pump pulse in a pump–probe experiment excites a subset of the molecules with a probability proportional to the projection of the transition dipole moment along the laser polarization axis ("excitation-induced alignment"). Therefore, time-resolved pump–probe diffraction measurements are generally carried out with partially aligned molecular ensembles.

Ideally, gas-phase diffraction measurements would use *perfectly* aligned molecular samples where the projection of the aligned bond lengths onto the detector plane is the same for all molecules in the sample.[3] In fact, a perfectly aligned sample of diatomic molecules can be considered a quasi one-dimensional object. The far-field electromagnetic field distribution (or Fraunhofer diffraction pattern) from a one- or two-dimensional object is given by the Fourier transform of the field distribution at the object. Therefore, if one could measure the *field* of the diffracted electrons or X-rays from a perfectly aligned diatomic sample, one could simply Fourier transform the measured distribution to yield the molecular structure (with the array of aligned molecules serving as the object). However, experimentally one measures *intensities* of electrons or X-rays by counting the number of particles incident on a spatially resolved detector. Since the intensity is proportional to the square of the field distribution, a Fourier transform of the far-field *intensity* distribution yields the autocorrelation of the molecular structure (as given by the convolution theorem — see Section A.1). There is no direct way to deconvolve the autocorrelation since the convolution is not a one-to-one operation, and thus one must use iterative phase-retrieval algorithms to recover the molecular structure, even for a perfectly-aligned sample.

Experimentally, the molecular ensemble can never be perfectly aligned, and the Fourier transform of the measured intensity pattern yields the autocorrelation of the molecular structure convolved with the angular distribution. The relationship between the measured diffraction pattern and the molecular structure becomes even more complicated for polyatomic molecules, since their three-dimensional structure can no longer be considered a one- or two-dimensional mask function. While more challenging than a perfectly aligned sample, extraction of the molecular structure from partially aligned polyatomic samples has been demonstrated (see discussion in Section 14.3.2 for an example aided by multiple measurements and learning algorithms).

Figure 9.5 helps visualize how one retrieves structure from the diffraction patterns of aligned molecules [51]. A strong-field, off-resonant femtosecond laser pulse creates impulsive (field-free) alignment of diatomic N_2 in the electronic ground state. The time delay between the aligning pump pulse and a UED probe is scanned to record diffraction patterns at various times throughout the *rotational* wave packet revival dynamics induced by the aligning pump pulse. The figure shows diffraction patterns from times corresponding to the maximum (panel a) and minimum (panel e) degrees of alignment as measured by $\langle \cos^2(\theta) \rangle$. Panels (c) and (g) plot the two-dimensional Fourier transforms of the respective diffraction data (the right column shows the corresponding simulations).

9.2.4 The Incoherent Case

We now proceed with the analysis, assuming the two spatial coherence conditions are *not* simultaneously satisfied. In general, each molecule consists of several atoms, and it is the total scattering *intensity* that is measured at the spatially resolved detector. From Equation 9.3, the intensity at a position θ on the detector from a *single* molecule can be written as the coherent sum of scattering amplitudes $f_n(\theta)$ from the individual atoms $(n = 1, \ldots, N)$ within the molecule:

---

[3]Note that, even for an aligned sample, the diffraction from different molecules does not add coherently due to the random *positions* of molecules in the gas-phase sample.

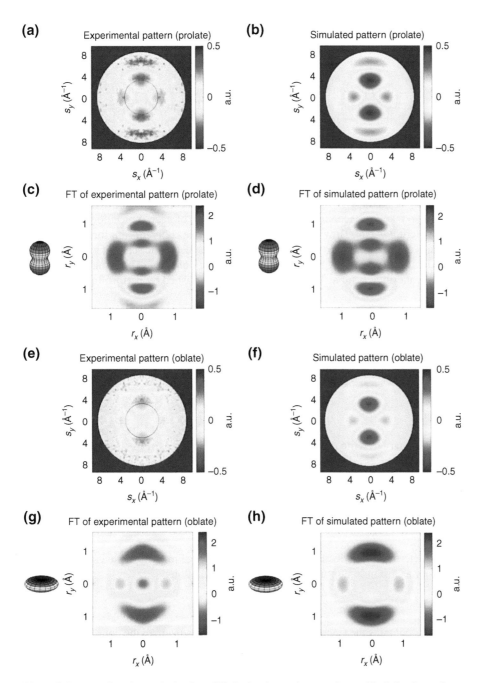

Figure 9.5   UED from impulsively aligned N_2 in the electronic ground state. The left column shows measurements and their two-dimensional Fourier transforms, whereas the right column shows the corresponding simulations. Figure from Ref. [51].

$$I_{\text{molecule}}(\theta) = \frac{I_0}{R^2} \left| \sum_{n=1}^{N} f_n(\theta) e^{i\mathbf{s}\cdot\mathbf{r}_n} \right|^2$$

$$= \frac{I_0}{R^2} \sum_{i=1}^{N} \sum_{j=1}^{N} f_i(\theta) f_j^*(\theta) e^{i\mathbf{s}\cdot\mathbf{r}_{ij}}, \tag{9.4}$$

where $\mathbf{r}_n$ is the position of the nth atom (relative to the detector), I_0 a constant, and R the distance from the molecule to the detector. The quantity $\mathbf{r}_{ij} \equiv \mathbf{r}_i - \mathbf{r}_j$ represents the vector distance between atoms i and j.

While the case of a partially aligned molecular sample is difficult to treat analytically, it can be easily treated numerically by weighting a discrete sampling of all angles. Here we consider the case of an unaligned sample, which can be treated mathematically in a straightforward manner. For an unaligned gas-phase sample of molecules randomly oriented with respect to each other, the incoherent sum over the ensemble must also take into account the random orientation of each molecule with respect to the scattering vector. How the atoms contribute to the coherent sum for each molecule depends on the orientation of the molecule with respect to the scattering vector, and we need to average the $e^{i\mathbf{s}\cdot\mathbf{r}_{ij}}$ factor over all possible orientations of the molecule relative to the scattering vector $\mathbf{s}$. This averaging is accomplished by performing the integral:

$$\frac{1}{4\pi} \int_0^{2\pi} d\beta \int_0^{\pi} e^{i\mathbf{s}\cdot\mathbf{r}_{ij}} \sin\alpha \, d\alpha = \frac{1}{2} \int_0^{\pi} e^{isr_{ij}\cos\alpha} \sin\alpha \, d\alpha, \tag{9.5}$$

where α and β are the polar and azimuthal angles of the molecules relative to the scattering vector $\mathbf{s}$. Evaluating the integral yields

$$\frac{-e^{-isr_{ij}} + e^{+isr_{ij}}}{2isr_{ij}} = \frac{\sin(sr_{ij})}{sr_{ij}}. \tag{9.6}$$

It is customary to write the scattering amplitude and intensity as a function of s instead of θ, where $s = 2k_o \sin(\theta/2)$ (see Equation 9.1). The orientationally-averaged version of Equation 9.4 thus becomes (keeping I_0 as the generic constant out front)

$$I_{\text{averaged}}(s) = \frac{I_0}{R^2} \sum_{i=1}^{N} \sum_{j=1}^{N} f_i(s) f_j^*(s) \frac{\sin(sr_{ij})}{sr_{ij}}. \tag{9.7}$$

When determining molecular structure, it is beneficial to write the averaged scattering intensity as a sum of atomic and molecular contributions:

$$I_{\text{averaged}}(s) \equiv I_A(s) + I_M(s), \tag{9.8}$$

where I_A contains the (incoherent) sum of scatterings from all the constituent atoms in a molecule, while I_M incorporates pairwise interference due to the coherent scattering from different nuclei in a given molecule. Practically speaking, $I_A(s)$ contains those terms in the double sum where the indices i and j are equal (i.e. the *intensity* from a single atom), while $I_M(s)$ contains all the other terms where the indices i and j are not equal. Specifically, I_A is simply

$$I_A(s) = \frac{I_0}{R^2} \sum_{i=1}^{N} f_i(s) f_i^*(s) = \frac{I_0}{R^2} \sum_{i=1}^{N} |f_i(s)|^2. \tag{9.9}$$

Note that $I_A(s)$ has no dependence on the $\mathbf{r}_i$'s, and therefore contains no information about the molecular structure. On the other hand, the molecular contribution to the scattering intensity is written as

$$I_M(s) = \frac{I_0}{R^2} \sum_{i=1}^{N} \sum_{j \neq i}^{N} f_i(s) f_j^*(s) \frac{\sin(sr_{ij})}{sr_{ij}}. \tag{9.10}$$

Since (unexcited) molecules in a gas-phase sample are randomly oriented, the resulting diffraction pattern will have cylindrical symmetry about the electron propagation direction (the only angular dependence is on the scattering angle). The atomic scattering intensity produces a signal that decays monotonically with scattering angle, while it is the interferences contained in I_M that provide the structural information.

We assume the complex scattering amplitude can be written as a product of a real amplitude and a complex phase:

$$f_i(s) = |f_i(s)| e^{i\eta_i(s)}, \tag{9.11}$$

where $\eta_i(s)$ is the scattering phase for the ith atom. In this case, the molecular scattering intensity reduces to

$$I_M(s) = \frac{I_0}{R^2} \sum_{i=1}^{N} \sum_{j \neq i}^{N} |f_i(s)| |f_j(s)| \cos[\eta_i(s) - \eta_j(s)] \frac{\sin(sr_{ij})}{sr_{ij}}. \tag{9.12}$$

For static molecules with well-defined atomic positions $\mathbf{r}_n$, Equation 9.12 describes the molecular contribution to the diffraction pattern. However, in general the molecules will be vibrating, and therefore the atomic positions $\mathbf{r}_n$ take on a range of values.[4] We therefore sum over all possible positions by integrating over $P(r_{ij})dr_{ij}$, where $P(r_{ij})$ is the probability distribution of internuclear distances r_{ij}. Putting these into Equation 9.12 we have for the molecular scattering intensity:

$$I_M(s) = \frac{I_0}{R^2} \sum_{i=1}^{N} \sum_{j \neq i}^{N} |f_i(s)| |f_j(s)| \cos[\eta_i(s) - \eta_j(s)]$$
$$\times \int dr_{ij} P(r_{ij}) \frac{\sin(sr_{ij})}{sr_{ij}}. \tag{9.13}$$

9.2.4.1 Homonuclear Diatomic Diffraction Pattern

We now explicitly consider the case of diffraction from a homonuclear diatomic molecule such as N_2 or I_2. Since the two atoms are identical, $f_i(s) = f_j(s) \equiv f(s)$ (and $\eta_i(s) = \eta_j(s) \equiv \eta(s)$). The phase dependence goes away, and the double sum over scattering amplitudes can be written out explicitly:

$$\sum_{i=1}^{2} \sum_{j \neq i}^{2} |f_i(s)| |f_j(s)| = (|f_1(s)| |f_2(s)| + |f_2(s)| |f_1(s)|) = 2|f(s)|^2. \tag{9.14}$$

In addition, the distance between atoms r_{ij} in a diatomic is simply the internuclear separation R. In fact, the probability of interatomic spacings $P(r_{ij})$ is exactly the magnitude squared of the vibrational wave function: $P(r_{ij}) \equiv P(R) = |\chi(R)|^2$. This reduces Equation 9.13 to the following:

---

[4]Note that even for the case of a molecule in its ground vibrational state, quantum mechanics implies a distribution of possible $\mathbf{r}_n$ values.

$$I_M(s) = \frac{I_0}{R^2} 2 |f(s)|^2 \int \frac{\sin(sR)}{sR} |\chi(R)|^2 dR. \tag{9.15}$$

It is usual to define the "modified molecular scattering intensity," sM, for presenting data and extracting molecular information:

$$sM(s) \equiv s \frac{I_{\text{averaged}}(s) - I_A(s)}{I_A(s)} = s \frac{I_M(s)}{I_A(s)}, \tag{9.16}$$

where $I_{\text{averaged}}(s)$ is the orientationally-averaged pattern one measures in the lab. For a homonuclear diatomic where $I_A(s) \propto |f(s)|^2$, this form of scattering intensity removes the leading constants and the atomic scattering factors:

$$sM(s) = \int \frac{\sin(sR)}{R} |\chi(R)|^2 dR. \tag{9.17}$$

The modified scattering intensity $sM(s)$ is in reciprocal space (it's a function of scattering vector s), while we are interested in determining the interatomic distribution in real space, or $|\chi(R)|^2$. Note that Equation 9.17 is simply the sine transform of the quantity $|\chi(R)|^2/R$. Therefore, we can determine the probability density in terms of the (inverse) sine transform of $sM(s)$ on the other side of the equation:

$$\frac{|\chi(R)|^2}{R} = \int_0^\infty sM(s) \sin(sR) ds. \tag{9.18}$$

For comparison to experiments, one typically inserts an artificial damping term to minimize contributions from the edges of the pattern and only integrates out to the maximum scattering vector measurable in the diffraction pattern. With these additions (and moving the R to the other side), we arrive at the expression for the nuclear probability density in terms of the measurable quantities:

$$|\chi(R)|^2 = R \int_0^{s_{\max}} sM(s) \sin(sR) e^{-\kappa s^2} ds, \tag{9.19}$$

where κ is the damping constant and $s_{\max}$ the largest scattering range the detector can accommodate. This is almost identical to the definition of the radial distribution function, $f(r)$:

$$f(r) = \int_0^{s_{\max}} sM(s) \sin(sR) e^{-\kappa s^2} ds. \tag{9.20}$$

9.3 DIFFRACTION DATA

9.3.1 Simulated Diffraction Data

Having developed a mathematical description of the technique, we are ready to present experimental results. For the sake of clarity, we begin by showing *simulated* diffraction data from a diatomic molecule. Figure 9.6 shows a simulation of electron diffraction for an unpumped, randomly aligned sample of diatomic N_2 in its ground electronic state. The upper-left panel shows the raw diffraction pattern as measured on the detector ($I_{\text{averaged}}(s)$). With no pump pulse present, the random alignment of molecular axes in the gas-phase sample leads to cylindrical symmetry about the propagation direction of the probe electrons. The upper-right panel plots the modified molecular scattering

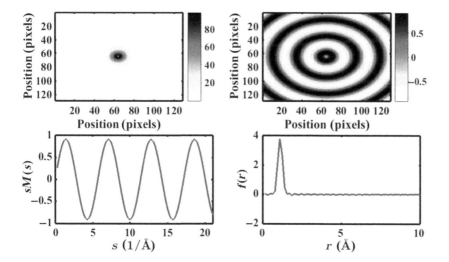

Figure 9.6 **(See color insert.)** Simulations of electron diffraction from diatomic nitrogen molecules with an electron beam energy of 100 keV. The top left panel is the raw diffraction data, while the top right shows the modified molecular scattering intensity, $sM(s)$. The bottom left panel shows an azimuthal average of $sM(s)$, and the bottom right panel plots the radial distribution function, $f(R)$. Simulation courtesy of Alexander Johnson and Martin Centurion.

intensity, $sM(s)$. Here one can see the periodicity in the signal that yields information on the interatomic spacing.

Given the symmetry, the patterns are typically integrated in the azimuthal direction to generate the diffracted intensity as a function of the distance away from the axis, s. The lower-left panel plots the azimuthally averaged, modified molecular scattering intensity $sM(s)$, which shows a clear sinusoidal oscillation. This signal is sine-transformed to form the radial distribution function, $f(r)$ (shown in the lower-right panel). As expected, one sees a narrowly peaked distribution around the equilibrium separation of N_2 in the electronic ground state.

9.3.2 Experimental Setup

The experimental arrangements are quite similar for both electron and X-ray diffraction. A pump pulse initiates dynamics by launching a vibrational wave packet in an excited electronic state of the molecule. Like other techniques, the pump is typically a pulse of light whose central frequency is tuned to resonance with an excited state. The wave function on the excited state is nonstationary and begins to evolve.

The probe is a short pulse of either energetic electrons or X-rays. The probe electrons or X-ray photons scatter (diffract) off the molecules and are collected by a spatially-resolved detector. The information recorded by the detector is essentially the two-dimensional diffraction pattern from the molecular sample and can be processed to extract time-dependent bond lengths as described in Section 9.2.

Whereas optical spectroscopy measurements are generally state-specific, in diffractive experiments, the measured signal contains contributions from all populated quantum states of the molecule. Typically, this means that one needs to subtract out the contribution from the unexcited, ground-state molecules. In such "difference diffraction" patterns, the static diffraction pattern (without or before the pump pulse) is subtracted from the diffraction pattern after excitation by the pump. Similar to differential measurements in transient absorption, this technique helps highlight small, pump-induced changes in the signal. In addition, subtracting the static pattern automatically removes the atomic scattering portion of the signal, leaving only the molecular contribution.

Mathematically, the difference diffraction intensity is given as the difference between the intensity with both pump and probe as compared to that with probe alone: $\Delta I(t,s) = I_{pu+pr}(t,s) - I_{pr}(s)$. Assuming an excitation fraction of ε to the excited electronic state (typically 10% or less), one can write the difference diffraction intensity as

$$
\begin{aligned}
\Delta I(t,s) &= I_{(pu+pr)}(t,s) - I_{pr}(s) \\
&= [\varepsilon I_{ES}(t,s) + (1-\varepsilon)I_{GS}(s)] - I_{GS}(s) \\
&= \varepsilon\left(I_{ES}(t,s) - I_{GS}(s)\right),
\end{aligned}
\tag{9.21}
$$

where I_{GS} and I_{ES} represent the diffraction intensity signals due to only the ground or excited electronic states.

Focusing on UED experiments in particular, the pulse of electrons (or "bunch") is typically generated by accelerating photoelectrons produced by focusing an ultrafast optical pulse onto a target. In this way, the pump–probe delay can be easily controlled by adjusting the time between the pump pulse and the optical pulse generating the electron bunch (just like in all-optical experiments). The energy of the accelerated electrons is an important factor in UED experiments. A minimum energy requirement, of course, is that the electrons have a deBroglie wavelength smaller than the separation between atoms. This requirement is easily met with an electron energy of only 1 kV, corresponding to a deBroglie wavelength of 0.38 Å (just under one atomic unit).

However, two other considerations make it attractive to work at significantly higher energies despite the lower cross section. First, the repulsion between electrons (known as "space charge") results in a broadening of the electron pulse in time. Since the electrons repel each other, the electron pulse duration tends to become longer as the pulse propagates, thereby compromising the time resolution. The degree of spreading decreases significantly with increasing electron energy as the electrons become relativistic.[5] Second, higher energies lead to a better match between the group velocities of the electron pulse and the (optical) pump pulse, since increasing the electron energy pushes their velocity closer to the speed of light. This leads to less temporal smear in a finite-thickness sample and improved time resolution. However, it is technically challenging to work at high electron energies, and thus one must make a compromise between experimental difficulty and time resolution.

9.3.3 Electron Diffraction Results

Figure 9.7 shows pump–probe UED data from diatomic iodine [52]. As with other techniques, the pump is a visible, femtosecond laser pulse that creates a wave packet on the B state of I_2. The probe is a relativistic electron bunch accelerated to an energy of 3.7 MeV. The measured instrument response time for the experiment, including pulse durations, velocity mismatch, and jitter, is 230 fs FWHM.

Panels (a) (experiment) and (b) (simulation) of Figure 9.7 plot the modified scattering intensity difference as a function of both scattering vector s and pump–probe delay. The modified scattering intensity difference is given by

$$
\Delta s M(t,s) = s\frac{\langle \Delta I(t,s)\rangle}{I_{atomic}(s)},
\tag{9.22}
$$

---

[5]We note that the other way to mitigate the space charge is to work with a low number of electrons per pulse. This of course leads to a reduction in the number of diffracted electrons per pulse, which can be compensated for by working at a higher repetition rate.

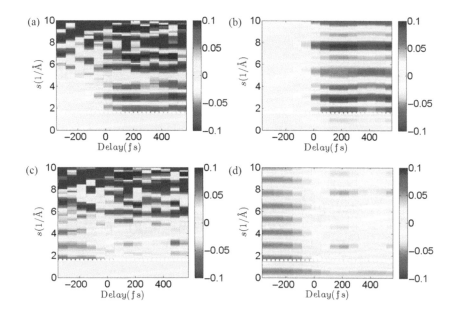

Figure 9.7 Time-resolved UED scattering from excited I_2 as function of scattering vector and pump–probe delay. The color axis is the modified scattering intensity difference, $\Delta sM(t,s)$. Total experimental (a) and simulated (b) signal, as well as excited-state experimental (c) and simulated (d) signal. Figure from Ref. [52].

where $\langle \Delta I(t,s) \rangle$ is the azimuthally averaged difference intensity and $I_{\text{atomic}}(s)$ is the atomic scattering intensity (which comes from simulation). Based on Equation 9.21, $\Delta sM(t,s)$ contains contributions from both the ground and excited states. The excited-state signal is the one of interest, and it can be obtained by adding the ground-state only contribution without the pump present (the excitation fraction ε is found by comparison between experiment and simulation). Panels (c) and (d) show the modified scattering signal from the excited state only ($sM_{\text{ES}}(t,s)$) for both experiment and simulation.

If the excited-state wave function were static, one would expect to see straight, horizontal bands after time zero in panel (c) (like the lower-left panel of Figure 9.6 stretched over all times). However, one clearly notices a modulation in time that corresponds to changes in the excited-state wave function. By putting $sM_{\text{ES}}(t,s)$ into Equation 9.19, one can extract the excited-state wave function (and average bond length) at all times. Panel (a) of Figure 9.8 plots the average bond length, $\langle R(t) \rangle$, as a function of time using the excited-state diffraction signal (markers are experimental data, dashed line is simulation). The wave packet oscillation in the excited state can be clearly seen. Panel (b) shows the extracted nuclear probability density, $|\chi(R)|^2$, using Equation 9.19 for both experiment (left) and simulation (right). Here both the oscillation and spreading of the wave packet are evident. In the simulation, the dashed lines are simulations with the experimental spatial resolution (0.07 Å), while the solid lines also include effects due to the finite temporal resolution (230 fs). The resolutions in space and time are just sufficient to resolve the expected changes in the wave function.

9.3.4 X-ray Diffraction Results

For comparison, we now discuss diffraction in diatomic iodine using an ultrafast X-ray probe [53]. Similar to the UED experiment, the pump is a visible femtosecond pulse that excites a time-dependent wave packet on the B state of I_2. The probe is now a 9.0 keV, 40 fs X-ray pulse. The diffraction pattern of scattered X-rays is measured on a spatially resolved detector at each pump–probe delay, and the processed data is plotted

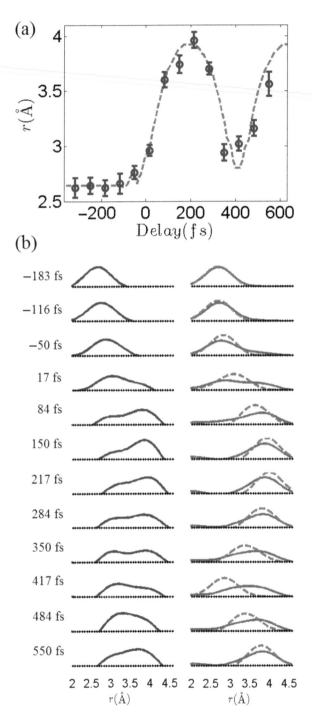

Figure 9.8 Time-resolved UED bond length (a) and probability density (b) from excited I_2. In panel (b), the left side is experimental, while the right side is from the model. See text for details. Figure from Ref. [52].

to resolve time-dependent changes in the atomic positions. Using the completeness of Legendre polynomials as angular basis functions, one can expand the scattered X-ray intensity in terms of Legendre polynomials:

$$I(Q,\theta,t) = A(Q,t)\left[1 + \sum_{n=1}^{3}\beta_{2n}(Q,t)P_{2n}(\cos\theta)\right], \qquad (9.23)$$

where Q is the momentum transfer, $A(Q,t)$ is the portion of the signal that does not depend on angle, and the $\beta_{2n}(Q,t)$ are the coefficients. The pump pulse can only transfer one unit of angular momentum, and the signal one measures is an intensity (proportional to an amplitude squared). Therefore, one expects the β_2 term to dominate the time-dependent diffraction signal. Fitting the data to Legendre polynomials and extracting components corresponding to specific terms in the expansion is akin to lock-in detection in the time domain (see Section B.2.2). In our diffraction results, the signal of interest only occurs at a specific "angular frequency" (corresponding to one or two specific Legendre polynomials), and thus focusing on that component allows one to filter out fluctuations in the signal that correspond to higher- or lower-order polynomials.

Figure 9.9 plots $\beta_2(Q,t)$ (the $n = 1$ term from Equation 9.23) as a function of both momentum transfer (Q) and time. The coefficient $\beta_2(Q,t)$ contains the primary time-dependent information, and a modulation can be seen in the data. To quantify the temporal behavior, Figure 9.10 plots the (inverse) Fourier transform of the data in Figure 9.9 along the momentum transfer coordinate. Like in Section 9.2.3, the Fourier transform of an intensity distribution yields an autocorrelation in the other domain. Because the data of Figure 9.9 is in reciprocal space, Figure 9.10 is in real space and represents an autocorrelation of the charge distribution of the molecule. The electron density in I_2 is primarily around the heavy iodine nuclei, and the excited-state charge distribution can be thought of as the atomic separation in the B state. Figure 9.10 displays two primary features. The first is a slow modulation in time corresponding to oscillation of the wave packet on the B state at an average internuclear distance of

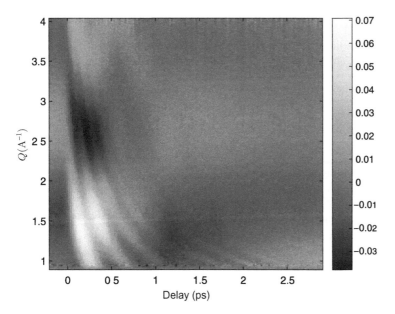

Figure 9.9 (See color insert.) Time-resolved X-ray scattering data from excited I_2. Plot shows the β_2 parameter as a function of both the momentum transfer magnitude, Q, and time, t. Figure from Ref. [53].

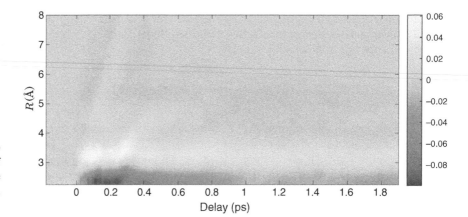

Figure 9.10 (See color insert.) (Inverse) Fourier transform of the data in Figure 9.9 along the momentum-transfer coordinate. Figure from Ref. [54].

just over 3 Å. The second feature is a "jet" of charge density that rapidly moves out to large separations. This corresponds to dissociating probability density, whose outgoing kinetic energy is consistent with the excitation wavelength used in the experiment (the authors note this dissociation is consistent with a transition to a repulsive B′ state).

9.4 ADVANTAGES AND DISADVANTAGES: WHY CHOOSE UED OR X-RAYS?

A primary advantage of probing with electron or X-ray diffraction is that it provides a more direct measure of time-dependent changes in molecular structure than optical probes. This is particularly important for larger molecules (see Part V). Although one must invert the diffraction data to determine molecular geometry, the extracted information provides a time-resolved picture of how the relative atomic positions change during a reaction. This is in contrast to optical probes that essentially measure the time dynamics of transitions between molecular states. Determining the actual molecular changes that correspond to these transitions typically takes significantly more theoretical modeling than is required to interpret results from diffractive measurements. In this way, electron and X-ray experiments are more "direct" measurements of time-dependent structural changes. Related to this is the fact that diffractive measurements are more generally applicable across molecules, as they do not require any specific molecular resonances for the probe process.

On the other hand, it is more challenging to achieve good time resolution in diffractive experiments. As mentioned, the charge densities for electrons must be low enough so that the time resolution is not degraded due to spreading of the electron bunch from Coulomb repulsion. These low charge densities lead to low count rates and long data collection times. In addition, with nonrelativistic electrons the group-velocity mismatch between the optical pump pulse and the electrons as they cross in the interaction region further lowers the time resolution. There are methods for overcoming these issues, including tilting the pulse front of the pump laser or using relativistic electrons. However, these modifications further complicate what is already a more complex experiment than most optical arrangements. For X-rays, it is challenging to produce sufficient numbers of photons in a short pulse duration. This limits experiments primarily to large-scale, accelerator-based facilities.

While X-ray and electron diffraction work the same way in principle, the fact that incoming electrons are diffracted by the nuclei as well as the electrons in a molecule leads to differences in practice. In general, X-rays diffract from the (electron) charge density of a molecule:

$$f(\vec{Q},t) = \int d^3x \rho(\vec{x},t) e^{i\vec{Q}\cdot\vec{x}}, \tag{9.24}$$

where $\vec{Q}$ is the momentum transfer and ρ the electron charge density. The scattering of X-rays from protons is negligible due to the large proton mass as compared to an electron; the heavier proton is accelerated almost $2,000$ times less than the electron, and thus radiates a negligible amount of energy as compared to an electron in the same electromagnetic field.

For heavy atoms such as our example of iodine, the total electron density is strongly peaked around the nuclei, and X-ray data should not differ markedly from that of UED. However, for molecules with lighter atoms, the electron density is more diffuse, making it a challenge to track nuclear positions. This is especially true for the hydrogen atoms in a molecule, where the *only* electron is used for bonding. Because of this, UED is generally more sensitive to atomic positions for lighter elements. This is discussed in detail in Chapter 14 (e.g., see Figure 14.8).

Due to the fact that there are generally many more electrons than atoms in a molecule, diffractive measurements are better suited for measuring nuclear dynamics than electron dynamics. For example, consider trying to follow the motion of a valence electron in a molecule: any change in the desired diffraction signal comes from the motion of one electron, while the total scattering signal comes from all the electrons. In a molecule such as benzene with six carbon atoms, the motion of one electron leads to a tiny change in the background diffraction signal from the 41 other electrons. Even in systems with a small number of electrons, one runs up against the fast timescale for electron motion. It is extremely challenging to put a sufficient number of photons or electrons into a probe pulse with short-enough duration to measure attosecond electron dynamics.

As noted in Figure 9.1, electrons interact more strongly and have cross sections several orders of magnitude larger than X-rays. This means UED is better suited to thin samples, since one can get reasonable count rates of diffracted electrons even though the sample size is limited. On the other hand, using electrons with thick samples can become problematic, as higher cross sections lead to a significant probability of multiple scattering events while transiting the sample (thereby degrading the signal). Multiple scatterings are rarely a concern for X-ray experiments. For reviews of ultrafast electron and X-ray diffraction experiments, see [55, 56, 57, 50].

This concludes our discussion of time-resolved spectroscopy in one dimension. In Part IV, we move on to modeling dynamics in multiple dimensions.

PART IV

Quantum Dynamics in Multiple Dimensions

CHAPTER 10

Explicit Approach to N-D Dynamics

10.1 INTRODUCTION

In Sections II and III, we focused on describing and measuring wave function dynamics in one dimension. While this captures the primary aspects of many systems, it is always an approximation to consider dynamics in actual molecules one dimensional. Whether it is a large molecule with many degrees of freedom, or a diatomic molecule not completely isolated from its environment, in general, one must consider multidimensional dynamics (it is an "N-dimensional problem"). If the dynamics are fully *separable*, the one, N-dimensional problem factorizes into N, separate one-dimensional problems, and the 1D treatment of Part II would be exact. However, in real systems, there will always be some amount of coupling between the different degrees of freedom, and this is what is really interesting to measure since it is difficult to describe theoretically.

For example, in large molecules, there is vibrational/vibrational coupling between different vibrational modes. In addition, while it is typically a good approximation to neglect rotations when studying vibrational dynamics, for large rotational energies, there will be vibrational/rotational coupling ("centrifugal distortion"). Finally, in regions where the Born–Oppenheimer approximation breaks down (see Section 2.5.1), there is coupling between the electronic and vibrational degrees of freedom. Thus, even if the initial motion is along one degree of freedom, the ensuing dynamics will always become multidimensional, and it is simply a question of how long this takes to happen.

Part IV aims to describe these multidimensional dynamics and analyze the primary implications. Chapter 10 develops a numerical model of dynamics that treats each degree of freedom explicitly, while Chapter 11 considers implicit approaches to the dynamics. Three important features we consider in detail are energy flow driven by mode coupling, decoherence, and nonadiabatic coupling between different potential energy surfaces (PESs). Our models illustrate that the first two are closely connected, while for the third, we discuss how nonadiabatic coupling is related to the breakdown of the Born–Oppenheimer approximation and the noncrossing rule in more than one dimension. We begin by considering vibrational modes of a polyatomic molecule as a representative multidimensional system. These can be approximated as a factorizable system for short times ("normal modes"), but will inevitably exhibit the effects of mode coupling for long times.

10.2 MODE COUPLING IN TWO DIMENSIONS

The simplest description of a polyatomic molecule pictures the atoms as masses connected by springs. In the limit of small displacements from equilibrium, the spring restoring force is linear with displacement, corresponding to a locally harmonic potential.[1] From the study of coupled harmonic oscillators in classical mechanics, the most efficient way to solve the differential equations governing the motion is to use a "normal-mode basis," in which the normal modes are linear combinations of atomic displacements. In an ideal case, these modes form eigenmodes of the potential such that motion along a given mode is independent of the others. Of course, in a real molecule, there will always be *some* degree of coupling between these modes, since the potential energy as a function of normal mode coordinates will never be exactly harmonic. Thus, motion away from equilibrium inevitably leads to sampling anharmonic terms in the molecular potential (proportional to $x^n y^m$ for $n + m > 2$). The key point is that the cubic and higher-order terms in the potential that are necessarily present (and become more important as one moves away from equilibrium) result in a force along one coordinate due to displacement along another - this is mode coupling.

We begin with a general PES for a system of two coupled vibrational modes:

$$V(Q_1, Q_2) = \frac{1}{2}k_1 Q_1^2 + \frac{1}{2}k_2 Q_2^2 + \frac{1}{2}k_{12}\left(Q_1^2 Q_2 + Q_1 Q_2^2\right), \tag{10.1}$$

where Q_1 and Q_2 represent the two normal mode coordinates, k_1 and k_2 are the harmonic constants for modes 1 and 2, and k_{12} the constant that describes the amount of mode coupling. Note that in general there are higher order terms in Q_1 and Q_2, and we have simply truncated the expansion to third order. Were it not for the term containing k_{12}, a wave function with initial displacement along coordinate Q_1 would only ever oscillate along this coordinate, and there would be no coupling into the second mode (the situation would be separable into two, independent one-dimensional problems).

We note that in Equation 10.1, the lowest-order anharmonic coupling term in the potential is cubic. Bilinear terms proportional to $Q_1 Q_2$ also lead to an apparent coupling between modes. However, this does not represent a true anharmonic coupling, since it can be removed by performing a rotation (linear transformation) of the coordinates. In other words, one just chose the wrong combination of coordinates to begin with. The first anharmonic terms in the potential that describe true coupling in an otherwise normal-mode basis near equilibrium are cubic in their overall coordinate dependence.

Figure 10.1 illustrates the essential features of quantum wave function evolution in the anharmonic 2D potential of Equation 10.1, showing snapshots of the modulus of the wave function at different times. The 2D wave function is initially displaced from the minimum along only one coordinate, as seen in panel (a). Panel (b) shows the wave function one-half period later, where the only hint at coupling into the other coordinate is the apparent slight tilt of the shape. Panel (c) shows the wave function at a sufficiently long time delay such that the motion has spread significantly into the other coordinate. Panel (d) shows the wave function one-half period after this.

---

[1]We note that the potential energy as a function of position for any system near equilibrium is always locally harmonic. In a Taylor series expansion of the potential energy, the constant term is unimportant and the linear term is zero at equilibrium. Thus, the first nonzero term is the quadratic (harmonic) one, and for small displacements away from equilibrium, this term will dominate the next (cubic) term.

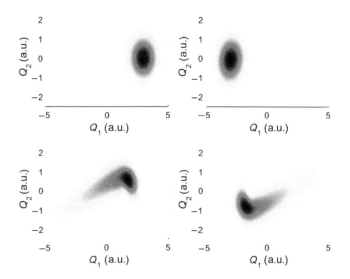

Figure 10.1 (See color insert.) Snapshots of a wave function evolving on the 2D anharmonic potential of Equation 10.1. Panel (a) shows the initial wave function displaced along only one coordinate. Panel (b) shows the wave function one-half period later. Panel (c) shows the wave function many oscillations later when motion along one coordinate has coupled into motion along the other. Panel (d) shows the wave function one-half period after panel (c).

The flow of energy between modes evident in Figure 10.1 can also be seen by considering Hamilton's equations for this potential. The full Hamiltonian for this system is given as

$$H = T + V = \frac{P_1^2}{2\mu_1} + \frac{P_2^2}{2\mu_2} + \frac{1}{2}k_1 Q_1^2 + \frac{1}{2}k_2 Q_2^2 + \frac{1}{2}k_{12}\left(Q_1^2 Q_2 + Q_1 Q_2^2\right), \quad (10.2)$$

where P_i is the momentum along coordinate Q_i, and μ_i is the reduced mass corresponding to motion along coordinate Q_i. Hamilton's equation for the momentum in the second coordinate is

$$\dot{P}_2 = -\frac{\partial H}{\partial Q_2} = -k_2 Q_2 - \frac{1}{2}k_{12}\left(Q_1^2 + 2Q_1 Q_2\right), \quad (10.3)$$

where it can be seen that displacement along Q_1 leads to a change in momentum (a force) along Q_2. At its most basic level, this is mode coupling.

10.2.1 Mode Coupling: Quantum vs. Classical

Given that mode coupling is dictated by the Hamiltonian in both classical and quantum mechanics, it is natural to ask what differences there might be between the classical and quantum pictures (like we did for one-dimensional wave function propagation in Section 5.5.2.2). To answer this question, we calculate the dynamics two different ways: quantum mechanically using the TDSE and classically using Hamilton's equations of motion. In both cases, we solve the problem numerically for the 2D potential given in Equation 10.1. We consider the case of initial displacement along only one of the modes, which corresponds to setting the initial amplitudes of the two oscillators distributed around $Q_1 = 1$ Å and $Q_2 = 0$ Å with (Gaussian) widths corresponding to the initial ground-state wave function spread (there is a similar distribution of initial momentum values whose width is dictated by the initial spatial distribution).

Figure 10.2 shows the amplitude of oscillation in both modes for the quantum and classical cases. As expected for both cases, one sees an oscillation of amplitude, and therefore energy, between the two modes in addition to the dephasing. As is typical

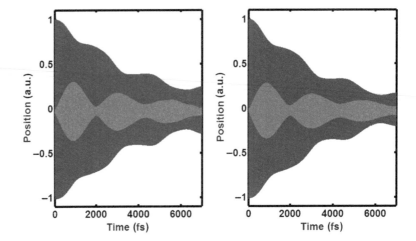

Figure 10.2 (See color insert.) Quantum (left) and classical (right) solutions for two modes anharmonically coupled as in Equation 10.1. Graph shows the first moments $\langle Q_1(t) \rangle$ (outer, black) and $\langle Q_2(t) \rangle$ (inner, gray). Note that the fast harmonic oscillations (period of $\sim 20\,\text{fs}$) under the envelope are not resolved due to the finite line thickness in the plot.

for weakly anharmonic systems near a potential energy minimum, it requires many oscillations of the individual modes before there is significant amplitude exchange between the two. In other words, the period of mode coupling is typically much longer than the period of individual modes (this is why the fast sinusoidal oscillations in each mode are not resolved in the plots and instead appear as a solid fill). Note that the energy coupling in the classical case follows that of the quantum case very closely.

In the time-independent picture, the initial wave packet can be represented in terms of a finite number of eigenstates, and these eigenstates will eventually rephase after a finite amount of time. However, when the number of states is large, the timescale for rephasing can be much longer than is experimentally relevant. As is discussed in detail in Section 10.3, for large number of modes the coherence leaves the initial mode and does not return on any experimentally accessible timescale. As we saw in Section 5.5.2.2, the concept of dephasing is incorporated into the classical simulations by considering a range of initial conditions in classical phase space. However, the timescale for "rephasing" in the classical result is determined by the choice of numerical sampling and does not quantitatively predict the quantum behavior.

The agreement between classical and quantum calculations is not accidental. Rather, it is a consequence of the fact that coupling between different degrees of freedom plays a much larger role than any uniquely quantum aspects of the dynamics. If the anharmonic terms in the potential are sufficiently large, mode coupling happens before any purely quantum aspects of the motion, such as wave packet rephasing, can take place.

10.2.2 Strength of the Coupling

For the coupling to be effective, the frequencies of oscillation must be nearly degenerate (as we shall see, it actually requires $mk_1 \approx nk_2$ for some integers m and n). Otherwise, driving of one mode by the other quickly shifts from being in phase to out of phase before significant amplitude transfers to the second mode. The detuning of the two modes can, to some extent, be offset by the strength of coupling: if you couple energy between the two modes strongly enough, you can beat the time for the modes to drift out of phase. The overall effectiveness and timescale for coupling therefore depends on both the intrinsic coupling strength (k_{12} in Equation 10.1) and the detuning $(k_2 - k_1)$, similar to off-resonant Rabi oscillations of a two-level system (see Section 3.2.3). For a given coupling strength, the amplitude of oscillation in the second mode falls off quickly with detuning, with the speed of the falloff depending on the coupling strength in an expected way: a strong coupling strength allows for greater detuning while still effectively coupling the two modes. In the simulations, we generally chose a coupling strength (k_{12} in Equation 10.1) that was a few percent of the harmonic constants (k_1, k_2), and a detuning between the modes of a few percent.

Since the model was designed to demonstrate the mechanism behind mode coupling, the coupling strength was intentionally kept low such that the coordinates Q_1 and Q_2 were an approximate normal-mode basis. In an actual molecule, the coupling strengths could be significantly larger, such that energy flows more readily between the two coordinates. In addition, one would expect Morse-like PESs along a given coordinate in a real molecule; when the potential rolls over to an asymptote, the mode spacings become closer together. Therefore, wave packets that undergo larger-amplitude excursions (and are necessarily composed of more closely spaced eigenstates) will more likely have near-degeneracies with another mode. Furthermore, as the number of degrees of freedom grows in larger molecules, one expects the mode coupling process to happen more readily due to the larger density of states; for a sufficiently large number of degrees of freedom, there will always be energy levels with a small enough detuning to facilitate strong coupling. In Section 10.3.1, we consider how this naturally leads to the idea of intramolecular vibrational energy redistribution.

10.2.3 Eigenstate Model of Mode Coupling

In this section, we develop an analytic treatment of mode coupling in the quantum picture by considering the four lowest eigenstates of the two-mode problem outlined earlier. The PES described by Equation 10.1 is essentially an anharmonic bowl whose center is located at the origin ($Q_1 = Q_2 = 0$). Far from equilibrium, the coupling between modes becomes large, and there can be significant energy transfer even when the normal mode frequencies are nondegenerate. Near equilibrium, however, the mode coupling is usually quite weak, and effective transfer between modes only takes place when there is a near-degeneracy between integer multiples of the mode frequencies. For simplicity, we assume that the frequency of mode one is approximately twice that of mode two.[2] Figure 10.3 depicts the four lowest, two-mode eigenstates, where E_{ij} represents the *uncoupled* energy of a state with i quanta of energy in mode 1 and j quanta of energy in mode 2. Note that in this representation, the states $|n_1 n_2\rangle = |ij\rangle$ are assumed to be eigenstates of the uncoupled potential (Equation 10.1 with $k_{12} = 0$).

The full Hamiltonian for the two-dimensional system of modes 1 and 2 is given by Equation 10.2. Since we are interested in the effect of coupling on the two near-degenerate states, it is simplest to consider only a subsystem comprising the two eigenstates with uncoupled energies E_{10} and E_{02} whose detuning from resonance is given by 2Δ (circled in Figure 10.3). In matrix form, the Hamiltonian for this two-state subsystem can be written in a way that explicitly shows the effect of coupling:

$$H = \begin{vmatrix} E_{10} & V' \\ V' & E_{02} \end{vmatrix}, \tag{10.4}$$

---

[2]This situation occurs in real systems when, for example, a stretch vibration is nearly degenerate with two quanta of a bend. An excellent example is CO_2, which has a near degeneracy of two quanta of the bend mode (667 cm^{-1}) with one quantum of the symmetric stretch mode (1337 cm^{-1}). The strong coupling between these modes, known as "Fermi resonance," is well studied (e.g. see [58]).

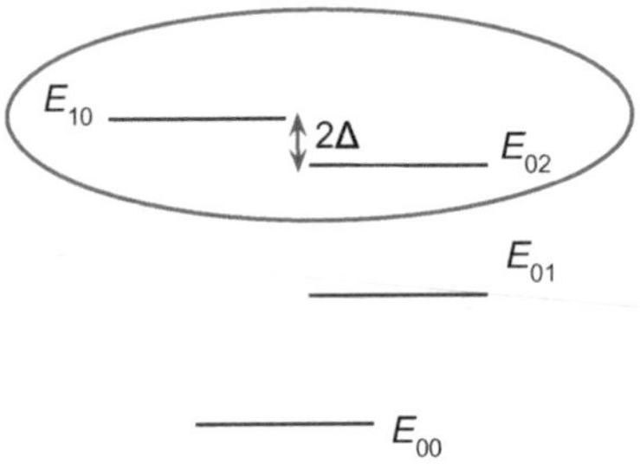

Figure 10.3 Four lowest (uncoupled) energy levels for a two-mode vibrational system, where one of the mode frequencies is approximately twice the other.

where V' is the "coupling potential" given by

$$V' = \left\langle 10 \left| \frac{1}{2}k_{12}\left(Q_1^2 Q_2 + Q_1 Q_2^2\right) \right| 02 \right\rangle. \tag{10.5}$$

For the purposes of our discussion, we consider only the second term in Equation 10.5 (one can show that the first term doesn't contribute to the matrix element calculation). We evaluate V' by expressing Q_1 and Q_2 in terms of the quantum raising and lowering operators:

$$Q_i = \sqrt{\frac{1}{2\mu_i \omega_i}}\left(a_i^\dagger + a_i\right), \tag{10.6}$$

where $a_i^\dagger$ raises the vibrational quantum number of mode Q_i by one unit (and a_i lowers by one unit). The product $Q_1 Q_2^2$ in the potential results in eight total terms, each containing products of three raising and lowering operators:

$$V' = \left\langle 10 \left| \frac{1}{2\sqrt{8\mu_1 \omega_1 \mu_2^2 \omega_2^2}} k_{12} \left(a_1^\dagger + a_1\right)\left(a_2^\dagger + a_2\right)^2 \right| 02 \right\rangle$$

$$= \frac{1}{2\sqrt{8\mu_1 \omega_1 \mu_2^2 \omega_2^2}} k_{12} \langle 10| a_1^\dagger a_2^\dagger a_2^\dagger + a_1^\dagger a_2 a_2 + a_1^\dagger a_2^\dagger a_2 + a_1^\dagger a_2 a_2^\dagger \tag{10.7}$$

$$+ a_1 a_2^\dagger a_2^\dagger + a_1 a_2 a_2 + a_1 a_2^\dagger a_2 + a_1 a_2 a_2^\dagger |02\rangle.$$

The different combinations of operators in Equation 10.7 correspond to different changes in mode occupation numbers. For example, one of the terms is $a_1^\dagger a_2^\dagger a_2$, which implies gaining one quantum of vibration in mode one and no net change in occupation of mode two. There is another term of the form $a_1^\dagger a_2 a_2$, or gaining one quantum of vibration in mode one and losing two quanta from mode two. In the eigenstate picture, the seemingly simple coordinate product $Q_1 Q_2^2$ provides for a whole host of population transfer possibilities between the uncoupled eigenstates.

In the normal-mode basis $\langle nm|n'm'\rangle = \delta_{nn'mm'}$, and therefore the only term in Equation 10.7 that returns $|10\rangle$ when acting on $|02\rangle$ is $a_1^\dagger a_2 a_2$. Therefore, none of the other terms contribute to population transfer between the two states of interest, and our expression for V' reduces to

$$V' = \frac{1}{4\sqrt{\mu_1 \omega_1 \mu_2^2 \omega_2^2}} k_{12}, \tag{10.8}$$

which is proportional to the anharmonic coupling strength k_{12}.

To find the energies of the *coupled* subsystem, we diagonalize the Hamiltonian in Equation 10.4, arriving at the following two eigenvalues:

$$E'_{1,2} = \frac{E_{10} + E_{02}}{2} \pm \sqrt{\Delta^2 + (V')^2}, \tag{10.9}$$

where $\Delta = \frac{E_{10} - E_{02}}{2}$. If we redefine the zero energy by subtracting off the average energy of the states from the total Hamiltonian, the eigenenergies simplify to

$$E'_{1,2} = \pm\sqrt{\Delta^2 + (V')^2}. \tag{10.10}$$

The eigenstates with the coupling included can be expressed in the uncoupled basis as (see Section A.6.1):

$$|1'\rangle = \sin\theta\,|02\rangle + \cos\theta\,|10\rangle \tag{10.11a}$$
$$|2'\rangle = \cos\theta\,|02\rangle - \sin\theta\,|10\rangle, \tag{10.11b}$$

where the mixing angle θ is defined by

$$\tan(2\theta) = \frac{V'}{\Delta}. \tag{10.12}$$

The evolution of the system can be tracked easily in the coupled basis, where the time dependences of the true eigenstates are dictated by their phase evolution, which themselves are determined by their eigenenergies. Since the initial condition is given in the uncoupled basis, one first needs to rewrite this in the new basis using Equation 10.11. In particular, if the system is prepared in the initial, uncoupled state $|10\rangle$:

$$\psi(t=0) = |10\rangle = \cos\theta\,|1'\rangle - \sin\theta\,|2'\rangle, \tag{10.13}$$

the time-dependent state of the system is given by

$$\psi(t) = \cos\theta\,|1'\rangle e^{-iE'_1 t} - \sin\theta\,|2'\rangle e^{-iE'_2 t}. \tag{10.14}$$

Coupling to the other mode is determined by taking the projection of Equation 10.14 onto the uncoupled state $|02\rangle$. Writing $|02\rangle$ in the coupled basis from Equation 10.11 and using the orthonormality of the coupled states yields

$$\langle 02\,|\,\psi(t)\rangle = \left(\sin\theta\,\langle 1'| + \cos\theta\,\langle 2'|\right)\left(\cos\theta\,|1'\rangle e^{-iE'_1 t} - \sin\theta\,|2'\rangle e^{-iE'_2 t}\right)$$
$$= \sin\theta\cos\theta e^{-iE'_1 t} - \sin\theta\cos\theta e^{-iE'_2 t}$$
$$= \frac{1}{2}\sin(2\theta)\left(e^{-iE'_1 t} - e^{-iE'_2 t}\right). \tag{10.15}$$

Squaring this leads to a *probability* of being in state $|02\rangle$ given by

$$P_{|02\rangle}(t) = \frac{1}{2}\sin^2(2\theta)\left[1 - \cos\left((E'_1 - E'_2)t\right)\right]$$
$$= \sin^2(2\theta)\sin^2\left(\frac{(E'_1 - E'_2)}{2}t\right) \tag{10.16}$$
$$= \frac{V'^2}{V'^2 + \Delta^2}\sin^2\left(\frac{(E'_1 - E'_2)}{2}t\right). \tag{10.17}$$

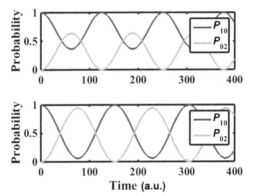

Figure 10.4 Eigenstate populations of $|10\rangle$ and $|02\rangle$ as a function of time for two different detunings. Top panel: $E_{02} = 0.97E_{10}$. Bottom panel: $E_{02} = 0.99E_{10}$. In both cases, the coupling strength $V' = 0.02$.

Figure 10.4 shows the probability of being in states $|10\rangle$ and $|02\rangle$ for short times after $t = 0$ for two different detunings. Consistent with the discussion earlier, energy flow between modes is very sensitive to both the detuning from degeneracy and the coupling strength. While the change in probability for being in one mode or the other is proportional to the coupling squared in the limit of large detuning, it only goes to 1 in the limit of zero detuning. Population transfer to other vibrational eigenstates is negligible, as these are the only two eigenstates coupled by raising/lowering operator terms that come close to conserving energy.

This detuning dependence is intuitive from a classical, time-domain perspective: the direction of energy flow between modes depends on their relative phase. Since this phase evolves at the detuning frequency, the energy transfer switches sign rapidly for large detuning, with the net result of very little energy transfer. Note that these results are quite general, applicable to any coupled, two-state system with a Hamiltonian of the form in Equation 10.4. In particular, the results for two, coupled vibrational modes are equivalent to those seen in Equation 3.21 for two electronic states coupled by an applied field.

10.3 MODE COUPLING IN HIGHER DIMENSIONS

In Section 10.2, we considered explicit coupling of modes in a system with two degrees of freedom. In larger molecules one cannot handle every degree of freedom explicitly, and it is helpful to consider a specific subset of vibrational modes. For example, some of the modes may be far from any resonance and play a negligible role in the dynamics. Alternatively, the molecule may not be well isolated from its environment (e.g. in a liquid solvent). While interactions with the environment can affect the system, what happens to the degrees of freedom within the environment is not important. In both of these situations, it can be useful to treat the situation as a "system" plus a "bath." The bath could be additional vibrational/rotational modes in the molecule, or part of the environment. In either case, the system contains the few degrees of freedom in the molecule that are of interest (and must be treated explicitly), while the remaining degrees of freedom in the bath are weakly coupled to the system and treated as a whole (either explicitly or implicitly).

Solvent	τ_{IVR} (fast)/ps	τ_{IVR} (slow)/ps	τ_{IET} /ps
CCl_4	7±1	⋯	50±5
$CDCl_3$	7±2	⋯	44±8
$(CD_3)_2CO$	5±3	⋯	16±6
Gas phase	<7[a]	400±200	⋯

[a]The fast risetime of the gas phase signal increases with probe wavelength from 2±1 ps to 7±3 ps, see Fig. 4.

Figure 10.5 Table illustrating differences in IVR and IET times for gases and liquids. The fast and slow IVR timescales correspond to different coupling strengths between various modes: the rate of energy flow from the initially excited mode to an intermediate set of modes can be different from the rate to the final mode(s). In the liquid phase, energy transfer to solvent molecules through collisions (IET) competes with IVR and can occur before energy is fully redistributed within a given molecule. Table from Ref. [59].

As we will see, coupling to a bath with many degrees of freedom leads to a decay of motion in the system mode(s). Such a decay is typically referred to as either intramolecular vibrational energy redistribution (IVR) or intermolecular energy transfer (IET), depending on whether the bath represents other modes in the same molecule (IVR) or in the environment (IET). Figure 10.5 shows a table with typical dephasing times for both gas- and liquid-phase relaxation in CH_3I [59]. There are essentially no collisions in the gas phase, and so IVR is the only mechanism. In the liquid phase, interactions with solvent molecules place an upper limit on the coherence time.

In principle, a Hamiltonian that encompasses all possible intra- and intermolecular interactions could accurately predict IVR, IET, and other similar mechanisms. However, since the Hilbert space grows exponentially with the number of degrees of freedom in quantum mechanics, it becomes practically impossible to account for all modes explicitly.[3] In this chapter, we continue to explicitly treat all degrees of freedom classically, with only a subset considered the "system." In Chapter 11, we extend the analysis to consider implicit approaches that model the process as an interaction between the system and a composite bath.

10.3.1 Classical Simulation in N-D

As a specific example, we consider a many-atom molecule with a large number of coupled vibrational modes. Although all modes are treated explicitly in the simulation, for discussion purposes, we divide the N-dimensional problem into a system-bath model by the scheme depicted in Figure 10.6. Initial excitation is localized in a single degree of freedom known as the system mode; in an actual molecule, the system mode could be a vibrational wave packet in a quasi-normal mode, accessible via an electric-dipole transition from the ground state. Modes 1, 2, and 3 represent the three vibrational modes in the molecule most strongly coupled to the system mode. The rest of the modes constitute the "bath" and represent other vibrational modes in the molecule for which one would expect a reasonable amount of coupling. As a whole, the couple scheme is meant to model how vibrational energy might flow in a large, polyatomic system.

---

[3]This is known as the "exponential wall" associated with computing quantum dynamics with many degrees of freedom.

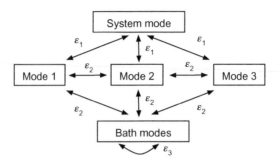

Figure 10.6 Coupling scheme for an N-D molecule with one system mode, three additional modes coupled to the system mode and to the bath, and N−4 bath modes coupled to each other.

As shown in Figure 10.6, modes 1, 2, and 3 are anharmonically coupled to the system mode (through the potential energy) with a coupling strength ε_1. These three modes are also coupled to each other, as well as to N−4 bath modes, with coupling strength $\varepsilon_2 < \varepsilon_1$. All bath modes are additionally coupled to each other with coupling strength $\varepsilon_3 < \varepsilon_2$ (ε_1 is typically 20–25% of the harmonic constant, while ε_2 and ε_3 are 5–10%). The frequencies of the modes are within approximately 10% percent of each other (the frequencies of the bath modes are randomized over an interval near the system-mode frequency). While the results focus on the system mode, all modes, including the bath modes, are treated explicitly in the simulation.

We begin with an initial displacement in the system mode and integrate the classical Hamilton's equations for a given bath size. To simulate the quantum-mechanical nature of a delocalized wave function, 200 classical "trajectories" are run for initial conditions nearby in phase space. The energy remaining in the system mode as a function of time is shown in Figure 10.7 for three different bath sizes: total number of modes equal to 15, 25, and 50. For all three bath sizes, we see a rapid decay of energy from the system mode as the energy flows into the other degrees of freedom. While some residual beating, or partial relocalization of energy, is observed at later times, the long-term behavior shows a loss of energy from the system mode. While the simulation uses classical Hamilton's equations, in the quantum picture "decoherence" is due to a dephasing of eigenstates such that energy leaves the system mode and is not relocalized on a timescale relevant to the experiment.

Figure 10.7 Energy remaining in the system mode as a function of time for three different total dimension numbers: 15 (top curve moving out to large times), 25 (middle curve), and 50 (bottom curve). The period of the system mode was 20 fs.

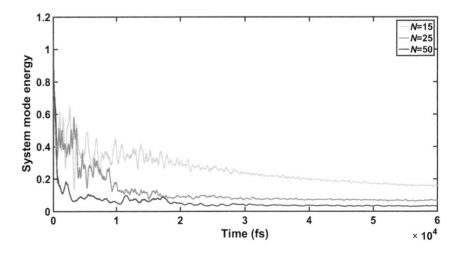

The decay of the system mode shown in Figure 10.7 is similar to the rapid, radiation-less decay of an excited atom in a dense gas (induced by collisions as opposed to spontaneous emission). Associated with this rapid, collision-induced decay is a broadening of the atomic line width in the frequency domain. Similarly, one finds that vibrational transition frequencies in molecules are broadened when the mode in question interacts with a bath of surrounding molecules (as in IET) or other, anharmonically coupled vibrations in the same molecule (as in IVR).

10.4 CONICAL INTERSECTIONS AND INTERNAL CONVERSION

In Chapter 2, we introduced the idea of a conical intersection (CI) as a consequence the noncrossing rule being violated in more than one dimension. The proximity of two adiabatic electronic states near a CI facilitates nonadiabatic coupling between states, allowing a nuclear wave function prepared on one of the two states to cross over to the other. In this section, we introduce a simple model to illustrate this idea.

In an atom, ground and excited electronic states are typically coupled only by radiation. Without an externally applied field, excited states decay via spontaneous emission with a timescale on the order of 10^{-8} s (or longer). In molecules, electronic and nuclear coupling often leads to faster, nonradiative "internal conversion" from one electronic state of the system to another. Of particular interest is how these dynamics allow multidimensional systems to rapidly convert substantial amounts of electronic energy into vibrational energy.

To model dynamics in the vicinity of a CI, we consider a simple system consisting of two electronic states and two vibrational coordinates. Away from the CI, we assume the PES for both electronic states are harmonic along both vibrational coordinates: $H^{HO} = \frac{1}{2}k_1 Q_1^2 + \frac{1}{2}k_2 Q_2^2$, where Q_1 and Q_2 are the two vibrational coordinates and k_i the effective spring constant. These two PESs are assumed to cross at a CI, and the Hamiltonian describing the system in the immediate vicinity of the CI can be written as

$$H^{CI} = (s_1 Q_1 + s_2 Q_2) \begin{vmatrix} 1 & 0 \\ 0 & 1 \end{vmatrix} + \begin{vmatrix} gQ_1 & hQ_2 \\ hQ_2 & -gQ_1 \end{vmatrix}, \tag{10.18}$$

where Q_1 and Q_2 are the two coordinates describing the so called "branching plane" (the plane where the degeneracy is lifted linearly), and s_i, g, and h are parameters that define the potentials and coupling in the vicinity of the CI [60, 61].

Figure 10.8 plots the adiabatic potentials corresponding to the Hamiltonian given in Equation 10.18, added to a two-dimensional harmonic oscillator for each surface.[4] The parameters for the potentials are $k_1 = 0.15$, $k_2 = 0.15$, $s_1 = 0.1$, $s_2 = 0.5$, $g = 0.5$, and $h = 0.09$. Figure 10.9 shows a side view of the potentials, where the point of degeneracy at the CI can be seen. Both figures show the adiabatic PESs in the vicinity of the CI located near $(Q_1, Q_2) = (0, 0)$. We solve the TDSE for the Hamiltonian in Equation 10.18 added to the harmonic oscillator potentials on each surface. The wave packet starts on the upper adiabatic surface at $(Q_1, Q_2) = (10, 0)$, and subsequently

---

[4]We note that propagating the wave function on the potentials without the addition of the harmonic oscillator potentials leads to a rapid spreading of the wave function, which obscures the coupling we are interested in observing near the CI.

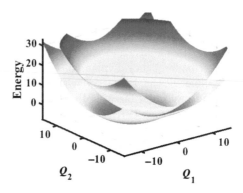

Figure 10.8 Adiabatic PESs for the Hamiltonian of Equation 10.18 plus two-dimensional harmonic oscillator potentials on each surface. All quantities are in atomic units.

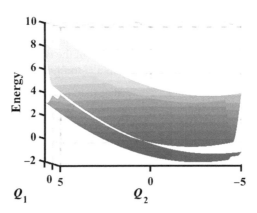

Figure 10.9 Side view of adiabatic PESs of Figure 10.8.

rides down the upper PES toward the CI. Along the way, it is coupled to the lower adiabatic PES by the TDSE with the Hamiltonian in Equation 10.18.

Figure 10.10 shows the results of the calculation. The top panel plots the expectation values of the vibrational wave packet along the two coordinates as a function of time in the upper PES (V_b) (similar for lower PES (V_a) in middle panel). The bottom panel plots the probabilities for being in the two adiabatic states as a function of time. Note that while most of the population transfer occurs when the wave packet in the upper PES approaches the CI near $(Q_1, Q_2) = (0, 0)$, there is population transfer to the lower PES away from the CI as well. While nonadiabatic coupling is greatest at the degeneracy, it is nonzero elsewhere, allowing the wave packet to move between surfaces before and after the wave packet passes through the CI. In addition, the population transfer is nonmonotonic. This underscores the point that the process is coherent, and population transfer can go in either direction depending on the relative phase of the wave function on the two surfaces.[5]

---

[5]Passage through a CI involves a coherent superposition of electronic states, and therefore electronic dynamics, because the motional (Landau-Zener) coupling between potentials is not limited to just the CI or seam. The coupling happens in a region around the intersection, so that as a vibrational wave packet propagates toward and through a CI or seam, it has some probability of switching between potentials. There is phase coherence between the portions on the different potentials, corresponding to a coherent superposition of electronic eigenstates.

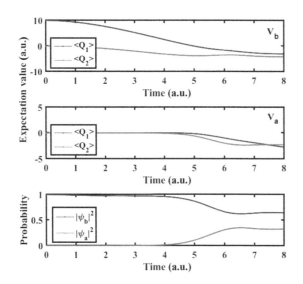

Figure 10.10 (See color insert.) Expectation values of Q_1 and Q_2 as functions of time on upper PES V_b (top panel) and lower PES V_a (middle panel). Bottom panel plots adiabatic-state probabilities as a function of time.

10.5 MULTIELECTRON DYNAMICS

We began this chapter by focusing on vibrational dynamics in multiple dimensions and the effect of coupling between modes. In Section 10.4, we tracked nuclear dynamics across multiple electronic states using a model for a conical intersection. In this section, we explicitly consider electron dynamics in a multielectron system.

10.5.1 Doubly Excited States and Autoionization

Electron dynamics become especially interesting when correlations between electrons play an important role. The simplest example of a correlated-electron system is the two-electron helium atom. As discussed in Section 2.2, the electronic wave function for the two-electron system could be written as a product of one-electron orbitals were it not for the $1/r_{12}$ electron–electron interaction term in the Hamiltonian:

$$\hat{H} = \frac{p_1^2}{2} + \frac{p_2^2}{2} - \frac{Z}{r_1} - \frac{Z}{r_2} + \frac{1}{r_{12}}, \tag{10.19}$$

where $Z = 2$ in helium. If one ignores the last term, the problem is reduced to two, single-particle problems, and stationary states of the system could be expressed in terms of hydrogen-like wave functions.

Including the electron–electron interaction term in the Hamiltonian leads to *correlated* electron dynamics, where the position and momentum of one electron affects the other. In this case, one cannot describe their motion independent of each other or write stationary state of the system in terms of a single, product state. Nevertheless, the single-electron, hydrogen-like orbitals can serve as an initial basis set for describing two-electron helium wave functions, where the electron–electron interaction acts as a perturbation (we followed this approach in Section 2.2.3, and will do so again in Section 10.5.2).

A nice example of correlated electron dynamics involves the decay of doubly excited states whose total energy is above the first ionization potential of the system (the states are in the "single-electron continuum"). For such states, there is an additional decay mechanism besides spontaneous emission known as autoionization. In this process, one electron drops down to occupy a more tightly bound orbital, while the other electron is ejected from the atom and emitted into the continuum (the total electronic energy remains conserved). Specifically, in helium, a doubly excited state corresponds to having both electrons occupy hydrogen-like orbitals that are less-tightly bound than the orbitals occupied in the ground state of the system. In other words, there are two "holes" in the $1s$ orbital.

If the two electrons were completely independent of each other with no overlap of their wave functions (i.e. uncorrelated), the doubly excited state could be written in terms of a separable product of one-electron wave functions that would be an eigenstate of the system. However, the fact that the electrons interact means that the doubly excited state is not an eigenstate of the full, multielectron system, and it can decay through autoionization if its total energy is above the first ionization potential. This process is illustrated in Figure 10.11 for the doubly excited $2p^2$ state. For larger systems where there are more electrons, the removal of a single core electron can result in a similar decay, where one electron drops from a higher-lying occupied orbital to fill the core hole, while another electron is ejected from the atom/molecule, conserving energy. This is known as Auger decay.

The decay of a doubly excited state can be considered a consequence of the coupling between bound and continuum states: the interaction between the two electrons couples the bound, doubly excited state of the atom to a state corresponding to an atomic ion and a continuum electron. In this framework, autoionization is similar to spontaneous emission from an excited atom. In spontaneous emission, there are two states that are uncoupled in the absence of an atom-field interaction: the first state corresponds to an excited atom with no photons in the field, while the second state has an unexcited atom with one photon in the field. The unavoidable coupling between the atom and the field leads to what looks like a decay of a bound state to the continuum with the emission of a photon.

The reason one typically considers spontaneous emission a *decay* process (where the photon is not reabsorbed) is due to the fact that, in the continuum, the spacing between electromagnetic modes goes to zero. Therefore, the time required for the modes to rephase in such a way that the photon is well localized at the site of the atom is essentially infinite. One can approximate the continuum by quantizing the radiation field in a box and taking the limit of a very large box so that the spacing between modes goes to zero. However, if one considers shrinking the box (as would apply if the atom were contained in a cavity), the spacing between electromagnetic modes of the cavity could be large enough such that only one or two modes are populated by the emitted photon. In contrast to an irreversible decay into the continuum, in this case the atom can reabsorb the photon and undergo Rabi oscillations. The case of autoionization/Auger decay is similar: if the electron ejected from the atom were emitted into a finite size box, there would be a finite spacing between the "continuum" states. After a time corresponding to the inverse of the energy-level spacing, the electron wave function would rephase at the location of the atom and could repopulate the doubly excited state of the atom from which it came.

Before moving on, we note a connection to the example of non-Born–Oppenheimer coupling between diabatic states of a diatomic molecule (see Sections 2.5.2–2.5.3). Consider a crossing between bound and dissociative diabatic potential curves, such as that occurs in NaI (see Section 7.2.5). The terms that we ignored in the Hamiltonian during the Born–Oppenheimer approximation led to a coupling between the two

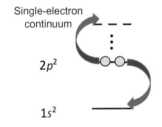

Figure 10.11 Autoionization of the doubly excited $2p^2$ state of atomic helium. One electron drops down to the 1s orbital, while the other is ejected from the atom. The total energy of the system is conserved.

states. This coupling causes population in the bound potential to "leak out" into the continuum in a process known as predissociation. As with autoionization, there is a coupling between bound and continuum states that leads to a decay of the initially prepared bound state.

10.5.2 Timescale for Correlated Electron Dynamics

In this section, we consider the basic timescale associated with correlated electron dynamics. To explore this idea, we carry out a simple perturbative calculation based on Fermi's golden rule that estimates the autoionization lifetime of a doubly excited state in helium. In particular, we consider decay of the $2p^2$ ^{1}S doubly excited state, which can be excited via two-photon absorption from the $1s^2$ ground state [62]. The notation $2p^2 = 2p2p$ indicates the hydrogen-like orbitals associated with the two electrons, while the ^{1}S label indicates the total orbital angular momentum of the state. The electron–electron interaction term in the Hamiltonian acts as the perturbation driving the transition from the $2p^2$ initial state to a final state with one electron in the $1s$ orbital and the other in the continuum with energy ε and angular momentum l $(1s, \varepsilon l)$. Note that for an initial total orbital angular momentum of zero, the final continuum state of interest is $l = 0$.

We begin with Fermi's golden rule for the (initial) transition rate between the two states:

$$\Gamma = 2\pi \left| \langle V_{\text{int}} \rangle \right|^2 \rho(\varepsilon), \tag{10.20}$$

where $\langle V_{\text{int}} \rangle$ is the electron–electron interaction matrix element between the initial and final states, and $\rho(\varepsilon)$ is the density of states at energy ε. The primary work is in evaluating the matrix element, which can be expressed in terms of the electronic orbitals as

$$\langle V_{\text{int}} \rangle = \left\langle \psi_{1f} \psi_{2f} \left| \frac{1}{r_{12}} \right| \psi_{1i} \psi_{2i} \right\rangle, \tag{10.21}$$

where $\psi_{1i}(r_1, \theta_1, \phi_1)$ and $\psi_{1f}(r_1, \theta_1, \phi_1)$ are the initial and final orbitals of electron 1 in terms of its coordinates (similarly for electron 2 in terms of its coordinates), and r_{12} is the separation between the two electrons: $r_{12} = |\vec{r}_1 - \vec{r}_2|$. The matrix element contains two, three-dimensional spatial integrals that cannot be separated into products due to the r_{12} factor.

A major advantage of choosing hydrogen-like orbitals as our basis states is that we can use known properties of the radial and angular wave functions. However, determining the wave function for the ionized electron in the continuum is more complicated. While the details are addressed later, here we simply note that the problem remains spherically symmetric in the continuum, and so the angular portion of the wave function is a spherical harmonic. Notationally, we represent the continuum function as $\psi_{\varepsilon,l} = R_{\varepsilon,l} Y_{lm}$, where ε is the energy of the free electron and l its angular momentum.

We write out the full spatial integral with $|2p, 2p\rangle$ as the initial state, $|1s, \varepsilon l\rangle$ as the final state, and an s-wave spherical harmonic $(l = 0)$ in the continuum. This yields

$$\langle V_{\text{int}} \rangle = \int_1 \int_2 R_{10}(r_1) Y_{00}(\theta_1, \phi_1) R_{\varepsilon,0}(r_2)$$
$$\times Y_{00}(\theta_2, \phi_2) \frac{1}{r_{12}} R_{21}(r_1) Y_{10}(\theta_1, \phi_1)$$
$$\times R_{21}(r_2) Y_{10}(\theta_2, \phi_2) r_1^2 r_2^2 dr_1 dr_2 d\Omega_1 d\Omega_2. \tag{10.22}$$

We follow our approach from Section 2.2.3, expressing the $1/r_{12}$ factor in terms of spherical harmonics:

$$\frac{1}{r_{12}} = \sum_l \frac{r_2^l}{r_1^{l+1}} P_l \left(\cos \theta_{12} \right) \quad \text{for } r_2 < r_1 \tag{10.23a}$$

$$= \sum_l \frac{r_1^l}{r_2^{l+1}} P_l \left(\cos \theta_{12} \right) \quad \text{for } r_1 < r_2, \tag{10.23b}$$

where the Legrendre polynomial $P_l \left(\cos \theta_{12} \right)$ is

$$P_l \left(\cos \theta_{12} \right) = \frac{4\pi}{2l+1} \sum_m Y_{lm}^* \left(\theta_1, \phi_1 \right) Y_{lm} \left(\theta_2, \phi_2 \right). \tag{10.24}$$

Specifically, we have for the angular integrals only (moving the sums outside the integrals):

$$\sum_l \frac{4\pi}{2l+1} \sum_m \int_1 \int_2 Y_{00} \left(\theta_1, \phi_1 \right) Y_{00} \left(\theta_2, \phi_2 \right)$$
$$\times Y_{lm}^* \left(\theta_1, \phi_1 \right) Y_{lm} \left(\theta_2, \phi_2 \right) Y_{10} \left(\theta_1, \phi_1 \right) Y_{10} \left(\theta_2, \phi_2 \right) d\Omega_1 d\Omega_2. \tag{10.25}$$

The two Y_{00}'s multiplied by each other simply bring out a factor of $1/4\pi$, canceling the leading factor. Grouping the integrals in terms of the electron coordinates yields:

$$\sum_l \frac{1}{2l+1} \sum_m \int_1 Y_{lm}^* \left(\theta_1, \phi_1 \right) Y_{10} \left(\theta_1, \phi_1 \right) d\Omega_1$$
$$\times \int_2 Y_{lm} \left(\theta_2, \phi_2 \right) Y_{10} \left(\theta_2, \phi_2 \right) d\Omega_2. \tag{10.26}$$

Both of these integrals yield $\delta_{l1} \delta_{m0}$; one of these picks out $l = 1, m = 0$ from the two sums, while the other then yields one via normalization. Putting this all back together in Equation 10.22, we are left with the radial integral that must be evaluated in two pieces due to the different functional forms of $1/r_{12}$ for $r_2 < r_1$ and $r_1 < r_2$:

$$\langle V_{\text{int}} \rangle = \frac{1}{3} \left[\int_0^\infty dr_1 \int_0^{r_1} dr_2 r_1^2 r_2^2 R_{10} \left(r_1 \right) R_{\varepsilon,0} \left(r_2 \right) \frac{r_2}{r_1^2} R_{21} \left(r_1 \right) R_{21} \left(r_2 \right) \right.$$
$$\left. + \int_0^\infty dr_1 \int_{r_1}^\infty dr_2 r_1^2 r_2^2 R_{10} \left(r_1 \right) R_{\varepsilon,0} \left(r_2 \right) \frac{r_1}{r_2^2} R_{21} \left(r_1 \right) R_{21} \left(r_2 \right) \right]. \tag{10.27}$$

Simple analytic expressions for the bound states are available, and we make use of these in calculating the radial integrals in Equation 10.27. This is not the case for the continuum-wave function, and we must now consider it in detail. Two issues arise when applying the formalism of Fermi's golden rule (Equation 10.20) to this calculation: the density of states in the continuum is infinite, and the continuum-wave function is not normalizable in the traditional sense (space normalization). A solution to both these problems is to discretize the continuum by assuming that the atom is enclosed in a (large) box, leading to finely discretized continuum states with a finite density of states. We calculate the discretized continuum-wave function by numerically solving the time-independent Schrödinger equation for a hydrogenic atom enclosed in a spherically symmetric box. The energy of the continuum state is given by subtracting the first ionization energy from that of the $|2p, 2p\rangle$ state: $\varepsilon = 62.5\text{eV} - 24.6\text{eV} = 37.5\text{eV}$ [62].

One could start the numerical integration outside of the box where the radial wave function is exponentially decaying to zero, and then integrate in toward smaller values of r. However, in this approach the details of the wave function near the core ($r = 0$) are very sensitive to the energy of the continuum-wave function, and thus the value of the matrix element depends critically on the exact energy of the free electron. We instead start the integration at small r and integrate outward. The box is constructed with a radial Heaviside function that switches at the box radius. We choose 550 Bohr for the radius but note that the size is not critical; while normalization of the continuum-wave function clearly depends on the size of the box, this is exactly canceled by the dependence of the density of states on the (same) box size. Figure 10.12 shows the radial wave functions $R(r)$ involved in the calculation. The $2p$ wave function (R_{21}) enters into the calculation twice, while the continuum ($R_{\varepsilon,0}$) and 1s (R_{10}) wave functions enter only once.

The density of states in the continuum at energy ε is estimated by calculating the density of states for a spherical well (infinite potential outside of a given radius, zero inside) while neglecting the Coulomb potential. The energy levels of the infinite spherical well are given in atomic units as

$$E_{nl} = \frac{z_{nl}^2}{2a^2},\qquad(10.28)$$

where a is the radius of the well and z_{nl} are the nth zeros of the lth-order spherical Bessel function, j_l. For a box size of 550 Bohr, an l value of 0, and an energy of 37.5 eV, the z_{n0} value is approximately 913. The next zero is about $z_{n0} = 918$. If we approximate the density of states by the inverse of the energy level spacing

$$\frac{1}{\Delta E} = \frac{2a^2}{z_{n+1}^2 - z_n^2},\qquad(10.29)$$

we find a density of states at this energy of roughly 66 (atomic units).

Numerical evaluation of the integrals in Equation 10.27 yields $|V_{int}| = 0.0008$. Combining this matrix element with the density of states in Fermi's golden rule (Equation 10.20) produces a rate of 9 THz, corresponding to a lifetime of approximately 110 fs. This is in reasonable agreement with the measured value of 140 fs given the simple approach to our calculation [63]. Part of the discrepancy is due to the nature of the initial wave functions used for the calculation: we chose uncorrelated hydrogenic wave functions that cannot accurately represent the true initial (correlated) state.

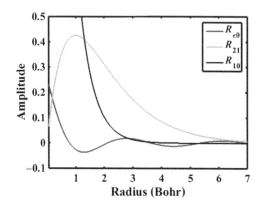

Figure 10.12 Wave functions involved in the decay of the $|2p,2p\rangle$ doubly excited state of helium. The vertical scale cuts off R_{10} in order to show details of $R_{\varepsilon 0}$.

A number of points are worth mentioning. The rate depends directly on an integral that involves products of radial wave functions (along with r or $1/r^2$ operators). The rate will vary substantially with the first and second moments of these wave functions. For example, if one of the electrons were initially in a weakly bound orbital instead of the $2p$ orbital, the wave function overlap would be significantly smaller. This would lead to a slower rate and a correspondingly longer lifetime. In fact, the integral would be zero if the two wave functions involved did not overlap at all (i.e. if one were zero everywhere the other was nonzero). Furthermore, the integral involves a continuum wave function whose oscillation frequency depends on the continuum energy. Thus, the rate depends on the continuum energy in a manner similar to the dependence of the transition dipole moment between bound and continuum states, falling off with increasing continuum energy as rapid oscillations in the continuum wave function lead to a vanishing integral over the slowly-varying, bound-state wave function. Finally, this calculation can serve as the basis for estimating timescales of correlated electron dynamics across different systems. For example, in heavier atoms, the initial and final bound states are more localized near the core, so one expects a larger matrix element and shorter lifetime (consistent with measured and calculated values of a few femtoseconds).

10.5.3 Sudden Removal of an Electron

In Section 10.5.2, we calculated a typical timescale for correlated electron dynamics using as an example autoionization of a doubly excited state in helium. In this section, we consider correlated electron dynamics resulting from the sudden removal of a single electron from a many-electron system. In particular, we focus on the rearrangement of the remaining electrons after one electron has been removed.

As before, it is easiest to approximate a multielectron atom using hydrogen-like orbitals. This is reasonable if each orbital describes an electron experiencing a spherically symmetric core whose total positive charge is shielded by the other electrons. For example, removal of the most weakly bound electron should not affect the inner electrons significantly (think Gauss' law). In this case, hydrogen-like orbitals are a good description, and ionization of a valence electron leaves the system in a very close approximation to an eigenstate of the cation. This means there will be little, if any, subsequent dynamics in the cation after ionization.

However, the limits of the hydrogenic-orbital approximation become quite evident when one considers the abrupt removal of an inner electron from the system (such as might occur with X-ray absorption). Removing a deeply bound electron leads to a substantial modification of the potential seen by the outer electrons due to the change in screening, and one could expect significant rearrangement of the remaining outer electrons after ionization. In other words, the orbitals chosen for the neutral atom do not provide a good basis set for describing this particular cation with only a single configuration.

The removal of an inner electron inevitably leads to an *electronic* wave packet that evolves on a timescale associated with the differences in energies between cationic states. This timescale tends to be attoseconds, and one can think of these dynamics as "attosecond charge transfer," where the remaining electron density moves around the atom or molecule in response to the sudden change in potential. If the total energy of the system is above the next ionization continuum, there is a possibility for Auger decay (Auger decay is the ultimate charge migration, with some charge actually leaving the system). In this sense, Auger decay is not some mysterious jump that suddenly occurs, but rather a natural consequence of multielectron dynamics, where the total energy of the N−1 electron system after removal of an inner electron is above the N−2 continuum.

A simple analogy to molecular vibrations in an excited PES helps illustrate such multielectron dynamics. Consider launching a vibrational wave packet on an excited electronic state of a diatomic molecule (e.g. B state in I_2). Since the minima for the ground and excited-state PES are generally not at the same location, the wave packet in the excited state is displaced from equilibrium and subsequently evolves in time. The electronic transition from the X to B state is analogous to the removal of a core electron, and the subsequent vibrational dynamics (or "relaxation") are equivalent to the valence electron rearrangement as a consequence of the impulsively-generated core hole.

Figure 10.13 shows simulations of attosecond charge transfer in three different systems (atom, small molecule, large molecule) [64]. For this discussion, we concentrate on the atomic case and consider removing (ionizing) an inner $2p$ electron from a krypton atom (as might occur with absorption of an X-ray photon). As mentioned, the removal of this inner electron prepares a time-dependent, N−1 electron wave packet in the remaining cation (rather than an eigenstate). This is shown in Figure 10.13 using the concept of "hole occupation number" for the $2p$ electron hole. Immediately after removal of the $2p$ electron, the hole occupation number is one in this orbital and zero in all other orbitals (not shown). If this configuration were an eigenstate of the cation, the hole occupation number for the $2p$ orbital would stay constant at one (indicating no subsequent dynamics). However, this configuration is not an eigenstate, and the hole occupation number changes on an attosecond timescale as the charge redistributes itself around the atom and other orbital holes become occupied. All three systems show a similar response over the first 50 attoseconds as the electron charge distribution reacts to the sudden change in potential. Disparities in dynamics occurring after 50 attoseconds reflect differences in the potentials for the three systems.

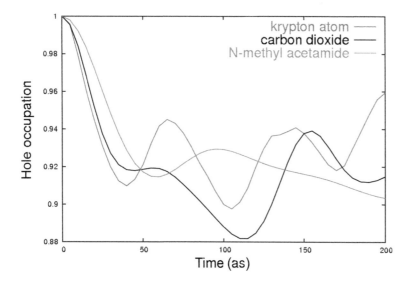

Figure 10.13 (See color insert.) Attosecond charge migration following core electron removal in three different systems. All three show initial dynamics on an approximately 50 attosecond timescale. Figure from Ref. [64].

10.6 HOMEWORK PROBLEMS

1. Numerically, solve Hamilton's equations (Equation 10.30) for the case of a two-dimensional, coupled anharmonic oscillator. Examine how the mode coupling varies for different relative mode frequencies and anharmonic potential strengths. Use both the xy^2 and x^2y terms in the Hamiltonian. Discuss how one can interpret the dependence of the mode coupling on how close the two frequencies are (the "detuning dependence").

$$\dot{q}_j = \frac{\partial H}{\partial p_j} \qquad \dot{p}_j = -\frac{\partial H}{\partial q_j} \qquad (10.30)$$

2. Numerically, calculate two-dimensional wave packet dynamics in the following potential:

$$V(x,y) = \frac{1}{2}m_x\omega_x^2 x^2 + \frac{1}{2}m_y\omega_y^2 y^2 + \frac{1.5 \times 10^{-6}}{2}\sqrt{m_x m_y}(xy^2 + x^2y). \quad (10.31)$$

Use the following parameter values in atomic units: $m_x = m_y = 100$, $\omega_x = 0.0075$, and $\omega_y = 0.0076$.

Start the wave packet with an initial displacement of 3 a.u. along the x-direction. Calculate and plot the expectation values $\langle x \rangle$ and $\langle y \rangle$ for a total time of 500 fs.

3. Numerically, calculate the two-dimensional dynamics for a wave packet approaching a conical intersection. Consider the potential energy to be the sum of the potential for a minimal Hamiltonian describing a conical intersection, V^{CI}, plus a simple harmonic potential for two different electronic states, V^{HO}. The sum of V^{CI} and V^{HO} gives the diabatic potentials. Plot the adiabatic potentials. Make a movie showing the wave packet on both PESs as a function of time. Plot the probability of being on the ground and excited states as a function of time.

Use the following for V^{HO} and V^{CI}:

$$V_1^{HO}(x,y) = V_2^{HO}(x,y) = \frac{1}{2}k_x(x-x_0)^2 + \frac{1}{2}k_y(y-y_0)^2 \qquad (10.32a)$$

$$V^{CI} = (s_1x + s_2y)\begin{bmatrix} 1 & 0 \\ 0 & 1 \end{bmatrix} + \begin{bmatrix} gx & hy \\ hy & -gx \end{bmatrix} \qquad (10.32b)$$

Choose parameter values of

$$k_x = 0.1, \quad k_y = 0.1 \qquad (10.33)$$
$$x_0 = 0, \quad y_0 = 0 \qquad (10.34)$$
$$s_1 = 0.1, \quad s_2 = 0.5 \qquad (10.35)$$
$$g = 0.5, \quad h = 0.09 \qquad (10.36)$$

Use $m = 1$ and calculate for $t = 0$–10. Try two different initial conditions of a Gaussian wave packet at $(x,y) = (10,0)$ or $(0,10)$. Use a width of 1.5 $(1/e)$ for both x and y. Hint: make use of the analytic approach to rewrite $e^{-iV\Delta t}$.

4. Calculate the autoionization lifetime for the $2p^2\ ^1D$ state of atomic helium. You can follow the calculation of the $2p^2\ ^1S$ lifetime given in Section 10.5.2.

CHAPTER 11

Implicit Approaches to N-D Dynamics

11.1 INTRODUCTION

In the previous chapter, we considered multidimensional dynamics where all the degrees of freedom were treated explicitly. As the number of dimensions in the system gets very large, it becomes intractable to solve the complete problem in such a manner. At that point, it is natural to divide it into two subsystems: one where each degree of freedom is considered explicitly (the "system"), and one where all degrees of freedom are aggregated and not treated individually (the "bath"). A common example is a molecule in solution, where the molecule is the system and the surrounding solvent molecules the bath. Another is an atomic gas with collisions, where any given atom can be the system and the surrounding atoms the bath. No matter the physical situation, for an appropriate choice of system and bath, the coupling between various degrees of freedom in the system is much stronger than between the system and bath. It is then reasonable to calculate only the system properties explicitly using quantum mechanics, while treating the bath in a simpler, often classical, way.

As we saw with explicit calculations in Chapter 10, energy in an initially excited system degree of freedom (e.g. one mode of a large molecule) tends to flow to the bath and its many degrees of freedom over time. This spread of energy is accompanied by an apparent loss of coherence in the system mode. In this chapter, we consider *implicit* approaches for multidimensional dynamics that do not track all the bath modes explicitly. We will need to develop a simplified model for the bath and how it interacts with the system. Our approach is facilitated by using the density matrix description, which allows one to consider partially-coherent systems and develop equations of motion that include system-bath coupling.

11.2 DENSITY MATRIX REPRESENTATION

An ensemble of (identical) atoms or molecules is said to be in a pure state if each particle in the ensemble can be characterized by the same wave function $|\psi\rangle$. In any real situation, there will be interactions between molecules, or between molecules and the outside world, that may lead to different wave functions for different molecules in the ensemble; such impure ensembles are known as "mixed states." Describing mixed states is important when one cannot treat all degrees of freedom explicitly and instead must separate the problem into a system and bath. In the next section, we define the density operator for working with such mixed states.

11.2.1 The Density Operator

A pure state is completely defined by its wave function, $|\psi(t)\rangle$, where we have suppressed the spatial dependence for compactness. The expectation value of any operator A is given by

$$\langle A(t)\rangle = \langle \psi(t)|\hat{A}|\psi(t)\rangle. \tag{11.1}$$

We can represent $|\psi(t)\rangle$ in a basis of energy eigenstates $|n\rangle$ as

$$|\psi(t)\rangle = \sum_n c_n(t)e^{-i\omega_n t}|n\rangle = \sum_n a_n(t)|n\rangle, \tag{11.2}$$

where the a_n coefficients incorporate the time-dependent phase evolution (in the absence of any coupling between states, the c_n are constant). Using the (more compact) a_n notation, Equation 11.1 becomes

$$\langle A(t)\rangle = \sum_{n,m} a_n(t)a_m^*(t)\langle m|\hat{A}|n\rangle \equiv \sum_{n,m} a_n(t)a_m^*(t)A_{mn}. \tag{11.3}$$

To eventually describe mixed states, we introduce the density operator formalism of J. von Neumann. For a pure state, the density operator $\hat{\rho}$ is defined as the outer product of the state vector $|\psi(t)\rangle$ with itself:

$$\rho \equiv |\psi(t)\rangle\langle\psi(t)| = \sum_{n,m} a_n(t)a_m^*(t)|n\rangle\langle m| \equiv \sum_{n,m} \rho_{nm}(t)|n\rangle\langle m|, \tag{11.4}$$

where the density matrix elements ρ_{nm} are given by the following product of eigenstate coefficients:

$$\rho_{nm}(t) = a_n(t)a_m^*(t). \tag{11.5}$$

This allows us to rewrite Equation 11.3 as

$$\langle A(t)\rangle = \sum_{n,m} \rho_{nm}(t)A_{mn} = \mathrm{Tr}\left[\rho(t)\hat{A}\right], \tag{11.6}$$

where the last step comes from the definition of matrix multiplication. Equation 11.6 says that the expectation value of an operator can be expressed as the trace of the product of the density matrix and the operator.

For a pure state, the density matrix formalism does not add anything to the description of the dynamics. However, it can be very useful in describing mixed states, for which it is impossible to write down a single state vector describing all molecules. To write the density matrix for a mixed state, we must average our earlier expression over all molecules in an ensemble:

$$\langle \rho_{nm}(t)\rangle_{\mathrm{ensemble}} = \langle a_n(t)a_m^*(t)\rangle_{\mathrm{ensemble}}. \tag{11.7}$$

The off-diagonal elements of the density matrix depend on the relative phases between eigenstates describing the state of each molecule in the ensemble. For a mixed state, each molecule in the ensemble has a different phase between eigenstates, and thus the ensemble average leads to all the off-diagonal elements vanishing. This allows us to define the density matrix for a mixed state by

$$\rho \equiv \sum_n P_n \, |\psi_n(t)\rangle \, \langle \psi_n(t)| \, , \tag{11.8}$$

where the fractional populations P_n satisfy the normalization condition

$$\sum_n P_n = 1. \tag{11.9}$$

The same expression for the expectation value of any operator as given in Equation 11.6 applies.

11.2.2 Evolution of the Density Operator

To describe dynamics in the density matrix formalism, we must determine how the density operator evolves in time. We start with a pure state, later carrying out the ensemble average to describe a mixed state using specific physical models. From the definition of $\rho(t)$, its evolution is given by

$$\frac{d\rho}{dt} = \frac{d}{dt}\left(|\psi\rangle \langle \psi|\right) = \frac{d\,|\psi\rangle}{dt}\langle\psi| + |\psi\rangle\frac{d\,\langle\psi|}{dt}. \tag{11.10}$$

From the Schrödinger equation, we know how the state vector $|\psi\rangle$, as well as its adjoint $\langle\psi|$, evolve in time:

$$\frac{d\,|\psi\rangle}{dt} = -iH\,|\psi\rangle\,; \quad \frac{d\,\langle\psi|}{dt} = +i\,\langle\psi|H. \tag{11.11}$$

Putting these together, we find

$$\frac{d\rho}{dt} = -iH\,|\psi\rangle\,\langle\psi| + i\,|\psi\rangle\,\langle\psi|\,H = -i\,[H,\rho]. \tag{11.12}$$

This result is known as the Liouville-von Neumann equation.

11.3 MODELS OF DEPHASING

In Section 10.3.1, we partitioned an N-dimensional problem into subsystems with varying degrees of coupling between the different parts. We tracked the flow of energy out of a single mode (the "system") and into the remaining modes (the "bath"), finding that the energy essentially did not return to the system mode due to the many degrees of freedom in the bath. Throughout that simulation, we considered all degrees of freedom explicitly. As the size of the problem increases, it becomes computationally beneficial to develop approaches that do not require tracking all degrees of freedom explicitly, but instead only the few "system" modes (in other words, an *implicit* approach for the bath). In this section, we introduce two simple, physical models for describing an ensemble of interacting atoms or molecules. The models consider a given particle in the ensemble to be the system, while the surrounding particles constitute the bath. Both models consider the system-bath interaction in terms of collisions with neighboring particles (frequent and mild in the case of a liquid, rare and dramatic in the case of a gas). The goal is to derive a mathematical description of how the system dynamics depend on these interactions.

The first model describes collisional effects in a gas-phase ensemble of atoms (or molecules). The basic idea is illustrated on the left side of Figure 11.1. Assume that an applied field has induced coherent electronic oscillations in the gas-phase atoms. The phase of these electron oscillations can be disrupted by collisions with other atoms or molecules, even when the collisions are elastic (i.e. populations remain unchanged). Each collision interrupts the oscillating electron, essentially rephasing the atomic dipole. With collisions occurring at random times for the different atoms, the phase of the oscillating dipole in one atom is randomized as compared to the others in the ensemble. This model leads to a number of observed effects, including collision broadening of line shapes and decoherence of the collective dipole (the ensemble-averaged displacement) as a function of time (see homework).

A second model considers how vibrational modes of a molecule in solution are affected by interaction with the surrounding solvent molecules. The various vibrational modes in a given molecule serve as the system, while the solvent molecules represent the bath. The right side of Figure 11.1 depicts the model (for clarity, only a single vibrational mode is considered). As neighboring solvent molecules move with respect to the system molecule, they apply forces that push or pull on the potential for each vibrational mode of the system molecule. This time-dependent change in the vibrational potential modulates the vibrational frequency of the mode. As we describe in the next section, this model leads to a decay of the vibrational coherence (the ensemble-averaged displacement) as a function of time.

11.3.1 Vibrations in Solution

In this section, we describe the second model in detail. In particular, we consider a molecule in solution where interaction with neighboring molecules leads to a modulation of the vibrational frequency for a given mode of the molecule [65]. The simplest time-dependent state describing vibrations along a single mode of a molecule is a two-eigenstate superposition given by the vibrational wave function

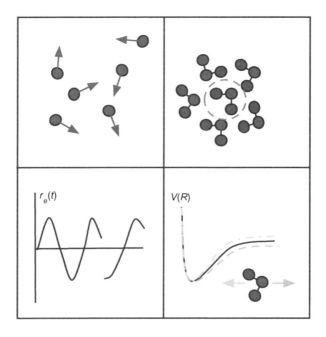

Figure 11.1 Models of collisional dephasing in an atomic gas (left) and vibrational dephasing due to solvent interactions (right). In an atomic gas, collisions between atoms interrupt the phase of the atomic dipole, causing the electronic displacement to be reset to a random value at the time of collision. In a liquid, the motion of neighboring molecules modifies the potential for a given vibrational mode in a time-dependent manner.

$$\chi(R,t) = c_1 e^{-i\omega_1 t}\chi_1(R) + c_2 e^{-i\omega_2 t}\chi_2(R) = a_1(t)\chi_1(R) + a_2(t)\chi_2(R). \qquad (11.13)$$

where the c_n coefficients are constant in the absence of any coupling between the states. The Hamiltonian matrix for this system (with no coupling between the states) is simply the energy eigenvalues along the diagonal:

$$H = \begin{bmatrix} \omega_1 & 0 \\ 0 & \omega_2 \end{bmatrix}. \qquad (11.14)$$

The density matrix for this two-state subspace (our "system") is

$$\rho = \begin{bmatrix} \rho_{11} & \rho_{12} \\ \rho_{21} & \rho_{22} \end{bmatrix} = \begin{bmatrix} a_1(t)a_1^*(t) & a_1(t)a_2^*(t) \\ a_2(t)a_1^*(t) & a_2(t)a_2^*(t) \end{bmatrix}. \qquad (11.15)$$

Using the Liouville-von Neumann equation (Equation 11.12), we arrive at

$$\frac{d\rho}{dt} = \begin{bmatrix} 0 & -i\rho_{12}\omega_{12} \\ -i\rho_{21}\omega_{21} & 0 \end{bmatrix}, \qquad (11.16)$$

where $\omega_{12} \equiv \omega_1 - \omega_2$ represents the frequency difference between the two eigenstates (similarly, $\omega_{21} \equiv \omega_2 - \omega_1$). In our model, interactions with the solvent molecules cause the shape of the potential to fluctuate in time. This means the frequency difference between states will also fluctuate in time: $\omega_{12} \to \omega_{12}(t)$. This modifies our equation of motion for the density matrix to read

$$\frac{d\rho}{dt} = \begin{bmatrix} \dot{\rho}_{11} & \dot{\rho}_{12} \\ \dot{\rho}_{21} & \dot{\rho}_{22} \end{bmatrix} = \begin{bmatrix} 0 & -i\rho_{12}\omega_{12}(t) \\ -i\rho_{21}\omega_{21}(t) & 0 \end{bmatrix}. \qquad (11.17)$$

This differential equation is separable (by term), and we can integrate it to solve for $\rho(t)$ (given the symmetry of the two terms, we simply focus on ρ_{12}):

$$\rho_{12}(t) = \rho_{12}(0)e^{-i\int_0^t \omega_{12}(t')dt'}. \qquad (11.18)$$

The exact form of $\omega_{12}(t)$ is unknown, as this comes from the random fluctuations of neighboring molecules. Presumably, each molecule in the ensemble will have a slightly different interaction with its neighbors, and thus a slightly different $\omega_{12}(t)$. However, we are interested in describing measurements on a molecular ensemble, and we should rewrite Equation 11.18 incorporating the average over all molecules:

$$\rho_{12}(t) = \rho_{12}(0)\left\langle e^{-i\int_0^t \omega_{12}(t')dt'}\right\rangle, \qquad (11.19)$$

where $\langle \ldots \rangle$ represents the average over the ensemble (we assume the ensemble has been coherently prepared such that $\rho_{12}(0)$ is the same for all molecules). We can separate $\omega_{12}(t)$ into a sum of the average frequency ω_{12} and the fluctuating portion $\alpha(t)$ (whose average over time is, by definition, zero): $\omega_{12}(t) = \omega_{12} + \alpha(t)$. Plugging this into Equation 11.19 yields

$$\rho_{12}(t) = \rho_{12}(0)e^{-i\omega_{12}t}\left\langle e^{-i\int_0^t \alpha(t')dt'}\right\rangle. \qquad (11.20)$$

We next expand the exponential in a Taylor series and move the ensemble average inside the time integral:

$$\left\langle e^{-i\int_0^t \alpha(t')dt'}\right\rangle = \left\langle 1 - i\int_0^t \alpha(t')dt' - \frac{1}{2}\int_0^t dt'\int_0^t dt''\alpha(t')\alpha(t'') + \ldots\right\rangle$$
$$= 1 - i\int_0^t \langle\alpha(t')\rangle dt' - \frac{1}{2}\int_0^t dt'\int_0^t dt''\langle\alpha(t')\alpha(t'')\rangle + \ldots \qquad (11.21)$$

This last move is possible since both the ensemble average and time integral are simply adding up contributions from different molecules and different times; the ordering of the contributions doesn't matter. The linear term in Equation 11.21 vanishes by definition of $\alpha(t)$. Keeping to second order, we are left with

$$\rho_{12}(t) = \rho_{12}(0)e^{-i\omega_{12}t}\left(1 - \frac{1}{2}\int_0^t dt' \int_0^t dt'' \langle\alpha(t')\alpha(t'')\rangle\right).\tag{11.22}$$

This can rewritten as (see homework)

$$\rho_{12}(t) = \rho_{12}(0)e^{-i\omega_{12}t}\left(1 - \int_0^t dt' \int_0^{t'} dt'' \langle\alpha(t'')\alpha(0)\rangle\right).\tag{11.23}$$

At this point, we are ready to address the form of the frequency fluctuations. Although the random fluctuations described by $\alpha(t)$ are not known explicitly, the quantity $\langle\alpha(t)\alpha(0)\rangle$ is the ensemble-averaged, frequency–frequency correlation function; it describes how the frequency of each given oscillator in the ensemble is correlated with the frequency of the same oscillator at a later time. A well-known guess, or ansatz, for this correlation function relevant to our physical situation of a molecule in solution was introduced by Kubo [66]. It posits that

$$\langle\alpha(t)\alpha(0)\rangle = \alpha_0^2 e^{-|t|/\tau},\tag{11.24}$$

where α_0 is the magnitude of the frequency fluctuations in solution and τ a correlation time. This ansatz says that when averaging over the ensemble, the interaction with the solvent (or environment in general) leads to an exponential decay in the correlation between the frequency at one time and some later time. Note that this is an elastic, or energy-conserving, interaction, which only affects the phases of the oscillators but not their amplitude/energy. Plugging Equation 11.24 into Equation 11.23 (dropping the absolute value since the integrals begin at zero) and integrating twice leads to the following expression for the coherence:

$$\rho_{12}(t) = \rho_{12}(0)e^{-i\omega_{12}t}\left[1 - \alpha_0^2\tau^2\left(e^{-t/\tau} + t/\tau - 1\right)\right].\tag{11.25}$$

We consider Equation 11.25 in two different limits. If the fluctuations in frequency occur very quickly, the correlation time τ in Equation 11.24 will be short compared to any timescale of interest. In this case, $t/\tau \gg 1$, and the only term that matters in the parentheses of Equation 11.25 is t/τ:

$$\rho_{12}(t) \approx \rho_{12}(0)e^{-i\omega_{12}t}\left(1 - \alpha_0^2\tau t\right) \approx \rho_{12}(0)e^{-i\omega_{12}t}e^{-\alpha_0^2\tau t}.\tag{11.26}$$

The time constant for the decay of the ρ_{12} coherence is known as the pure dephasing time and denoted as $T_2 = 1/(\alpha_0^2\tau)$. In terms of T_2, the equation for $\rho_{12}(t)$ can be written as

$$\rho_{12}(t) = \rho_{12}(0)e^{-i\omega_{12}t}e^{-t/T_2}.\tag{11.27}$$

We emphasize that Equation 11.27 is not valid for any individual molecule in the ensemble, but rather only when describing the ensemble as a whole. It expresses the fact that any coherent vibrational motion initiated in the ensemble will decay in time as a result of the varied interactions with the solvent. These interactions, which at the quantum level involve multidimensional dynamics among many degrees of freedom, are considered implicitly rather than explicitly in the model in order to simplify the calculation.

In Section 10.3, we performed a classical calculation that explicitly tracked the energy in a particular mode when coupled to many other degrees of freedom. Those results can be compared to the implicit model of this chapter by considering the oscillation amplitude of a particular mode (the "system mode"). In our implicit model, the expectation value of the mode coordinate, $\langle R \rangle$, in our two-state superposition is simply proportional to $\rho_{12}(t)$ (see homework). Figure 11.2 plots the oscillation amplitude of the system mode as a function of time for both the explicit calculation (jagged curve) and implicit Kubo model (smooth curve). The explicit calculation involved a total of 50 degrees of freedom, and the exponential decay of the implicit curve has been offset from zero to account for the fact that the energy in the system mode can only be shared among 50 modes at most. Our goal is to show the similar behavior between the explicit and implicit treatments (as opposed to extracting any quantitative results); therefore, the dephasing time T_2 in the implicit model has simply been fit by eye to the fast decay in the explicit calculation.

While the initial decay of system mode amplitude is similar between two treatments, note that the explicit treatment shows some degree of "population trapping" between times 200 and 900 periods. This is due to the finite number of modes in the explicit calculation, where it takes a longer time to reach the equilibrium limit. Only in the limit of a very large number of modes would we expect full agreement between the explicit calculation and the implicit model.

We now consider the opposite limit of Equation 11.25, where the fluctuations due to solvent interactions are much slower, and the correlation time τ is large compared to the experimental timescales. In this case $t/\tau \ll 1$, and the exponential in Equation 11.25 can be expanded to second order in t/τ:

$$\rho_{12}(t) = \rho_{12}(0)e^{-i\omega_{12}t}\left(1 - \alpha_0^2\tau^2\left(1 - \frac{t}{\tau} + \frac{1}{2}\frac{t^2}{\tau^2} + \frac{t}{\tau} - 1\right)\right). \tag{11.28}$$

The zeroth- and first-order terms cancel the other two terms in parentheses, and we are left with

$$\rho_{12}(t) = \rho_{12}(0)e^{-i\omega_{12}t}\left(1 - \frac{\alpha_0^2 t^2}{2}\right) \approx \rho_{12}(0)e^{-i\omega_{12}t}e^{-\frac{\alpha_0^2}{2}t^2}, \tag{11.29}$$

where the second step is valid in the short time limit. Note that in this limit, the decay of the coherence ρ_{12} is Gaussian in nature (as opposed to exponential).

Of course, in a real system the actual decay need not occur in either one of the analytic limits, and the specifics of the decay can differ. In fact, it is worth verifying that both

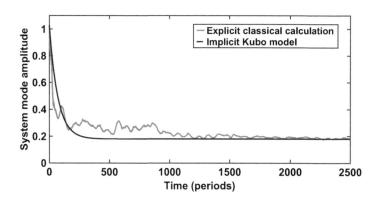

Figure 11.2 Oscillation amplitude of system mode as a function of time for explicit calculation of Section 10.3 with 50 modes (jagged curve) and implicit Kubo decay model (smooth curve).

our explicit and implicit models are relevant in actual systems. Figure 11.3 shows vibrational decay in three different experimental systems, along with corresponding illustrations of the behavior based on similar principles to our explicit and implicit models [67]. The top row in Figure 11.3 shows the case where the dephasing is dominated by weak coupling between only two modes. The middle row shows the case of intermediate coupling between many modes, while the bottom row illustrates strong coupling with many modes. In each case, the right column shows experimental data, while the left column shows illustrations of different expected decay behaviors.

One can compare the top row of Figure 11.3 to the envelope of the oscillations in the left panel of Figure 10.2, where we explicitly considered two modes with a weak anharmonic coupling. The middle and bottom rows of Figure 11.3 can be compared with the results shown in Figures 10.7 (explicit, many modes) and 11.2 (both explicit and implicit). The difference between the middle and bottom rows of Figure 11.3 is the strength of the coupling, whereas the difference between the various explicit results shown in Figure 10.7 is the number of modes participating. Both the coupling strength and the number of modes contribute to the shape of the decay curve, with a larger number of modes and stronger coupling each leading to a smoother and more monotonic decay. We will return to these models again when considering experiments that measure dynamics in large, multidimensional systems.

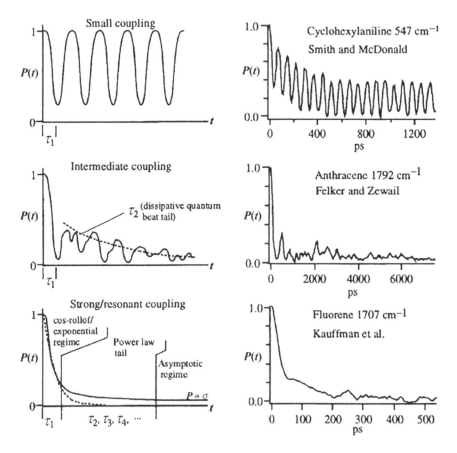

Figure 11.3 Decay (dephasing) of vibrational motion in three different molecular systems. Right column shows experimental data, while left column shows illustrations corresponding to the different types of expected decay behaviors. Figure from Ref. [67].

11.4 HOMEWORK PROBLEMS

1. Starting from the Liouville-von Neumann equation (Equation 11.12), derive the differential equations describing the time dependence of the density matrix for a two-level system coupled by an external field. Assume that the Hamiltonian is given in matrix form by Equation A.38:

$$H = \frac{1}{2} \begin{bmatrix} -\Delta & -\chi_{\text{Rabi}} \\ -\chi_{\text{Rabi}} & \Delta \end{bmatrix}, \tag{11.30}$$

where $\chi_{\text{Rabi}}(t)$ is the Rabi frequency and Δ the detuning. You should find the following for your result (note that these can also be derived using the standard wave function approach and the definition of the density matrix):

$$\dot{\rho}_{11} = -\frac{i}{2}\chi_{\text{Rabi}}\left(\rho_{12} - \rho_{21}\right) \tag{11.31a}$$

$$\dot{\rho}_{22} = +\frac{i}{2}\chi_{\text{Rabi}}\left(\rho_{12} - \rho_{21}\right) \tag{11.31b}$$

$$\dot{\rho}_{12} = +i\Delta\rho_{12} + \frac{i}{2}\chi_{\text{Rabi}}\left(\rho_{22} - \rho_{11}\right) \tag{11.31c}$$

$$\dot{\rho}_{21} = -i\Delta\rho_{21} - \frac{i}{2}\chi_{\text{Rabi}}\left(\rho_{22} - \rho_{11}\right) \tag{11.31d}$$

2. In Section 11.3, we briefly introduced a model (the first model discussed) that shows how elastic collisions in a gas lead to dephasing of the atomic dipoles in an ensemble. Assume that the collisions are both frequent (with a mean time of τ between them) and short compared to a Rabi oscillation period. In addition, assume that each collision will, on average, reset the oscillation of the atomic dipole such that $\rho_{21} = 0$ at the time of collision. Derive an expression for the off-diagonal element of the density matrix, $\rho_{21}(t)$, that is valid for the ensemble (including collisions).

Start by showing that for a collision occurring at time t_1, the following solution satisfies the differential equation from Equation 11.31 *and* ensures that $\rho_{21}(t = t_1) = 0$:

$$\rho_{21}(t)|_{t_1} = -\frac{\chi(\rho_{22} - \rho_{11})}{2\Delta}\left(1 - e^{-i\Delta(t-t_1)}\right). \tag{11.32}$$

Next, calculate $\rho_{21}(t)$ for the atomic ensemble with collisions by averaging over all times t_1. Make use of the fact that the fraction of atoms that had a collision between t_1 and $t_1 + dt_1$ is given by

$$df(t,t_1) = \frac{e^{-(t-t_1)/\tau}}{\tau}dt_1. \tag{11.33}$$

Finally, check that you get the same result for $\rho_{21}(t)$ if you modify the equation for $\dot{\rho}_{21}(t)$ to read (assuming constant χ, ρ_{22}, and ρ_{11}):

$$\dot{\rho}_{21} = -\left(\frac{1}{\tau} + i\Delta\right)\rho_{21} - \frac{i}{2}\chi_{\text{Rabi}}\left(\rho_{22} - \rho_{11}\right). \tag{11.34}$$

One obtains a similar differential equation for the other off-diagonal element. Interested readers can find a detailed treatment of this model in Ref. [9].

3. Consider the differential equations you derived for the density matrix in Problems 1 and 2. Keeping the collisional-decay terms in both the off-diagonal equations, numerically integrate the equation to find the density matrix elements as functions of time for an applied, time-varying electric field of constant amplitude. Plot $\rho_{11}(t)$, $\rho_{22}(t)$, and $i(\rho_{21}(t) - \rho_{12}(t))$. Use the following values for the parameters (all values are in atomic units):

   On-resonance: $\Delta = 0.0$ a.u.
   Coupling strength: $\chi_{Rabi} = 1.0$ a.u.
   Collision period: $\tau = 2.0$ a.u.

4. Show that Equation 11.23 follows from Equation 11.22.

5. Show that the expectation value of the mode coordinate, $\langle R \rangle$, in our two-state superposition of Section 11.3.1 is proportional to $\rho_{12}(t)$.

PART V

Measurements of Multidimensional Dynamics

CHAPTER 12

Incoherent Measurements in ND

12.1 INTRODUCTION

In Part III, we examined techniques for measuring one-dimensional dynamics. We will find that the approaches for multidimensional measurements are not all that different. However, for most one-dimensional systems (and certainly those in gas phase), time-resolved techniques typically do not provide information beyond what would be available using frequency-domain spectroscopy. In Part V, we will encounter systems where a time-domain approach provides more information, and does so more easily, than frequency-domain methods. The primary reason is the multidimensional character of the systems; whether intramolecular (e.g. dephasing into many vibrational coordinates) or intermolecular (e.g. interactions with a solvent), the large number of modes accessible to the system results in dynamics that time-resolved approaches are especially well suited to measure.

As with the examples of Part III, it is generally not possible to measure the full, (now multidimensional) wave function. Experiments are limited to some subspace that may include projections, lower-dimensional slices, or integration over certain coordinates. Since one only has access to a portion of the total information, determining what dynamics produced the observed signal can be challenging. Considering many degrees of freedom really brings this issue to the fore, and the full picture often comes to light only after piecing together insights gleaned from a variety of different experiments, along with comparisons to theoretical models.

Our goal in Part V is to describe measurements of multidimensional dynamics, guided by the topics introduced in Chapters 10 and 11. We follow the same general outline as Part III, sorting the techniques as incoherent measurements (Chapter 12), coherent optical measurements (Chapter 13), and coherent diffractive measurements (Chapter 14). However, given the wealth of dynamics available in multidimensional systems, in Part V the different sections often highlight the particular molecular process being measured (as opposed to a specific technique as in Part III). Some of the experimental approaches have been seen before, such as transient absorption and ultrafast electron diffraction. We will also consider new techniques like two-dimensional spectroscopy that are especially useful for multidimensional systems. We finish with Chapter 15, where we investigate a molecular ring-opening reaction using a suite of different experiments.

12.2 MEASURING STRUCTURAL DYNAMICS

12.2.1 Introduction: X-ray Absorption

As described in Section 3.1, the particular electrons in an atom or molecule that respond most strongly to an applied field are determined by the frequency of the field. For example, in experiments using infrared (IR) and optical wavelengths of radiation, the applied fields interact primarily with the valence electrons. On the other hand, X-ray fields typically address the core-level electrons. As shown in Figure 12.1, in a multidimensional molecular system, the valence electrons tend to be delocalized over two (or more) atoms, while the core electrons are confined to particular atoms. Thus, X-ray absorption has the potential to offer element specificity in a way that optical light cannot.

Of particular interest for time-resolved spectroscopy, the spectral positions of core-level absorptions depend on the atomic separations in a molecule. By monitoring how the frequencies of core-level absorptions shift in time after excitation by a pump pulse, absorption of an X-ray probe can provide a time-resolved picture of structural rearrangement. In this way, time-resolved X-ray absorption spectroscopy is similar to diffractive measurements using electrons or X-rays (see Chapter 9). As compared to optical methods, time-resolved X-ray absorption spectroscopy and diffractive techniques offer a more direct measure of how molecular structure changes in time. In this section, we discuss time-resolved, X-ray absorption in the biologically relevant system of myoglobin [69].

12.2.2 Experimental Setup

As usual, a pump pulse initiates dynamics in an excited electronic state. In this case, a 70-fs, 538-nm optical pulse induces dissociation ("photolysis") in carboxymyoglobin (MbCO). A time-delayed, 30-fs X-ray probe pulse follows the subsequent dynamics, either by changes in the transient X-ray absorption spectrum, or by X-ray fluorescence following absorption. In this particular implementation, the primary experimental signal is X-ray fluorescence collected at $90°$ relative to the incident beam direction. The fluorescence yield serves as a measure of the absorption probability, and in this way, the experiment is analogous to the incoherent technique of laser-induced fluorescence (as opposed to the coherent technique of transient absorption).

From a signal-to-noise perspective, whether it is more favorable to perform an absorption or fluorescence measurement depends on the size of pump-induced changes in probe transmission compared to intrinsic fluctuations in probe intensity. Crudely, a

Figure 12.1 X-ray absorption in a molecule with two different atoms. While the valence electrons tend to be delocalized over the molecule, the core electrons are confined to particular atoms. X-ray absorption at specific frequencies (as indicated by the arrows) provides element specificity. Horizontal lines represent energies for removing electrons from the associated orbitals. Figure from Ref. [68].

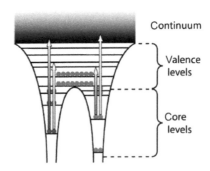

probe beam with 1% intensity fluctuations requires a sample density so that pump-induced changes in the probe are larger than 1%. Of course, signal averaging can help, but the basic idea remains the same: for samples where the pump-induced change is small, fluorescence measurements are often advantageous.

The X-ray absorption spectra for both reactant (MbCO) and product (dissociated Mb*) were measured to establish an appropriate probe photon energy for monitoring the dynamics. In particular, absorption should be sensitive to the reaction photolysis coordinate (CO dissociation), which given the structure of the molecule, is correlated with the distance between the iron atom and heme-group plane. Figure 12.2 shows the transient absorption spectra at two different time delays, as well as spectra with no excitation by the pump (bottom panel: normalized absorption, top panel: change in absorption). Movement of the iron atom relative to the heme plane is expected to shift the spectral position of the X-ray K-edge absorption (as seen in the bottom panel near 7.12 keV). A photon energy of 7123 eV (marked by left arrow in top panel) shows the largest change in X-ray absorption due to the action of the pump pulse. By monitoring the time-resolved change at this energy, one can map out structural dynamics following photolysis.[1]

12.2.3 Data and Interpretation

Figure 12.3 shows the time-resolved X-ray fluorescence signal when the energy of the absorbed probe photon is 7123 eV (the data are normalized to the fluorescence signal without the pump pulse). A sharp turn on of the fluorescence at time zero is followed by a slower approach to saturation, where the timescales for the two exponential fits are 73 fs and 400 fs. The changes in X-ray fluorescence (and therefore X-ray absorption) follow the structural modification that occurs after absorption of the pump pulse.

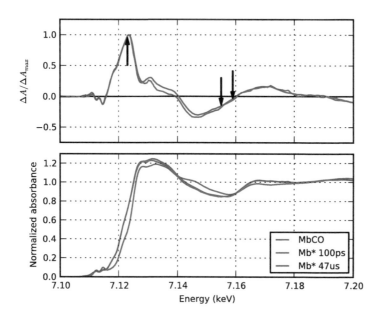

Figure 12.2 X-ray absorption spectra of MbCO before and after photolysis. The bottom panel plots normalized absorbance, while the top panel shows the change in absorption. Figure from Ref. [69].

---

[1]Experimentally, it was not possible to record full, time-resolved X-ray spectra; therefore, a few spectral positions were chosen and time scans recorded for each.

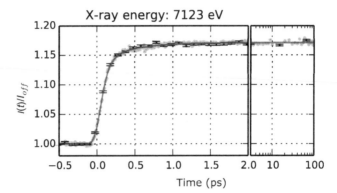

Figure 12.3 Time-resolved X-ray fluorescence for photolysis of MbCO with a probe photon energy of 7123 eV. Figure from Ref. [69].

These dynamics would be significantly more difficult to follow with optical absorption, as the connection between changes in molecular structure and optical absorption is much more complicated than with shifts in X-ray absorption. Nevertheless, detailed structural information is not available even with X-ray absorption, and the main information content from the measurements is the timescale for the iron atom to move. We revisit X-ray transient absorption on a different molecule in Section 15.2.4.

12.3 MEASURING ELECTRONIC STATE COUPLINGS

In this section, we consider two measurements of electronic state couplings in molecules. They both involve the process of "internal conversion," where a molecule undergoes rapid changes in electronic state character (shape of the electronic wave function) without spontaneous emission. The first experiment uses time-resolved photoelectron spectroscopy (TRPES) to time resolve changes in electronic state character, while the second uses fluorescence gating (FG) to measure nonradiative lifetimes of excited states.

12.3.1 Time-Resolved Photoelectron Spectroscopy

In Section 7.3 we discussed measuring dynamics in one-dimensional systems using TRPES. While TRPES can be a powerful tool for following dynamics in larger systems, the interpretation of TRPES in multiple dimensions is more complicated than in 1D and involves an understanding of electronic correlations. In particular, we will find that TRPES allows one to identify different excited electronic states via correlations between the intermediate and final states.

12.3.1.1 Experimental Setup

The experimental arrangement is similar to the one-dimensional case described in Section 7.3.2. Pump and probe pulses interact with a gas-phase molecular sample in a vacuum chamber. The pump initiates dynamics in an excited electronic state, and the time-delayed probe ionizes the molecules (the intensity of the probe pulse is kept low enough to ensure that one-photon ionization dominates the signal). An applied electric field across the interaction region collects the generated photoelectrons. The measured time of flight of the electrons is converted to a kinetic energy release upon ionization, and the kinetic-energy-resolved electron signal is plotted as a function of pump–probe delay.

12.3.1.2 Data and Interpretation

TRPES data from the molecule decatetraene (DT, $C_{10}H_{14}$) is shown in Figure 12.4 [70]. Panel (b) shows a clear change in the photoelectron energy distribution with pump–probe delay. While we expect this signal to contain information about excited-state dynamics following excitation by the pump, the variation in photoelectron energy accompanies a change in the electronic state character involving correlations between neutral and ionic states. Therefore, it is essential to have additional knowledge about the system to correctly interpret the experimental data.[2] In particular, a correct interpretation of the measurements in Figure 12.4 must account for a number of items, including the D_0 and D_1 state ionization energies (for ionization to the ground and first excited state of the cation), how the photoelectron energy relates to the change

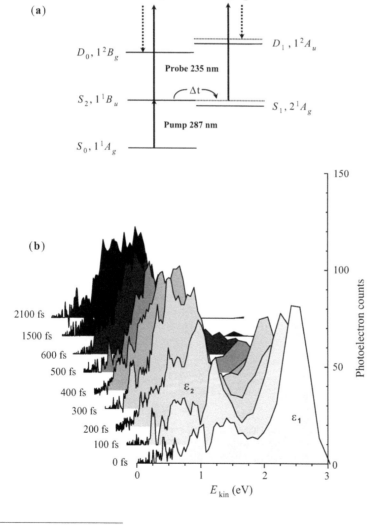

Figure 12.4 TRPES measurements in $C_{10}H_{14}$-DT. Top panel shows the energy-level structure, while the bottom panel plots the photoelectron spectrum as a function of pump–probe delay. Figure from Ref. [70].

---

[2]This is not to say that experiments in one dimension can be interpreted without any information about molecular potentials, but instead that it is clearly more complicated in multiple dimensions, especially when there are couplings between electronic states.

in ionization potential, correlations between ionic and neutral states, and the relative timescales for energy shifts of the initial excited state as compared to the peak in the photoelectron spectrum.

We begin with our result from Section 7.3.5, where Equation 7.11 gives the probability for finding a photoelectron with energy ε:

$$P(\varepsilon,t,\tau) = \left\langle \chi_{c,\varepsilon}^{(2)}(R,t) \,\middle|\, \chi_{c,\varepsilon}^{(2)}(R,t) \right\rangle$$
$$= \left\langle \chi_b^{(1)}(R,0)F(R,\varepsilon,t,\tau) \,\middle|\, F(R,\varepsilon,t,\tau)\chi_b^{(1)}(R,0) \right\rangle, \tag{12.1}$$

where

$$F(R,\varepsilon,t,\tau) \equiv \frac{-1}{2i} \int_0^t dt_2\, e^{i(V_I(R)-V_b(R)-\omega_2+\varepsilon)t_2}\, \mu_{cb} E_{02}(t_2), \tag{12.2}$$

and the τ dependence of F is contained in the arrival time of the probe field $E_{02}(t_2)$. From Equation 12.2 we see that $F(R,\varepsilon,t,\tau)$ is large when the photoelectron energy $\varepsilon = \omega_2 - (V_I(R)-V_b(R))$, which depends on the energy of both the initial neutral state and the final ionic state. As we saw in Section 7.3.4, if the wave packet moves from higher energy to lower energy on the potential, this can be reflected in a shifting photoelectron energy. However, evolution of the photoelectron spectrum can also reflect changes in the character of the electronic state on which the wave packet evolves. For the data shown in Figure 12.4, initial excitation takes the molecule to a state of primarily 1^1B_u character. As time progresses, this state changes to be primarily 2^1A_g character due to internal conversion, thereby changing the ionic state to which the neutral is correlated.[3]

Koopman's correlations are the most basic way to understand such neutral-ionic state correlations, and Figure 12.5 illustrates an example. Excitation by the pump pulse promotes an electron from the highest occupied molecular orbital (HOMO) to the lowest unoccupied molecular orbital (LUMO), leaving the molecule in a neutral state with an orbital configuration shown in the lower-left panel. For short time delays, the

Figure 12.5 An orbital energy diagram showing change in electronic character during internal conversion, as well as correlations between neutral and ionic states. The bottom portion shows orbital configurations for two different neutral states. During internal conversion, the character of the electronic state changes, corresponding to a change in orbital occupation. Ionization removes the most weakly bound electron and leaves the molecular cation in a state with the same orbital occupation as the neutral state below.

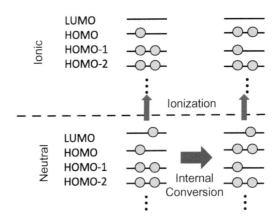

molecule remains in this configuration until the probe pulse arrives and ionizes the molecule. Removal of the most weakly bound electron from the LUMO via single-photon, weak-field ionization leads to the ground state of the cation with one electron missing from the HOMO. This is the ground state (D_0 - the lowest doublet state) in Figure 12.4 and corresponds to the orbital configuration shown in the upper-left panel of Figure 12.5. For longer time delays, the molecule changes character due to internal conversion, leaving a configuration with a missing electron (a "hole") in the HOMO-1 orbital (bottom-right panel of Figure 12.5). Probe ionization removing the LUMO electron now leads to the first excited state of the cation with an electron missing from HOMO-1 (upper-right panel of Figure 12.5, corresponding to D_1 in Figure 12.4). Note that the energy cost of removing an electron from a particular orbital depends on the nuclear coordinates (and can change during the internal conversion process).

Since D_1 and D_0 have different ionization potentials, the energy of a photoelectron associated with ionization to D_1 is different than ionization to D_0. Thus one can follow internal conversion dynamics by monitoring how the photoelectron spectrum changes with pump–probe delay. Returning to the experimental data of Figure 12.4, the photo-electron spectrum is dominated by a peak at 2.5 eV at short time delays and a peak near 1 eV for long time delays. Separate electronic state calculations assign these peaks as ionization to D_0 and D_1, respectively. The data therefore provide a measure of the timescale for internal conversion in the neutral molecule.

At intermediate delays, it is clear that one peak grows in while the other disappears. Note this is in contrast to a single peak shifting energy as seen in the one-dimensional results of Figure 7.14. Unlike our interpretation of Figure 12.4, one shifting peak is consistent with a wave packet simply riding down a single potential energy surface (PES), converting electronic potential energy to nuclear kinetic energy as it goes.

Before moving on, we briefly note that Koopman's correlations can be formalized and refined in terms of correlations based on Dyson orbital norms. Mathematically, the Dyson orbital is defined to be [71]:

$$\phi^D = \sqrt{N} \int \psi_i^N(1,\ldots,N)\psi_f^{N-1}(2,\ldots,N)dr_2\ldots dr_N, \tag{12.3}$$

where ψ_i^N is the initial, N-electron wave function of the neutral, and ψ_f^{N-1} is the final, $N-1$ electron wave function of the ion. The integral is over $N-1$ dimensions, leaving a one-electron function, or orbital, for ϕ^D. The magnitude of the Dyson orbital is large for neutral and ionic states that are Koopman-correlated as discussed earlier, while being very small for states that are not correlated. The matrix element used for calculating photoionization rates with Fermi's golden rule is expressed in terms of this Dyson orbital as

$$D_k = \left\langle \phi^D \left| -\hat{\boldsymbol{\varepsilon}} \cdot \mathbf{r} \right| \psi_k^e \right\rangle, \tag{12.4}$$

where $\hat{\boldsymbol{\varepsilon}}$ is the polarization direction of the light, $\mathbf{r}$ the dipole moment operator, and ψ_k^e the free (continuum) electron-wave function with momentum k.

12.3.2 Fluorescence Gating

We briefly introduced the technique of FG in Section 7.2.7. The basic idea is that time-resolving fluorescence from an excited electronic state allows one to determine the excited-state lifetime. In this section we discuss the application of FG to a series of uracil derivatives to study the effect of nonadiabatic coupling near conical intersections on excited state decay [72].

12.3.2.1 Experimental Setup

The experimental setup is the same as that shown in Figure 7.5. A femtosecond pump pulse at 267 nm promotes the solution-phase molecules to the first bright excited electronic state (S_1). The fluorescence is collected and focused into a nonlinear crystal where it overlaps with an 800-nm femtosecond probe pulse. Any fluorescence light in the crystal at the same time as the probe pulse undergoes sum-frequency generation, and the light produced is measured on a photomultiplier tube, with the integrated signal recorded as a function of pump–probe delay.

12.3.2.2 Data and Interpretation

Figure 12.6 shows the intensity of the sum-frequency-generation light as a function of pump–probe delay for uracil ($C_4H_4N_2O_2$) and four of its derivatives (along with a Gaussian representing the instrument response function of 330 fs full width at half maximum). The decays of uracil, 6-methyluracil, and 1,3-dimethyluracil can barely be distinguished from that of the fast instrument response. However, the decays of 5-methyluracil (thymine) and 5-fluorouracil are significantly longer (388 and 1736 fs, respectively). In looking at all the compounds, substitution at the C5-position (carbon next to the carbonyl, CO) on the uracil ring appears to be the primary driver of extended excited-state lifetime.

The FG data shown in Figure 12.6 only provide a timescale. Like a number of other techniques, comparison with calculation is required to draw meaningful conclusions regarding the dynamics responsible for the observed timescale. To interpret the data, the authors use electronic structure calculations to determine the shape of the excited (S_1) and ground (S_0) PESs in the region near a conical intersection between the two states. They use a reduced, two-dimensional subspace in which one of the coordinates is the out-of-plane motion of the C5 substituent. They find that the height of the

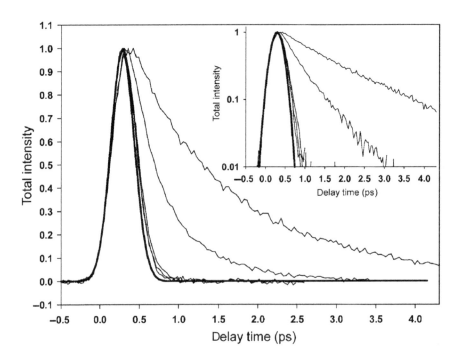

Figure 12.6 FG sum-frequency generation signal as a function of pump–probe delay (semilog plot on inset). Two of the uracil derivatives show an excited-state lifetime longer than the instrument response function. Figure from Ref. [72].

potential barrier between the minimum of the S_1 state and the position of the conical intersection follows the same trend as the lifetimes of the various uracil substitutions: the higher the barrier height, the longer the lifetime.

As seen in the inset of Figure 12.6, decay of the fluorescence signal is approximately exponential. We note that this exponential decay is related to the time it takes to reach the vicinity of the conical intersection, which is separate from exponential decay associated with radiative decay (and would be expected on a nanosecond timescale). The exponential decay of Figure 12.6 is typical of experiments measuring nonradiative relaxation, where there is a barrier to the region of nonradiative coupling. In contrast, for systems where there is no barrier to internal conversion (e.g. 1,3-cyclohexadiene in Chapter 15), the decay is not exponential.

The exponential decay associated with the barrier is not necessarily related to tunneling, as it is found even in systems where the energy of the wave packet is greater than the barrier height. In many cases, the timescale for the decay reflects the fact that the molecule needs to "pool" energy from several degrees of freedom to overcome the barrier. One can perform classical calculations for an ensemble of molecules in many dimensions with slightly different initial conditions, reflecting the spread of an initial wave function on the excited state in both coordinate and momentum space (similar to what we did in Section 10.3.1). The results show an exponential decay (in time) of the number of molecules in the ensemble whose trajectories have not passed over the barrier.

12.4 MEASURING INHOMOGENEOUS SYSTEMS

In this section we consider the effects of two different types of system inhomogeneities. In one case, single-molecule spectroscopy is used to separate the effects that different local environments have on identical systems in an ensemble. In the other example, it is variations in the system themselves that lead to the inhomogeneity. In both cases, one is interested in the dynamics of an ensemble in which there is variation in a particular property. In Section 13.3 we consider this idea in detail, introducing a new experimental method for measuring inhomogeneous systems.

12.4.1 Photoluminescence from a Quantum Dot

We first look at photoluminescence (PL) measurements in an important model system for solid-state physics: a quantum dot. A quantum dot is an engineered, atomic-like system on the surface of a bulk semiconductor. The dot can be made to have reasonably well-defined bound states with relatively weak coupling to the bulk (a solid-state analogy to an atom not fully isolated from the environment). The excited-state lifetime of the quantum dot is strongly influenced by the amount of coupling to its environment. For example, nonradiative decay mechanisms, mediated by defects in the dots and phonons in the bulk, can lead to more rapid decay than would occur via spontaneous emission. This is similar to an isolated molecule experiencing rapid, nonradiative decay from an excited electronic state to the ground state through nonadiabatic coupling.

The experiment uses time-resolved photoluminescence (TR-PL) to investigate the effects of ordering on quantum dot properties [73]. The setup is similar to that of FG discussed in Section 7.2.7. InAs quantum dots on an InGaAs surface are illuminated by an 80-fs pump pulse that promotes carriers to the excited state. The fluorescence lifetime of the dots is on the order of hundreds of picoseconds, which permits time-resolved measurements of the emitted light using a fast photodetector (note this implies that no probe pulse is required in these experiments). The intensity of the fluorescent light as a function of time after excitation by the pump serves as the experimental signal.

Figure 12.7 shows PL data from the InAs quantum dots. Panel (a) plots the time-independent fluorescence spectra for both ordered and disordered samples, while panel (b) shows the TR-PL measurements. The peaks in the spectra of panel (a) could be either homogeneously broadened (due to a short lifetime) or inhomogeneously broadened (due to nonuniformity of the quantum dot sizes). While it is not possible

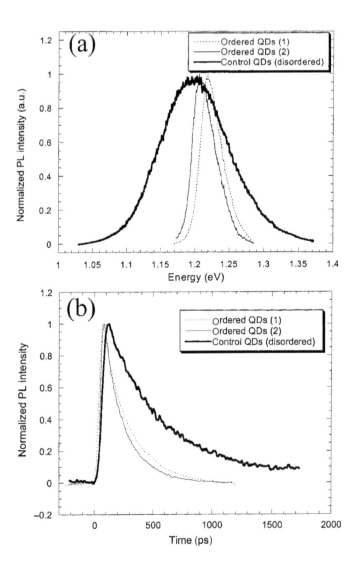

Figure 12.7 TR-PL from InAs quantum dots on a InGaAs surface. Panel (a) shows low-temperature PL spectra of InAs QD structures for both ordered and disordered structures. Panel (b) shows the TR-PL signals. Figure from Ref. [73].

to discriminate between these two possibilities based on PL spectra alone, the TR-PL data in panel (b) allows one to draw a conclusion. If the large spectral width of the peak in the disordered sample of panel (a) were due to homogeneous broadening, one would expect a shorter lifetime than the ordered dots. However, it is clear that the PL lifetime in the disordered dots is significantly longer than that for the ordered samples, indicating that it is the inhomogeneous broadening that is responsible for the width of the PL spectrum. Note that the measured PL decay times for both the ordered and unordered sample are shorter than those associated with radiative recombination, illustrating the role that defects and the bulk play in driving nonradiative decay and thus reducing the time over which fluorescence is emitted.

12.4.2 Single-Molecule Spectroscopy

In order to directly measure how the local environment affects the dynamics of a system in contact with a bath, one can attempt to separately measure each individual molecule in the ensemble. Looking at the molecules one at a time not only circumvents the inhomogeneities introduced by averaging over the ensemble, but also helps elucidate the role of the environment on a molecule-by-molecule basis. Of course, it is quite challenging to measure an ensemble one molecule at a time, and one generally needs control over the separation between molecules (single-molecule spectroscopy is typically carried out by diluting the sample and spin-coating it on a substrate). If the separation between molecules is large enough, one can isolate a single molecule on a surface in the focus of a microscope objective. In time-resolved measurements, this allows one to measure fluorescence from a single molecule as a function of pump–probe delay.

Figure 12.8 shows measurements on the large fluorophore DiD (1,1'-dioctadecyl-3,3,3',3'-tetramethylindodicarbocyanine) [74]. A pump pulse saturates the transition from the ground state of the molecule to the excited state, resulting in 50% probability

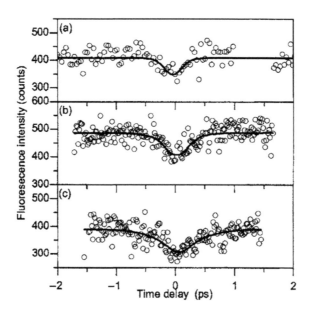

Figure 12.8 Single-molecule fluorescence yield as a function of pump–probe delay. The three panels correspond to three different individual molecules. Figure from Ref. [74].

for being in both the ground and excited states.[4] If the probe pulse arrives nearby in time to the pump, it can drive both stimulated absorption and stimulated emission; since the transition is saturated, there is no change in the excited-state population. However, if the wave packet on the excited state has moved away from the Franck–Condon (FC) region, light from the probe pulse will only be absorbed (not also emitted), and the fluorescence yield increases. Thus, this pump–probe arrangement measures how long it takes for the excited-state wave packet to move away from the FC region. The three panels in Figure 12.8 show results for three individual molecules on the surface. As the data illustrate, the time to move out of the FC region is different for different molecules in the ensemble, illustrating how the local environment affects the dynamics of each molecule differently.

12.5 MEASURING MULTIELECTRON DYNAMICS

As we have discussed, in a truly multidimensional problem, the total wave function cannot be written as a product of one-dimensional functions. Consider the example of the two-dimensional oscillator from Section 10.2. While the harmonic case certainly involves dynamics in multiple degrees of freedom, its wave function can be factored into a product of two, one-dimensional functions that are independent of one another. If the potential contains anharmonic terms, however, the true eigenstates cannot be written as a product of one-dimensional functions. In this situation the dynamics along the two coordinates are inherently coupled.

It is similar in the electronic case. If the electrons in a multielectron system can be considered independently, the total, multiparticle wave function can be reduced to a product of independent, single-electron wave functions, and the motion of each electron can be treated individually. However, when the electrons interact, the full, multiparticle wave function is no longer separable and is entangled between the many degrees of freedom. While electron correlations were safely ignored in the "one-electron" systems of Section 7.4, in the multielectron experiments of this chapter, correlated electron dynamics are essential for describing the behavior. As we did for the helium atom in Section 2.2, we continue to use products of one-electron wave functions (orbitals) as our basis set for expressing the full, N-electron wave function. While a finite sum of products is clearly an approximation for anything beyond the hydrogen atom, it serves as a convenient basis for including electron interactions.

12.5.1 Autoionization in an Atom

We begin with an example of correlated electron dynamics in a simple system [75]. The experiment is carried out in atomic barium (Ba), whose ground state consists of two valence electrons in the $6s$ orbital. A sequence of laser pulses excites both of these two electrons to coherent superpositions of high-lying orbitals (producing two Rydberg wave packets). If only one electron was excited, its energy would be slightly below the ionization threshold, and the electron would remain bound to the

---

[4]We note that, in principle, the pump pulse could invert the population on this transition. However, given the interaction of the molecules with the substrate and the vibrational dynamics on the excited state, coherence between the ground and excited state is not maintained for long. This leads to saturation of the transition during the interaction with the pump and a maximum population transfer of about 50%.

atom. However, excitation of both electrons leads to a doubly excited state whose total energy is above the single ionization limit (a "bound state embedded in the continuum"). It is the interaction between the two electrons that allows one of them to escape in a process called autoionization.

A completely rigorous description of the experiment would treat the excitation in terms of the full, multielectron wave function. Motivated by the earlier discussion, we instead consider the problem in terms of single-electron wave packets for the two valence electrons that subsequently interact due to the Coulomb potential. This is similar to the helium atom, where the Coulomb interaction between the two electrons was treated as a perturbation.

12.5.1.1 Experimental Setup

The excitation diagram is shown on the left side of Figure 12.9. The applied pump pulse is actually a pair of synchronous laser pulses that two-photon excite the atoms into a Rydberg wave packet just below the single-ionization threshold. This transition promotes one of the electrons to a superposition of high-lying Rydberg orbitals, while the other electron makes a small jump from the $6s$ to the $5d$ orbital. The probe is likewise composed of a pair of synchronous pulses, whose mutual time delay can be varied with respect to the pump. The probe interaction promotes the other valence electron into a second Rydberg wave packet (a superposition of Rydberg orbitals) at a total energy just below the double-ionization threshold. At this point, the electron–electron Coulomb interaction between the two excited electrons can lead to autoionization, producing a free electron and a singly charged barium cation. The experiment used field ionization to further ionize the cation, thereby creating a second electron and a dication. The final state of the system is measured by detecting the secondary electrons produced by quasi static field ionization in a manner sensitive to the state of the cation before the second ionization step.

12.5.1.2 Data and Interpretation

Prior to autoionization, both electrons are in coherent superpositions of principal quantum number states, producing a pair of "breathing" radial wave packets (see right side

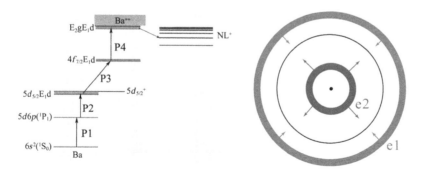

Figure 12.9 Left: Schematic diagram of two-electron wave packet generation and detection. The pump interaction incorporates laser pulses P1 and P2, while the probe involves pulses P3 and P4. The atom is left in a doubly excited state with an energy just below the double-ionization threshold. Subsequent autoionization leads to the ejection of an electron. Right: Representation of a two-electron wave packet in atomic barium. The electrons are in different "breathing" radial wave packets. Figure adapted from [75] (including supplemental material).

of Figure 12.9). Classically, the Coulomb interaction is strongest when the two electrons are closest. Quantum mechanically, the two wave packets "collide" when the different electron probability densities overlap in space and time. The time delay between the pump and probe pulses determines where in space the two wave packets first collide (the first collision results in the largest contribution to the autoionization yield). In a simple classical picture, when the wave packets collide at small radii, the electrons can exchange a significant amount of momentum with the core, leading to a more energetic autoionization electron and a more deeply bound state of the cation. On the other hand, if the electron wave packets collide far from the core, the final state of the cation is less deeply bound. Thus, the pump–probe delay controls both the amount of autoionization and the final energy of the cation.

Figure 12.10 shows the autoionization yield as a function of pump–probe delay for two different electron binding energies. Both signals undergo a modulation with a period of approximately 5 ps, matching the calculated wave packet oscillation period of the first excited electron (e1). The autoionization yield is enhanced when the second electron (e2) is launched either immediately after e1 or one round-trip later; in both cases, electron e1 is near the core at the time of collision.

12.5.2 Time-Resolved Auger Decay

As we saw in Section 10.5, Auger decay begins with a high-energy photon removing an electron from a core or other inner orbital of the system. If the true, N-electron eigenstates were simply products of one-electron orbitals, there would be no dynamics after removal of this inner-orbital electron. However, one electron missing from an inner orbital does not correspond to an eigenstate of the system, and the ensuing dynamics include a number of energetically allowed possibilities such, as X-ray fluorescence (spontaneous emission) or Auger decay.

In Auger decay, an electron occupying an outer orbital "falls down" into the inner-shell hole created by the incident X-ray, and rather than release the excess energy as a photon, the system emits a second electron. We considered the natural timescale for this type of electron–electron interaction in Section 10.5. In order to compare to experiment, we seek a measurement of the time it takes for the Auger electron to be emitted after the production of the original, core photoelectron. A technique known as "streaking" uses a strong-field laser interaction to measure, or gate, the Auger emission time. The experimental approach is somewhat different than those we have considered earlier, and we introduce it here.

Figure 12.10 Ionization yield as a function of pump–probe delay for autoionization in atomic barium. The two different panels show electrons with two different binding energies. Figure adapted from [75] (courtesy of author).

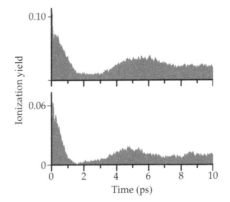

12.5.2.1 The Streaking Technique

The streaking technique uses an intense laser field present during the electron emission process, where the applied field shifts the momentum/energy of the electron measured by a time-of-flight or velocity-map-imaging measurement. As seen in Section 3.2, the momentum of an electron born into the continuum at time t will be modified by the presence of the streaking field by

$$\mathbf{p}'(t) = \mathbf{p}(t) - q\mathbf{A}(t), \tag{12.5}$$

where $\mathbf{p}$ is the momentum of the electron without the streaking field, q is the electron change $(-e)$, and $\mathbf{A}(t)$ is the vector potential of the streaking field at the time the electron is "born."[5] Since the (dynamical/kinematic) momentum $\mathbf{p}'$ is conserved even when the streaking field (and hence its vector potential) turns off, the final momentum that one measures after some delay τ is given by

$$\begin{aligned} \mathbf{p}'(t + \tau) &= \mathbf{p}'(t) \\ &= \mathbf{p}(t) + \mathbf{A}(t). \end{aligned} \tag{12.6}$$

Figure 12.11 illustrates a basic streaking measurement for ionization of a model atom represented by a Yukawa potential [76]. An extreme ultraviolet (XUV) attosecond pulse ionizes an electron in the presence of a few-femtosecond IR laser field (panel (a)). There is a controllable time delay τ between the XUV pulse and the peak of the vector potential for the IR probe pulse. Panel (b) shows both the measured electron momentum and the (negative) vector potential of the streaking field as functions of the time delay τ. The electron momentum shifts according to the time-dependent vector

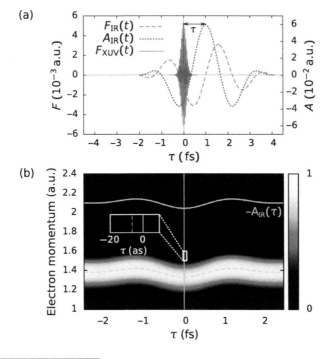

Figure 12.11 (See color insert.) Illustration of a streaking measurement using a combination of femtosecond IR and attosecond XUV pulses. Panel (a) shows the IR field $F_{IR}(t)$, its associated vector potential $A_{IR}(t)$, and the XUV attosecond field $F_{XUV}(t)$ as a function of time delay. Panel (b) shows a simulation of the streaking diagram, where both $A_{IR}(t)$ and the electron momentum are plotted as a function of time delay. The ionization delay is seen by the phase shift between the $A_{IR}(t)$ and the electron momentum $p(t)$ as a function of time delay. Figure from Ref. [76].

---

[5]This result can be derived by integrating the equation of motion for an electron in an oscillating electric field from the moment it is born until the end of the pulse.

potential of the streaking field at the time of the XUV pulse (which is when the electron is born).

Note that the nature of the shift depends on the optical period of the streaking field compared to the time it takes for the electron to be emitted. In our discussion, we assumed the photoemission took place in a time significantly shorter than an optical cycle of the streaking field. In that case, the momentum of the electron was shifted by the vector potential at the instant of birth and followed the behavior shown in panel (b) of Figure 12.11. It is instructive to consider the other limit, where the photoemission takes place over many optical cycles of the streaking field. In this case, each portion of the electron wave function has its momentum shifted by the vector potential at its particular time of emission. Thus, the photoelectron spectrum consists of electrons with a range of momenta/energies, and the mean energy is modified by the cycle-averaged energy shift: $(\mathbf{p}+\mathbf{A})^2/2$. Since the vector potential can add to or subtract from the outgoing electron momentum (both are vector quantities), the emitted photoelectrons are shifted both up and down in energy. Furthermore, the vector potential has the smallest derivative (with respect to time) at its maximum magnitude, and so a large percentage of the electron wave function experiences close to the maximal shift. This results in what appear to be sidebands on the photoelectron spectrum (as one would expect from a simple, perturbative picture of dressing the photoelectron with a photon from the streaking field).

This is the basic principle of a streaking experiment. Before moving on to discuss how it is used to measure Auger lifetimes, we note that streaking also highlights the concept of "ionization time delays." In panel (b) of Figure 12.11, a small time delay is seen between the oscillations of the vector potential and electron momentum. Qualitatively, the electron does not come out of the atom instantaneously due to the time it takes to "climb out of the potential." While there are many ways to quantify the phase and group delays of a wave function as it exits a particular potential, the observed effect on the electron wave function is typically described as a "Wigner-Smith" delay. The Wigner-Smith delay is defined as the group delay of the wave packet relative to a free wave packet (no potential) with the same asymptotic ($r \to \infty$) energy [77].

12.5.2.2 Auger Lifetime Measurement

We are now ready to describe an Auger lifetime measurement where the streaking pulse serves as the probe. As we will see, delaying this pulse relative to the XUV (or X-ray) pump pulse that removed the core electron allows one to measure the timescale for Auger decay. Figure 12.12 illustrates the approach in atomic krypton [78]. Attosecond XUV pulses with a central energy of 97 eV are produced via high-harmonic generation in a noble gas sample. Absorption of the XUV light removes a core electron from the atom (step a in panel a), producing an electron with a kinetic energy spectrum depicted by the "Core" peak along the vertical axis. The Auger process comes from the correlated electron dynamics in steps b and c: one electron falls down and fills the core vacancy (step b), while another electron is ionized with a resulting kinetic energy spectrum given by the "Auger" peak (step c). The probe pulse arrives with a variable time delay. In addition to producing a core hole, the pump pulse also produces additional, prompt photoionization of the atom from a valence orbital (step a').

Panel (b) illustrates the expected timescales for production of both the prompt photoelectron as well as the Auger electron. Since the Auger measurement depends on the temporal profile of the streaking field, detailed characterization of the probe pulse is

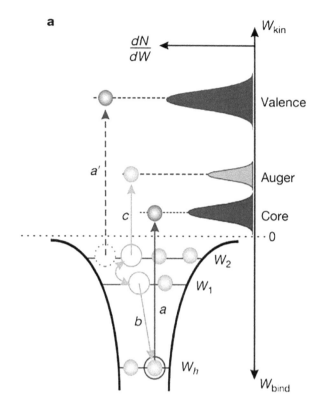

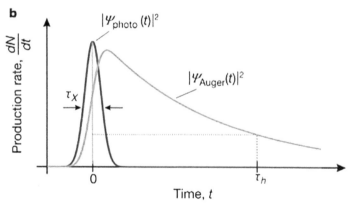

Figure 12.12 Illustration of Auger lifetime measurements using attosecond X-ray pulses and few-cycle, strong-field laser pulses for streaking. Panel a shows the excitation process and resulting photoelectron spectrum, while panel b depicts the expected timescales for electron production. Figure from Ref. [78].

essential. Fortunately, the "prompt" photoelectrons in the spectrum from direct ionization of valence orbitals by the pump pulse allow for characterization of the streaking pulse *in situ* (in addition to the Auger lifetime measurement).

The experimentally measured photoelectron spectrum as a function of pump–probe delay is shown in Figure 12.13. The spectrum at any given time delay contains a number of different peaks that can be interpreted with reference to the ground-state orbital configuration of krypton: $[Ar]3d^{10}4s^24p^6$. Ionization of a $4p$ electron by the XUV pulse leads to the highest energy peak in the spectrum of Figure 12.13. The width of this peak is determined by the XUV spectral bandwidth; since the bandwidth is larger

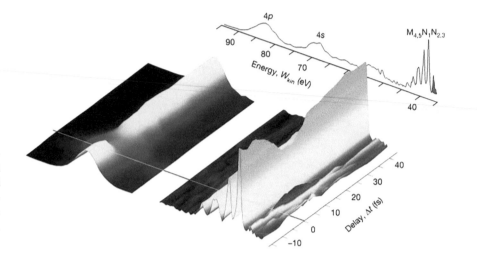

Figure 12.13 Photoelectron spectrum as a function of delay between an X-ray pump and a strong-field (streaking) probe. The peak highlighted in red contains the Auger lifetime information. Figure from Ref. [78].

than the fine-structure splitting, a single, broad peak that appears in the spectrum. There is also a peak corresponding to XUV ionization of a 4s electron.

Of greater interest are the series of peaks around 40 eV, as they correspond to Auger electrons. Figure 12.14 shows a more detailed view of the process. The binding energies of the 4s and 4p orbitals are split by the fact that s-orbitals penetrate the core to a greater extent than p-orbitals (which have no probability for finding the electron at $r = 0$). This lowers the energy of the 4s orbital since it sees a less-shielded core. There is also fine-structure splitting, including both spin–orbit and relativistic effects, that splits the 4p orbital into $4p_{3/2}$ and $4p_{1/2}$, and the 3d orbital into $3d_{5/2}$ and $3d_{3/2}$. This produces four distinct peaks, corresponding to removal of $3d_{5/2}$ or $3d_{3/2}$ electrons by the XUV pulse, followed by Auger emission of the $4p_{3/2}$ or $4p_{1/2}$ electrons (as depicted in Figure 12.14). Note that these peaks are due to secondary Auger ionization after the XUV pulse, and as such are always present, independent of streaking probe pulse.

The lowest-energy peak (shaded in red in Figure 12.13) corresponds to Auger electrons that have been dressed by the strong-field streaking pulse. In particular, cycle averaging over the period of the streaking pulse produces a sideband that shifts the red peak in the photoelectron spectrum away from the undressed Auger peak. This shift depends on the streaking field being present during Auger emission, and so it changes with pump–probe delay. In this way, the time dependence of the Auger process can be resolved.

The spectral integral of this red peak is shown as a function of pump–probe delay in the panel (a) of Figure 12.15. Panel (b) plots the spectral integral of the prompt photoelectron peak as a function of delay, permitting characterization of the streaking pulse. Comparison of the two peaks yields an extracted lifetime for this Auger process of approximately 8 fs. This time-domain result is consistent with an inferred lifetime from the width of the Auger peaks in the frequency-domain spectrum.

12.5.2.3 RABBITT Technique

Before moving on, we note that the Auger streaking experiment is closely related to another attosecond measurement technique known as Reconstruction of Attosecond

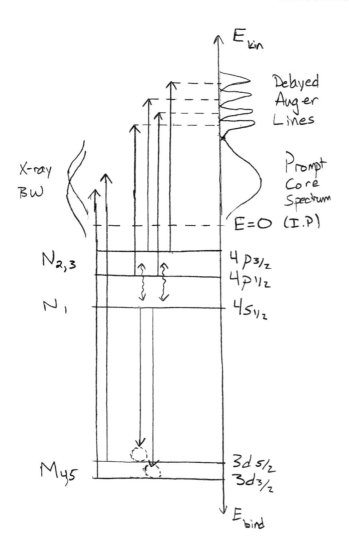

Figure 12.14 Diagram showing relevant atomic orbitals and energies for krypton Auger lifetime measurement. Note that the prompt X-ray ionization of the 4*p* and 4*s* levels (seen in Figure 12.13) is not shown in this diagram.

Beating By Interference of Two-photon Transitions (RABBITT) [79, 80]. The primary difference is that instead of a single attosecond pulse as used for streaking, RABBITT uses an attosecond pulse train produced by high-harmonic generation of the IR probe field. This leads to an XUV spectrum with a series of peaks at the odd-numbered harmonics (see Section B.4). In addition, this implies that ionization by the XUV pulse train in RABBITT takes place over several cycles of the streaking pulse, putting it in the sideband limit discussed in Section 12.5.2.1. The combination of the high harmonic spectrum with peaks at the odd harmonics and sideband production at the fundamental frequency of the streaking pulse leads to constructive and destructive interference of the ionization amplitude at photoelectron energies corresponding to ionization by even harmonics. In particular, ionization by the *n*th harmonic *plus* one photon of the fundamental interferes with ionization by the $(n+2)$th harmonic *minus* one photon of the fundamental (see panel (a) of Figure 12.17). This interference yields a phase difference that can be used to extract timing information about the ionization. We consider a RABBITT measurement of photoemission from a metal surface in Section 12.5.4.

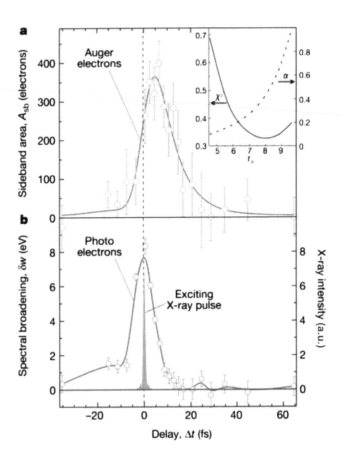

Figure 12.15 Panel (a) shows the integrated Auger sideband intensity as a function of pump–probe delay. Panel (b) shows the width of the direct ionization photoelectron peak as a function of delay. Figure from Ref. [78].

12.5.3 Charge Migration: Multielectron Wave Packets in Molecules

In Section 10.5.3 we discussed multielectron dynamics that occur after removal of an inner-orbital electron from a system. Theoretical models predicted an attosecond rearrangement of electron charge density that was quite general across different systems. In larger molecules, correlated electron dynamics are almost always important, and in this section we consider an experiment measuring such attosecond charge migration.

12.5.3.1 Experimental Setup

The experiment probes attosecond electron dynamics in the amino acid molecule phenylalanine [81]. An isolated attosecond XUV pulse (290 as duration, photon energy 15 to 35 eV) serves as the pump and impulsively ionizes the molecule. The probe is a 4-fs visible/near-IR pulse (central photon energy 1.77 eV) that induces a second ionization step in the molecule. The ions produced after the pump–probe interaction are collected using time-of-flight mass spectroscopy (see Section 7.3). In particular, the authors follow dynamics of the doubly charged immonium ion fragment produced by the probe pulse.

12.5.3.2 Data and Interpretation

Figure 12.16 panel (a) shows the measured doubly charged immonium fragment ion yield as a function of pump–probe delay in phenylalanine (lower plot has the exponential decay removed). Since XUV absorption does not lead to the formation of any

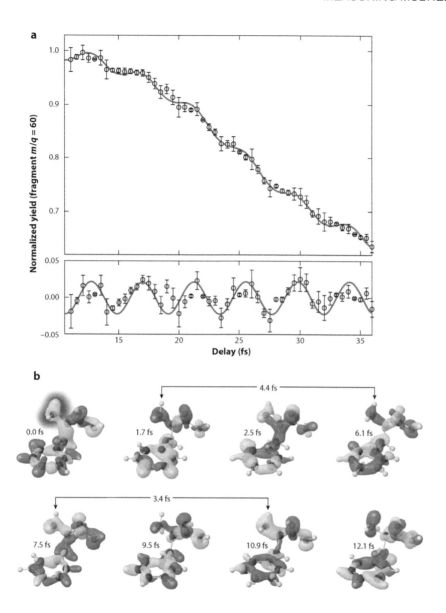

Figure 12.16 Pump–probe measurements illustrating charge migration in phenylalanine. Panel (a) shows the doubly charged immonium fragment ion yield as a function of pump–probe delay. Modulations in the fragment ion yield are highlighted in the lower plot of panel (a), where the exponential decay has been removed. Panel (b) shows calculated electronic charge densities in the molecule at different time delays. Oscillations with a similar period to the experimental signal are seen in the calculations. Figure from Ref. [82] (original data from [81]).

dications (and no dications are formed by the probe pulse alone), any dications in the time-of-flight mass spectrum must come from a combined action of the pump and probe pulses. Furthermore, since the dication yield depends on pump–probe delay, the second ionization step must be sensitive to dynamics in the cation. One can infer from the timescale of modulations in the dication yield (4.3 fs) that dynamics in the cation most likely involve electronic dynamics, as the shortest nuclear vibrational period in the molecule is about a factor of two slower.

Panel (b) depicts the *calculated* charge migration in the molecule after ionization with the attosecond pulse. The timescales associated with the electronic dynamics depicted in panel (b) are similar to the measurements, suggesting that the second ionization by the probe pulse measures electron dynamics in the cation occurring after the pump.

As in the case of an electronic wave packet in atomic potassium (see Section 7.4.3), ionization by the probe pulse depends on the changing position of the electrons (i.e. a time-dependent charge density distribution).

12.5.4 Photoemission Delays From Solids

In Section 12.5.2 we discussed streaking and RABBITT techniques for measuring the Auger lifetime in an atom. We also introduced the idea of ionization (Wigner-Smith) delays for a particular ionization channel. The RABBITT technique can be applied to measure delays in photoemission from solid surfaces, where the emission dynamics are more subtle than the atomic case. In this section, we discuss an experiment measuring photoemission from a silver metal surface [83].

12.5.4.1 Experimental Setup

The experimental setup is very similar to that of Section 12.5.2. An XUV pulse train initiates electron dynamics in the presence of a near-IR streaking field. The RABBITT apparatus simultaneously measures photoelectrons emitted from both argon atoms and the silver surface (Ag(111)). For both samples, the photoelectron spectrum is measured as a function of delay between the XUV pulse train and IR streaking field. The yields are compared in detail, allowing the authors to subtract off any contribution to the delays common to both systems.

12.5.4.2 Data and Interpretation

Panel (a) of Figure 12.17 depicts the RABBITT excitation scheme. The basic model of surface RABBITT involves three steps. Initial photoabsorption of valence electrons creates highly excited electrons in the solid near the surface (depicted by the blue arrows showing excitation by different odd-order, high-harmonic photons). These electrons then undergo transport inside the solid towards the surface. The final step involves "dressing" the photoelectrons at the surface of the metal with the IR probe pulse.[6] The IR-streaking field introduces sidebands at the even-order harmonic frequencies.

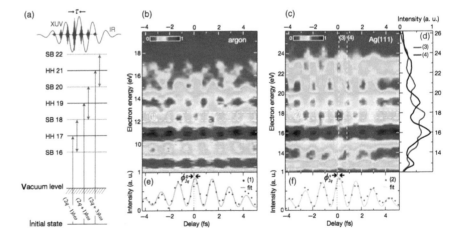

Figure 12.17 Attosecond RAB-BITT measurement of photoemission from a metal surface as compared to argon. Panel (a) depicts the excitation scheme and pulse sequence. Panels (b) and (c) plot the RABBITT photoelectron spectra for gas-phase argon and a silver surface, respectively. Figure from Ref. [83].

---

[6]The probe pulse essentially does not penetrate the metal and is most intense at the surface, allowing the authors to conclude that the streaking measurement takes place near the surface.

Panels (b) and (c) of Figure 12.17 show RABBITT photoelectron spectra for emission from the surface of silver (c) as compared to argon (b). Both measurements show peaks in the photoelectron spectrum corresponding to ionization by odd harmonics in the pump pulse, as well as sidebands at the even harmonics due to the IR streaking field. The intensity of the sidebands are modulated as a function of pump–probe delay.[7] The measured spectra are used to extract photoemission time delays for photoelectrons at different energies ($\varepsilon = h\nu - \phi$, where ϕ is the work function).

The extracted photoemission delays measured by surface RABBITT are compared to existing models for photoemission from different bands in a solid (including ballistic transport). The calculations show expected photoemission delays between 150 and 180 as for photon energies between 25 and 35 eV, while the measurements yield delays that vary substantially (from -120 as to $+120$ as) over that same range of photon energies. The measurements indicate that current theoretical descriptions of photoemission are likely incomplete and must be developed further.

This concludes our discussion of incoherent detection techniques for measuring multidimensional dynamics. In the next chapter we consider coherent optical approaches for measuring such dynamics.

---

[7]The electron yield at energies corresponding to the odd harmonics is also modulated because of depletion by the streaking field.

CHAPTER 13

Coherent Optical Measurements in ND

13.1 MEASURING VIBRATIONAL MODE COUPLING

In Part IV we discussed models for vibrational mode coupling in molecular systems, finding that the anharmonic nature of the potential energy surface led to energy flow between different modes of the system. In this section, we examine experimental approaches for measuring such mode coupling in both gas-phase and liquid-phase samples. Before proceeding, we note that an alternative technique known as "two-dimensional spectroscopy" is also well suited for studying such mode coupling (see Section 13.3). Given the rather significant shift in the interpretation associated with two-dimensional spectroscopy, we delay its discussion until after experimental approaches that follow more directly from Part III.

13.1.1 IR–IR Transient Absorption

13.1.1.1 Experimental Setup

We begin by considering a relatively simple pump–probe experiment designed to measure energy transfer out of a vibrational mode. The experimental arrangement is similar to the optical transient absorption experiments discussed in Section 8.2. In infrared (IR)-transient absorption, an IR pump pulse initiates vibrational dynamics in a particular mode of the molecule through a direct IR transition. Vibrational population transfer is monitored via the time-resolved, transient absorption spectrum of an IR probe pulse. The transmitted probe is typically spectrally resolved to provide multichannel information.

13.1.1.2 Data and Interpretation

Figure 13.1 shows spectrally resolved, IR-transient absorption data in liquid-phase HOD (partially deuterated water) in a solvent of D_2O [84]. The pump is a 45-fs IR pulse with a central frequency of 3400 cm^{-1} that excites the OH-stretch in HOD. The time-delayed probe is an IR pulse whose extremely broad bandwidth allows one to simultaneously record transient absorption data across a large range of probe frequencies. The change in absorption is plotted as a function of both pump–probe delay (τ_2) and probe frequency (ω_3). The region in the middle is not recorded due to the large absorption of the probe by the OD-stretch in the solvent.

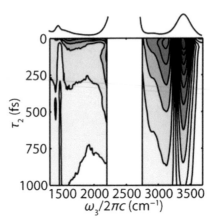

Figure 13.1 IR-transient absorption data showing mode coupling in HOD in a solvent of D_2O. Red colors correspond to larger transmitted probe intensity (less absorption as compared to when the pump is not present), while blue colors imply more absorption. Figure from Ref. [84].

In the O–H stretch region near 3400 cm^{-1}, one sees enhanced probe transmission due to the presence of the pump. This represents ground-state bleach from excitation of the O–H stretch by the pump pulse (removal of population from the ground state implies there is less absorption of the probe at this color). In our two-mode notation of Section 10.2.3, the pump excites the transition $|00\rangle \rightarrow |10\rangle$, adding one quanta of vibration to the first mode (the O–H stretch) and leaving zero quanta in the second mode (the HOD bend in this case).[1] Immediately to the red of this peak near 3100 cm^{-1} is a region of less probe transmission (enhanced absorption). This represents excited-state absorption of the probe on the $|10\rangle \rightarrow |20\rangle$ O–H stretch transition; the pump pulse has left population in the first-excited state that can now be promoted to the second-excited state. The reason for the red-shift from the $|00\rangle \rightarrow |10\rangle$ transition is the anharmonicity of the potential well for the O–H stretch mode (in fact, these types of measurements can also be used to measure anharmonicities).

There is also a small additional region of enhanced absorption centered at 2860 cm^{-1} that is seen as a shoulder on the 3100 cm^{-1} peak in Figure 13.1. It is this peak that most directly provides information about mode coupling in the molecule. By comparison with the known mode energies seen in Figure 13.2, this peak is identified as the $|10\rangle \rightarrow |12\rangle$ transition (excitation of two quanta of the HOD bend by the probe). This normally dark transition is excited due to the anharmonic coupling between the two modes, as one quanta of the stretch is close in energy to two quanta of the bend and the actual motion associated with each of the vibrational states corresponds to a mixture of stretching and bending. In this sense, the observation of this transition, along with its temporal dynamics, provides a direct measure of the coupling between the $|20\rangle$ and $|12\rangle$ states (a Fermi resonance); were it not for this coupling, further excitation would only follow the $|10\rangle \rightarrow |20\rangle$ pathway.[2]

13.1.1.3 Mathematical Description

The mathematical description for this technique builds directly on our treatment of transient absorption in Section 8.2. We follow our usual prescription of first writing

---

[1] We note that here and throughout this chapter, the indices describing the number of quanta in each mode do not correspond to "good quantum numbers" (i.e. they are not conserved), because of the anharmonic terms in the molecular Hamiltonian which the experiments aim to probe.

[2] In this system, it happens that the energies are such that the stretch-bend coupling is stronger for the overtone states ($|20\rangle \longleftrightarrow |12\rangle$) than the coupling between the first-excited states ($|10\rangle \longleftrightarrow |02\rangle$).

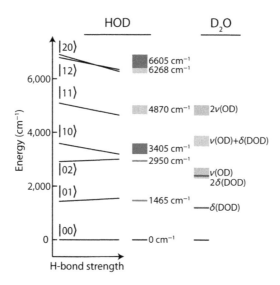

Figure 13.2 Energy-level diagram for the different vibrational modes of HOD showing shifts due to hydrogen bond strength. Black lines on the left indicate how the energies associated with different quanta of stretching and bending vary with hydrogen bond strength, while the shaded bars on the right illustrate the absorption spectrum for HOD and D_2O. Figure from Ref. [84].

down the wave function after interaction with the pump pulse, followed by determining the experimentally measured signal after the probe. Although the observed mode coupling was strongest in the overtone states ($|20\rangle \longleftrightarrow |12\rangle$), for simplicity in the mathematical model, we focus on coupling between the lowest states ($|10\rangle \longleftrightarrow |02\rangle$). The relevant energy levels and couplings for the model are shown in Figure 13.3.

We begin with Equation 6.31 describing the vibrational wave function on excited electronic state b after interaction with the pump pulse (connecting electronic states a and b):

$$\chi_b^{(1)}(R,t) = \frac{1}{i} \int_0^t dt_1 e^{-iH_b(t-t_1)} \left(-\mu_{ba}E_1(t_1)\right) e^{-iH_a t_1} \chi_a(R,0). \tag{13.1}$$

There are three primary modifications required to describe multidimensional, IR-transient absorption. First, while the field-induced transitions in Equation 6.31 are between different electronic states, in IR-transient absorption, the transitions are between two different vibrational states on a single electronic potential energy surface (PES). Second, there are now multiple vibrational modes, and so our vibrational wave function will be a function of at least two different coordinates. Finally, the Hamiltonian describing the system will need to include coupling between the two modes, since excitation in one of the modes results in energy flow to the other.

To accommodate IR transitions, we begin by rewriting Equation 13.1 in terms of pump excitation to the first excited state of a *single* vibrational mode corresponding to coordinate Q_1 (we add the second mode in the next step):

$$\chi_1^{(1)}(Q_1,t) = \frac{1}{i} \int_0^t dt_1 e^{-iH(t-t_1)} \left(-\mu_{10}E_1(t_1)\right) e^{-iH t_1} \chi_0(Q_1,0), \tag{13.2}$$

where $\chi_0(Q_1,0)$ is the initial $n=0$ vibrational eigenstate for mode Q_1, H is the Hamiltonian for mode Q_1, $E_1(t_1)$ is the pump field, and μ_{10} is the (IR) dipole moment connecting vibrational state 0 to state 1. In the case of a single mode with no coupling, χ_0 is a vibrational eigenstate, and the Hamiltonian propagator would simply pull out the time-dependent eigenenergy phase evolution of the individual state: $e^{-iH t_1}\chi_0(Q_1,0) = e^{-iE_0 t_1}\chi_0(Q_1,0).$

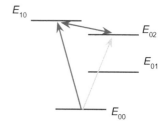

Figure 13.3 Vibrational energy-level diagram relevant for describing intramolecular vibrational energy redistribution. The pump and probe pulses are tuned to couple the common ground state E_{00} to the first-excited states E_{10} and/or E_{02} (shown by the two upward arrows). As discussed in Section 10.2.3, cubic anharmonic terms in the potential lead to efficient coupling between two quanta of one mode that are nearly degenerate with a single quantum for another (shown by the third arrow).

We next generalize Equation 13.2 to include two different vibrational modes Q_1 and Q_2, along with a term in the Hamiltonian that includes the coupling between the two modes. As in Chapter 10, we use two subscripts to keep track of the quanta of vibration in each mode, noting as mentioned earlier that these are not actually "good" quantum numbers because of the anharmonic coupling between modes. For example, χ_{01} corresponds to no quanta of excitation ($n = 0$) in mode Q_1 and one quanta ($n = 1$) in mode Q_2. We assume the two modes start in the (common) ground state χ_{00}. With this notation, the first-order wave function after the pump pulse can be written as

$$\chi^{(1)}(Q_1,Q_2,t) = \frac{1}{i} \int_0^t dt_1 e^{-iH(t-t_1)}$$
$$\times \left(-\mu_{mn,00}E_1(t_1)\right) e^{-iHt_1} \chi_{00}(Q_1,Q_2,0). \tag{13.3}$$

For the moment, we have kept this expression general so that a broadband pump pulse $E_1(t_1)$ could connect the common ground state $|00\rangle$ to either of the two excited states $|10\rangle$ or $|02\rangle$. For this reason, we write the dipole moment as $\mu_{mn,00}$, which connects the ground state to either of the excited states. In the case when one of the excited states is dark, the dipole moment would only include one of the modes (e.g. $\mu_{10,00}$ or $\mu_{02,00}$, where $nm = 10$ or 02). The Hamiltonian H includes kinetic terms for both mode 1 and mode 2, as well as the potential. The anharmonic potential V, including the coupling, comes from Equation 10.1:

$$V(Q_1,Q_2) = \frac{1}{2}k_1 Q_1^2 + \frac{1}{2}k_2 Q_2^2 + \frac{1}{2}k_{12}\left(Q_1^2 Q_2 + Q_1 Q_2^2\right). \tag{13.4}$$

Independent of the particular excitation path, the final state $\chi^{(1)}$ at time t is, in general, a superposition of $\chi_{10}^{(1)}$ and $\chi_{02}^{(1)}$ due to the coupling.

The coherent optical signal we measure comes from an induced polarization that is first order in the probe field (e.g. see Equations 6.39 and 8.1). In our notation for IR-transient absorption, this takes the form:

$$P_{\text{probe}}^{(1)} = \left\langle \chi^{(1)}(Q_1,Q_2,t) \left| \mu_{nm,00} \right| \chi^{(2)}(Q_1,Q_2,t) \right\rangle, \tag{13.5}$$

where $\chi^{(1)}$ is the vibrational wave function after interaction with the pump pulse (in general a superposition of excited $|10\rangle$ and $|02\rangle$ states), while $\chi^{(2)}$ is the wave function after interaction with both the pump and probe. Similar to the case of ground-state bleach for electronic transient absorption (see Section 8.2.3.2), one can consider the probe pulse bringing the excited-state population back down to the (common) ground state. In any case, it is the polarization of Equation 13.5 that leads to the observed signal.

Following our usual approach, we write Equation 13.5 in terms of the wave function after excitation by the pump ($\chi^{(1)}$). As in Section 8.2, we assume pump excitation occurs at time $t = 0$, which requires a redefinition of time zero as compared to Equation 13.3, where the ground state evolved in time before the pump arrived. Although the pump can, in principle, excite both modes, for concreteness we assume that the pump pulse selectively excites only the state $|10\rangle$. With this assumption, we write Equation 13.5 as

$$P_{\text{probe}}^{(1)} = \frac{-1}{i} \left\langle \chi_{10}^{(1)}(Q_1, Q_2, 0) \middle| e^{+iHt} \mu_{nm,00} \right.$$

$$\times \int_0^t dt_2 e^{-iH(t-t_2)} \left(\mu_{00,nm} E_2(t_2) \right) e^{-iHt_2} \middle| \chi_{10}^{(1)}(Q_1, Q_2, 0) \right\rangle. \quad (13.6)$$

Reading right-to-left, we interpret Equation 13.6 as follows. The pump-prepared state $|\chi_{10}^{(1)}(Q_1, Q_2, 0)\rangle$ evolves in time from $t = 0$ to $t = t_2$ under the Hamiltonian H. In addition to phase evolution of the eigenstate(s), the anharmonic term in the Hamiltonian serves to couple population from state $|10\rangle$ to $|02\rangle$ as described in Section 10.2.3. The resulting wave function after this evolution is a superposition of χ_{10} and χ_{02}. The probe pulse $E_2(t_2)$ then couples this excited-state wave function with the common ground state; the portion of the excited-state wave function that returns to the ground state leads to the polarization we wish to measure. There will, in general, be population remaining in the excited χ_{10} and χ_{02} states. However, the coherence established between these two states by the coupling will not lead to a polarization at the probe frequency.

The propagator $e^{-iH(t-t_2)}$ leads to a phase evolution in the common ground state $|00\rangle$ from time t_2 to final time t (in the ground state, there is no mixing between the two). It is this evolving ground-state wave function that generates a time-dependent polarization with the wave function on the left side generated by the pump and evolving (and mixing) until final time t. The $\mu_{nm,00}$ is the polarization dipole operator that connects these two states. The fact that the pump-excited $|10\rangle$ state has mixed with $|02\rangle$ means there will be a time-dependent polarization at the $|00\rangle \leftrightarrow |02\rangle$ frequency (in addition to the $|00\rangle \leftrightarrow |10\rangle$ frequency). Therefore, a broadband probe pulse shows changes in absorption at both these spectral positions and serves as a measure of the coupling, as observed in the data shown in Figure 13.1.

13.1.2 IR-Optical Transient Absorption

One of the challenges for experiments measuring vibrational energy flow in molecules is that while loss of energy from a particular mode is easy to observe, determining where the energy goes can be more difficult. Here we consider an approach using a combination of vibrational and electronic transitions to track energy flow. The basic idea is shown in Figure 13.4 for the molecule CH_2I_2 [85]. Schematic PESs for the ground and first-excited electronic states are shown as a function of two vibrational coordinates: a C–I stretch and a C–H stretch. For a laser pulse with central wavelength λ_{abs}, the Franck–Condon overlap with the excited-state surface is significantly enhanced when the vibrational wave function in the ground state is displaced along the C–I coordinate. One can think about this in terms of the relative curvatures of the two different PESs: for a probe wavelength tuned to the red side of the electronic absorption, only ground-state population with substantial displacement along the C–I coordinate is vertically resonant with the excited state. The critical aspect is that one can put energy into one vibrational mode (C–H) with an IR transition, while using an experimental probe that is sensitive to vibrational energy in a *different* mode (C–I). This allows one to track the flow of energy from one mode to the other.

13.1.2.1 Experimental Setup

As shown in Figure 13.4, an IR pump pulse with wavelength λ_{vib} is tuned to excite vibrational population in the C–H stretch mode. The pump pulse has a central wavelength of 1.7 μm (5880 cm^{-1}) and a bandwidth of 170 cm^{-1}. This central frequency

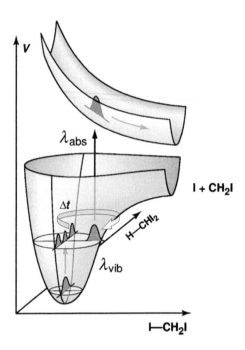

Figure 13.4 Diagram showing how an IR-pump, ultraviolet (UV)-probe experiment in CH_2I_2 can provide information on mode coupling. An IR pump pulse excites vibrations on the electronic ground state of the molecule. The vibrational dynamics are probed by a UV pulse that excites the molecules to a dissociative excited electronic state. Figure from Ref. [85].

corresponds to the first overtone of the C–H stretch: $\nu_{C-H} = 0 \to 2$. Any vibrational energy that flows from the initially excited C–H mode to the C–I mode will be detectable through increased absorption of the probe pulse at λ_{abs}. In this particular experiment, the probe pulse has a central wavelength of 290 nm and is tuned for preferential absorption to the excited electronic state when there is motion along the C–I vibration. The polarizations of the two pulses are set to the magic angle (54.7°) to suppress any rotational contribution to the signal. The change in absorption of the probe due to the pump is measured as a function of pump–probe delay.

13.1.2.2 Data and Interpretation

Transient absorption data for both gas-phase and liquid-phase CH_3I is shown in Figure 13.5 [86] (although Figure 13.4 shows the schematic PES for CH_2I_2, the idea is the same). Both gas and liquid phase show a fast, initial rise in probe absorption with a timescale of approximately 5–7 ps. This indicates vibrational energy coupling from the initially excited C–H coordinate to the C–I coordinate. Since this signal is present in both gas and liquid phases, it must be an *intra*molecular effect, or IVR (there are no solvent-molecule collisions in the gas phase).[3]

After the initial rise, the liquid-phase data shows a slower, 50-ps decay. This is attributed to solvent-mediated energy dissipation, or *inter*molecular energy transfer. As seen in Section 11.3.1, a measurement over an ensemble of molecules that experience different, fluctuating vibrational frequencies due to their local solvent environment produces a loss of vibrational coherence. In particular, this leads to a decay of the oscillating expectation value of the normal mode coordinate (the decay was exponential in the Kubo model). This is manifest in the liquid data of Figure 13.5, where the enhanced absorption decays due to solvent interactions. In fact, different solvents

---

[3]It was tested and shown that collisions in the gas phase do not play a role.

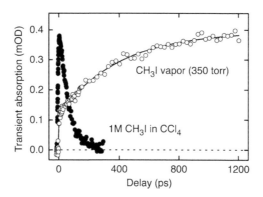

Figure 13.5 Transient absorption signal as a function of delay between an IR pump and UV probe for CH_3I in the gas and liquid phases. Figure from Ref. [86].

produce different timescales for decay, and these types of experiments can provide information on solvent–solute interactions.

In contrast, the gas-phase data shows a continuing rise in the transient absorption signal, albeit at a slower rate (with an approximately 400 ps timescale). In the gas phase, this must be an intramolecular effect, and the increase in signal corresponds to *further* excitation along the C–I coordinate. The authors analyze the gas-phase data with an approach similar to our discussion of N-dimensional coupling in Chapter 10. Specifically, they consider an order for coupling (or coupling "tiers") that describes how strongly any particular vibrational state of the molecule is coupled to the initially excited state. States that are strongly coupled will show fast (short) coupling timescales, while those with weak coupling to the original state participate only secondarily. This is similar to the approach of our model described in Figure 10.6: there is a strong coupling to one or more modes that progresses rapidly, while a weaker coupling to other modes plays a role in the long-time dynamics.

13.1.3 Femtosecond Stimulated Raman Spectroscopy

As noted, while it is relatively straightforward to observe signals that suggest mode coupling, determining precisely which modes participate and how is not always a simple task. For example, vibrational dephasing along a single coordinate can mimic a flow of energy to other modes, as a measurement of $\langle R(t) \rangle$ cannot distinguish between a loss of probability and simple delocalization of the wave packet along that degree of freedom. In addition, many modes can participate in dynamics after initial excitation (some of which are IR active and some of which are only Raman active), and it can be quite complicated to track the flow of energy on timescales associated with significant energy transfer.

In this section, we discuss a technique known as femtosecond stimulated Raman spectroscopy (FSRS) that takes a somewhat different approach. With a goal of making as direct a measurement as possible of vibrational mode coupling, FSRS tracks changes in the frequency of one mode as a function of displacement along another to probe the local anharmonicity of the PES. As its name implies, instead of IR light driving vibrational transitions directly, FSRS uses Raman scattering in the probe process.

13.1.3.1 Experimental Setup

The basic idea is shown in Figure 13.6 [87]. A pump pulse is used to initiate motion in one or more vibrational modes. This can be through an electronic transition to an

(a) **Time-resolved FSRS**

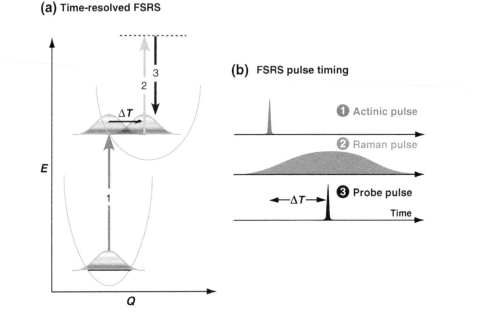

(b) **FSRS pulse timing**

Figure 13.6 (See color insert.) Timing and energy diagram illustrating the technique of FSRS. An "actinic" pump pulse initiates vibrational dynamics in an excited PES. Together, the Raman and probe pulses measure mode coupling. Figure adapted from [87].

excited PES (as shown in panel a of Figure 13.6), or through impulsive stimulated Raman scattering on the ground PES (as discussed in the following example). In either case, the role of the pump pulse is to initiate the vibrational dynamics. The probing process is accomplished using two pulses: a narrowband, long-duration "Raman pulse" and a broadband, short-duration "probe pulse." The detailed interaction of the two pulses in the probe process are best explained in terms of a specific example.

We consider an experiment designed to measure mode coupling in the ground electronic state of the molecule $CDCl_3$ [88]. A 50-fs pump pulse centered at 800 nm excites two, low-frequency vibrations using impulsive stimulated Raman scattering in the ground PES. The approximately 300 cm^{-1} bandwidth of the pump pulse is sufficient to impulsively excite the E and A_1 C–Cl bends at frequencies of 262 and 365 cm^{-1}. As the wave packet evolves along the vibrational coordinates for these two low-frequency modes (Q_{bend1} and Q_{bend2}), anharmonicity of the full potential leads to small changes in the vibrational frequency of the C–D stretch mode. In other words, for different expectation values along the C–Cl coordinate, the value of the C–D stretch frequency is slightly shifted due to the fact that the modes are coupled (they are not orthogonal). Determining how the C–D stretch frequency is affected by the C–Cl bend provides a measure of the coupling and information about the potential surface.

In principle, determining the C–D stretch frequency can be accomplished by measuring the frequency shift of a narrowband, Raman-scattered probe. However, a narrowband probe pulse necessarily implies a longer pulse in time, thereby averaging over many positions of the vibrational motion along Q_{bend1} and Q_{bend2}. Furthermore, the shift of the C–D frequency with displacement along the Q_{bend1} or Q_{bend2} coordinates is so small (a few parts per thousand or less), it is not possible to see it directly in the Raman spectrum of the probe.

The FSRS technique overcomes this problem by using a composite probe interaction consisting of a relatively intense, narrowband pulse (the "Raman pulse" with 2.4 ps duration), along with a weak, broadband pulse (the "probe pulse" with 20 fs duration).

As we shall see, the ultrafast probe pulse is used to heterodyne the Raman scattered light from the Raman pulse, allowing the authors to both time resolve the variations in the Q_{bend1} or Q_{bend2} coordinate, as well as measure the phase of the Raman scattered light.

13.1.3.2 Data and Basic Interpretation

Figure 13.7 shows the heterodyned Raman scattered light, or "FSRS spectra," in the C–D stretch region of the spectrum at a number of different delay times between the pump and probe pulses. The FSRS spectra are generated using a two-step procedure. First, spectra with and without the Raman pulse are divided by each other to isolate the stimulated Raman light from the broadband probe (the stimulated Raman line from the Raman probe appears on top of the probe spectra background). This step produces the "unpumped" spectra at the top of panel (a) in Figure 13.7, which shows the sharp Raman line of the C–D stretch at 2255 cm^{-1}. Second, FSRS spectra are recorded for cases both with and without the pump pulse present to isolate the effect of vibrational motion induced by the pump. The bottom trace in panel (a) shows a zoomed view of the difference between the pumped and unpumped traces at a time delay of 777 fs. One can see sideband peaks on both sides of the fundamental; these sidebands are spaced by 262 and 365 cm^{-1}, the frequencies of the two, low-frequency bend modes excited by the pump pulse.

Panel (b) of Figure 13.7 plots the FSRS difference spectra at a number of different delay times, showing how the phases of the heterodyned sidebands evolve in time. Each sideband transitions through a cycle, going from a peak, to a dispersive shape, to a dip, back to a dispersive shape, and finally returning to a peak again. The period for this cycle is exactly the vibrational period of the respective bend mode: 127 fs for the 262 cm^{-1} mode and 91 fs for the 365 cm^{-1} mode. By 5 ps, the vibrational coherence in the low-frequency mode is gone, and the sidebands are no longer present. Panel (c) shows simulated FSRS spectra at similar delay times based on a "coupled-wave theory" of FSRS, where the light fields propagate according to Maxwell's equations in a medium with an induced polarization. Before describing the basic principles of this model, we consider the experimental results qualitatively.

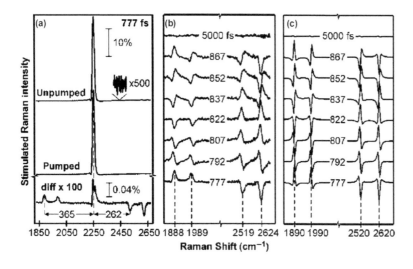

Figure 13.7 Heterodyned femtosecond stimulated Raman scattering from CDCl$_3$ illustrating coupling between the high-frequency C–D stretch mode and two low-frequency modes. See text for details. Figure from Ref. [88].

It is the behavior of the sidebands in Figure 13.7 that provides information about vibrational mode coupling. In particular, were there no coupling between the low-frequency C–Cl bends and the high-frequency C–D stretch modes, there would be no sidebands on the C–D Raman line. This can be understood by considering Figure 13.8 [87]. The top panel presents the conceptual picture, where the potential energy is plotted along two different coordinates: Q_{low} (corresponding to the low-frequency bending modes) and Q_{high} (corresponding to the high-frequency stretch mode). For low-amplitude motion, the potential energy curves along each direction resemble harmonic parabolas, with a narrow, steep parabola along Q_{high} and a wide, shallow parabola along Q_{low}. If the two modes were completely uncoupled, one could consider these 1D curves independently. However, at different points along the Q_{low} coordinate, weak coupling between the modes implies that the curvature (and hence frequency) of the high-frequency mode is slightly different.

The top panel of Figure 13.8 depicts this by showing three different Q_{high} potential curves at three different locations along the Q_{low} coordinate. Q_{high} has a slightly lower frequency of oscillation when the Q_{low} mode is at Q_0 (red), as compared to a slightly higher frequency of oscillation when the Q_{low} mode is at Q_2 (blue). Importantly, as Q_{low} oscillates back and forth in position, the *frequency* of Q_{high} will oscillate (as shown in the middle panel). It is this modulation in frequency of Q_{high} that leads to the sideband behavior in the bottom panel (and in the data of Figure 13.7). We note that the modulation of the Raman pulse by the coherently vibrating molecular sample can also be described in terms of a time-dependent index of refraction. The changing index of refraction imparts a time-dependent phase on the Raman pulse, visible as sidebands in its spectrum.

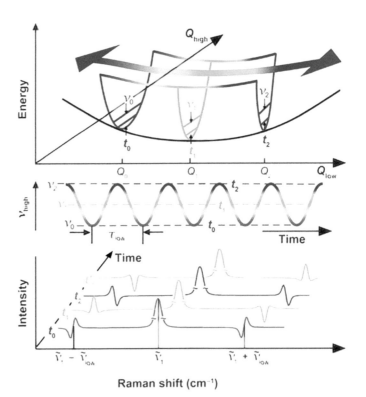

Figure 13.8 Illustration of how FSRS can track anharmonic mode coupling by measuring small frequency shifts in a high-frequency "reporter" mode as a function of displacement along lower frequency modes. See text for details. Figure adapted from [87].

13.1.3.3 Mathematical Description

Based on our treatment of Raman transitions in Chapter 3 (in particular Equations 3.49 and 3.50), as well as Equation 6.44 in the perturbative description of Chapter 6, the heterodyned field emitted along the probe direction is the sum of the probe field and the Stokes field generated by stimulated Raman scattering of the Raman pulse in the medium:

$$E_{\text{FSRS}}(t) = E_{\text{probe}}(t) + cNE_{\text{Raman}}(t)\left(\alpha_o + \frac{\partial\alpha}{\partial Q_{\text{high}}}Q_{\text{high}}(t)\right), \qquad (13.7)$$

where $Q_{\text{high}}(t)$ is the vibrational amplitude of the high-frequency C–D stretch, $E_{\text{probe}}(t)$ and $E_{\text{Raman}}(t)$ are the time-dependent fields of the probe and Raman pulses, N is the density of molecules per unit volume, α_o is the polarizability of the sample, $\frac{\partial\alpha}{\partial Q_{\text{high}}}$ is the change in polarizability along the stretch coordinate, and c represents an overall constant [88]. The vibrational amplitude in the high-frequency mode builds up as the Raman pulse propagates through the sample and scatters light into the Stokes beam. The combination of the Stokes light and the fundamental produces an optical beat at the frequency of the high-frequency mode, driving a vibrational coherence that can be described in time as

$$Q_{\text{high}}(t) = \cos(\omega_{\text{high}}t), \qquad (13.8)$$

where ω_{high} is the frequency of the high-frequency mode, and we have assumed unit amplitude. If ω_{high} were constant, one would see a single Raman line at the Stokes shift of ω_{high} after interaction with the Raman pulse.

However, due to the coupling between modes (described earlier and seen in the middle panel of Figure 13.8), the frequency of the high-frequency mode is modulated by the low-frequency mode according to

$$\omega_{\text{high}}(t) = \langle\omega_{\text{high}}(t)\rangle\left(1 + \varepsilon\sin(\omega_{\text{low}}t)\right), \qquad (13.9)$$

where ω_{low} is the frequency of the low-frequency mode and ε is a small constant. This modulation leads to the Stokes light from the Raman pulse acquiring sidebands at the frequency of the *low*-frequency mode. While the presence of sidebands indicates a coupling between the two modes, they come from scattering by the long-duration Raman pulse and thus average over many oscillations of the low-frequency mode. Therefore, the sideband intensity alone does not contain any time-dependent information.

But the phase of the sideband evolves at the frequency of the low-frequency mode. By interfering the sideband with a short probe pulse, one can recover the time-dependent phase.[4] As the authors discuss, this time-dependent phase provides insight into the mode coupling. In particular, the phase behavior of the sideband modulation is consistent with their origin from anharmonic coupling between modes (and inconsistent with other possibilities). To mimic the experimental data, the simulated FSRS spectrum is calculated from the Fourier transform of $E_{\text{FSRS}}(t)$, or $|F.T.\{E_{\text{FSRS}}(t)\}|^2$. The signal in the presence of the pump field is normalized by that without the pump and shown in panel (c) of Figure 13.7.

---

[4]Note that in this experiment, it is not actually necessary for the probe pulse to pass through the sample, as its function is simply to time-gate the sidebands of the Raman Stokes pulse. This is similar to the role of the probe in fluorescence gating, where the gating pulse is simply used to time resolve the fluorescence yield and does not need to interact with the sample.

This example discussed the case of two slow modes modulating the frequency of a fast mode. This is a particularly simple case of anharmonic coupling, where one vibrational mode has its frequency modified by motion along other coordinates. In the limit of many coupled modes with a large distribution of frequencies, the situation resembles the cases discussed in Sections 10.3 and 11.3.1.

Section 13.1 discussed three different approaches for studying vibrational mode coupling in polyatomic molecules. For reviews of this research field, see [89, 86, 87].

13.2 MEASURING ELECTRONIC STATE COUPLINGS

In the previous section we discussed coupling between different vibrational modes in a polyatomic molecule. As we saw in Chapter 10, another consequence of multidimensional dynamics is the existence of degeneracies between electronic PESs where population moves from one surface to another. Examples include internal conversion, where the electronic state character changes, and intersystem crossing, where the spin configuration changes. Transfer between the different configurations of the system is vibrationally mediated and happens most readily when the energies of the configurations are similar. The degeneracy point itself between two, two-dimensional PESs is known as a conical intersection (or conical seam in high dimensions). Note that in all cases, the dynamics move beyond the Born–Oppenheimer approximation, as one can no longer consider vibrational motion on distinct, adiabatic PESs. In this section we examine how the experimental techniques developed in Part III can be used to measure and understand nonadiabatic dynamics of electronic state couplings.

13.2.1 Transient Absorption

As we have seen, transient absorption can be used to track vibrational dynamics in either the excited or ground electronic state. Interpretation of transient absorption experiments is typically more complicated in multiple dimensions, as the technique does not always provide direct dynamical information. For example, in a multidimensional coordinate space there may be many locations where the difference in energy between two PESs corresponds to the photon energy of the probe. Therefore, absorption at a given frequency is not necessarily associated with the wave packet being at a specific location along a particular coordinate. Nevertheless, transient absorption, in conjunction with modeling, has been successfully used to infer multidimensional dynamics such as internal conversion. In this section we provide two examples of such measurements.

13.2.1.1 Isomerization in Retinal

We begin by considering a measurement of the primary stage of vision: isomerization of the 11-*cis* retinal chromophore to its all-*trans* form [2]. The retinal chromophore sits within the photoreceptor protein rhodopsin, which is located in the rods of the eye retina. Upon absorption of light, the retinal system experiences rotations of bond angles near the center of the molecule. This experiment uses ultrafast electronic transient absorption to track isomerization dynamics through a conical intersection between the ground and excited electronic states.

The experimental arrangement is similar to previous transient absorption experiments. The absorption is between two different electronic states, and the wavelengths of the pump and probe are in the optical region of the spectrum. The reaction proceeds very quickly (completed within 200 fs), and good time resolution is essential for resolving the dynamics. A 10-fs pump pulse centered at 500 nm promotes a portion of the ground-state wave function to the excited electronic state where it begins to evolve. Two different broadband probe pulses are used, covering spectral windows of 500–720 nm or 820–1020 nm. The transient absorption signal of the probe is spectrally resolved and measured as a function of pump–probe delay.

Figure 13.9 shows the experimental (left side) and simulated (right side) spectrally-resolved transient absorption signal plotted in terms of the differential transmission (DT), or $\Delta T/T$. Panels a and c show color maps over both wavelength and time, while panels b and d show temporal dynamics at particular wavelength slices. Two primary features are visible in panels a and c. The blue-shaded feature at early times represents a positive DT signal, implying that more probe light is *transmitted* with the pump present than without. On the other hand, the red-shaded feature appearing at later times indicates more *absorption* of the probe due to the pump. Both of the features extend across the entire bandwidth of the probe and appear to peak on the blue side of the spectrum (∼650 nm).

In addition, both features have a "spectral chirp," visible as a slope in the wavelength–time plot. With the blue feature, the peak of the enhanced transmission shifts to longer wavelengths as time progresses from 0 to 50 fs. With the red feature the opposite

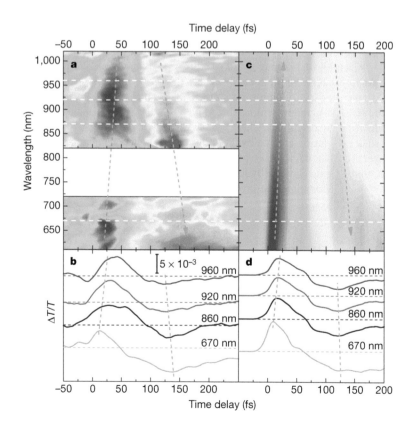

Figure 13.9 Optical transient absorption used to follow dynamics through a conical intersection in rhodopsin. Panel a shows the measured transient absorption spectrum as a function of pump–probe delay, while panel b shows lineouts at different probe wavelengths. Panels c and d show simulated measurements that can be compared with the data on the left. The color axis plots the differential transmission $(\Delta T/T)$. Figure from Ref. [2].

happens, and the peak enhanced absorption shifts to shorter wavelengths as time evolves from 100 to 150 fs. The experimental results agree quite well with calculations of the time-resolved absorption spectrum. The calculations use classical dynamics with surface hopping between quantum-mechanical PESs.

The two primary features of the data can be understood through the model of transient absorption developed in Section 8.2.3 and with the assistance of Figure 13.10. As mentioned, the blue feature in the data corresponds to more probe light transmitted at these wavelengths with the pump present; this represents either a ground-state bleach, where removal of ground state population results in less probe absorption, or "stimulated emission", where population in the excited state is stimulated back to the ground state, thereby adding photons to the probe field (both of the processes connect the ground and first-excited PESs). In this particular molecule, there is negligible ground-state absorption at these colors, indicating the signal is due to stimulated emission.

From Section 8.2.3.2, we begin with the polarization leading to changes in probe transmission connecting the ground and first-excited states (Equation 8.13). We alter the electronic PES labels to match those of Figure 13.10 and change the internuclear separation R to a generalized coordinate Q:

$$P_{\text{probe}}^{(1)} = \left\langle \chi_{S_1}^{(1)}(Q,0) \left| e^{+iH_{S_1}t} \mu_{S_1 S_0} e^{-iH_{S_0}t} F(Q,t,\tau) \right| \chi_{S_1}^{(1)}(Q,0) \right\rangle, \tag{13.10}$$

where

$$F(Q,t,\tau) \equiv \frac{-1}{2i} \int_0^t dt_2 \, e^{-i(V_{S_1}(Q) - V_{S_0}(Q) - \omega_2)t_2} \mu_{S_0 S_1} E_{02}(t_2), \tag{13.11}$$

and the τ dependence in F comes from the arrival time of the probe pulse $E_{02}(t_2)$. As earlier, the function $F(Q,t,\tau)$ deviates from zero for Q values where $\left(V_{S_1}(Q) - V_{S_0}(Q) - \omega_2\right)$ is roughly zero, indicating that the polarization is large whenever the wave packet is in the location where the probe pulse is resonant between the two PESs. As the wave function on state S_1 in Figure 13.10 moves down the slope out of the Franck–Condon region, different spectral portions of the probe spectrum come into resonance with the state S_0. Away from the Franck–Condon region, the excited-state population created by the pump pulse creates what appears to be a population inversion for the resonance at the probe wavelengths. Therefore, the polarization between states S_1 and S_0 actually leads to stimulated emission back to the ground state (putting more photons into the probe field than it contained before the sample). The

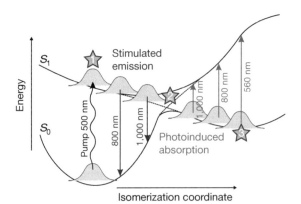

Figure 13.10 Cartoon PES model of wave packet propagation through a conical intersection in rhodopsin. See text for details. Figure from Ref. [2].

spectral chirp of the blue feature in Figure 13.9 allows one to map out the changing energy difference between states S_0 and S_1 along the isomerization coordinate.

The red feature in the data corresponds to enhanced absorption of the probe and represents the excited-state absorption component of the signal. We again draw on our results from Chapter 8, where we derived an expression for the polarization leading to excited-state absorption (Equation 8.4). Importantly, note that after passage through the conical intersection, the roles of the two states are reversed, and the first-order wave function created by the pump pulse now resides on state S_0:

$$P_{\text{probe}}^{(1)} = \left\langle \chi_{S_0}^{(1)}(Q,0) \left| e^{+iH_{S_0}t} \mu_{S_0 S_1} e^{-iH_{S_1}t} F(Q,t,\tau) \right| \chi_{S_0}^{(1)}(Q,0) \right\rangle, \qquad (13.12)$$

where again

$$F(Q,t,\tau) \equiv \frac{-1}{2i} \int_0^t dt_2 e^{+i(V_{S_1}(Q) - V_{S_0}(Q) - \omega_2)t_2} \mu_{S_1 S_0} E_{02}(t_2). \qquad (13.13)$$

The factor $F(Q,t,\tau)$ is large whenever $\left(V_{S_1}(Q) - V_{S_0}(Q) - \omega_2\right)$ is roughly zero, implying that a localized wave function moving along the isomerization coordinate picks out the Q values where this resonance condition is met. The absorption turns on in different portions of the probe pulse bandwidth at different times as the energy gap between the two states changes.

Considering the temporal dynamics of the two features together presents a simple, clear picture of the reaction depicted in Figure 13.10. After excitation with the pump, population on the excited state moves downhill on the potential. The chirp seen in the stimulated emission signal indicates a continual red-shift of the resonance frequency at which stimulated emission occurs as the wave packet approaches the conical intersection. After the wave packet moves through the conical intersection, population is now on the ground state and can participate in "excited-state" absorption (although the absorption is technically from the ground state, for consistency, we refer to enhanced absorption of the probe after pump excitation as excited-state absorption, or ESA). The chirp in the ESA signal indicates the relative slopes of the two potentials as the wave packet settles into the all-*trans* product minimum.

This experiment nicely shows how broadband transient absorption data can (and need to) be combined with calculations to understand ultrafast reaction dynamics. While the cartoon picture shown in Figure 13.10 is *consistent* with the data, there is no guarantee from these data alone that the observed signal actually corresponds to wave packet motion along the particular coordinate of interest. The simulations shown in panel c of Figure 13.9, along with additional experimental data (not shown), complete the full picture of the reaction. As with most pump–probe measurements, comparison with calculation or theory is required to interpret the experimental results and yield a detailed understanding of the underlying dynamics. On the other hand, it is only with direct experimental measurements that one can confirm calculations involving many degrees of freedom that necessarily make approximations to arrive at a result.

13.2.1.2 Relaxation in Nucleobases

Internal conversion plays an important role in the photoprotection of DNA and RNA bases. The fluorescence yield from photoexcited DNA and RNA bases such as cytosine, adenine, and uracil is about 10^{-4} due to rapid internal conversion after photoexcitation. This rapid decay helps prevent photodamage in these biologically important molecules, since the molecule spends a small fraction of the fluorescence lifetime in the reactive excited state.

In this section we consider an experiment in 9-methyladenine that uses three different transient absorption measurements to highlight a combination of multidimensional quantum dynamics, including electronic state coupling (internal conversion via conical intersections) and vibrational energy flow from the molecule to its surroundings (system-bath coupling) [4].[5] A schematic energy-level diagram is shown in panel (b) of Figure 13.11. In all of the measurements, a UV pump pulse at 267 nm launches a wave packet on the electronically excited $\pi\pi^*$ state (step 1 in panel b). Transient absorption of a time-delayed probe is then measured using one of the three different probe pulses: visible, UV, or IR.

We begin with excited-state absorption using a visible probe. The top portion of panel (a) in Figure 13.11 shows the ESA signal as a function of pump–probe delay. An initial rise is followed by a fast decay on a subpicosecond timescale. Electronic structure calculations indicate there are nonadiabatic pathways to the ground state via conical intersections (step 2 in panel b). Thus, it is natural to interpret decay of the ESA signal in terms of internal conversion enabled by nonadiabatic coupling between the initially excited $\pi\pi^*$ state and the ground state; as population leaves the excited electronic state, the ESA signal goes away.

Confirmation of this interpretation is provided by ground-state bleach (GSB) measurements using a UV probe (bottom portion of panel a). These data show a fast, initial ground-state bleach due to excitation by the pump: the sample absorbs less probe light after the pump since molecules promoted to the excited state no longer absorb in the UV. The UV probe absorption recovers as molecules excited to the $\pi\pi^*$ state decay back down to the ground state.

However, there is a clear difference in the timescales for decay of ESA and recovery of GSB, with the ESA signal decaying more rapidly independent of solvent. This suggests that the GSB recovery dynamics involve more than just fast internal conversion between electronic states. Indeed, the recovery of the GSB signal requires not only internal conversion back to the ground electronic state, but also that molecules in the ground state have favorable Franck–Condon overlap with the excited state at the probe frequency. Molecules that undergo internal conversion typically transfer their electronic energy to nuclear motion and have significant vibrational energy in the ground electronic state ("hot ground-state molecules"). Good Franck–Condon overlap (and recovery of the GSB signal) often necessitates "vibrational cooling" of the molecules, where the hot, ground-state molecules dissipate energy to the solvent through collisions with surrounding molecules (step 3 in panel b).[6] The Franck–Condon overlap of the hot ground state with a hot excited state is usually poor at the probe wavelength used to probe a cold ground state because the ground and excited state potential energy surfaces are typically not parallel. The role of solvent-mediated vibrational cooling in the process is supported by the dependence of the bleach-recovery timescale on the

---

[5]The derivative 9-methyladenine is used in place of adenine to speed up the process of vibrational cooling in the water solution (the internal conversion dynamics through the conical intersection are similar for the two molecules) [90].

[6]As described in Part IV, this transfer of vibrational energy to other modes arises from the large number of degrees of freedom involved, either within the molecule itself or in the surrounding solvent. This typically precludes a rigorous treatment of the entire quantum system, and it is common to break the large-dimensional system into a small subsystem (the methyladenine molecule in this case) coupled to a large-dimensional subsystem (the bath) that can be described classically.

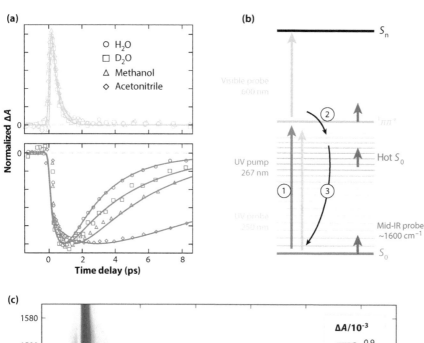

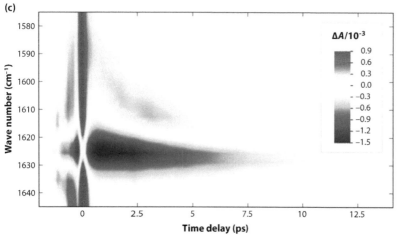

Figure   13.11 Experimental data for transient absorption in 9-methyladenine. Panel (a) plots transient absorption signals in various solvents for a visible probe (top, yellow) and a UV probe (bottom, blue). Panel (b) shows a schematic energy-level diagram, along with the pump excitation (step 1), internal conversion back down to the ground electronic state, (step 2), and vibrational cooling in the ground state (step 3). Panel (c) plots the spectrally-resolved transient absorption data with an IR probe. Figure from Ref. [4].

solvent (the vibrational cooling must be an *inter*molecular effect). In addition, the fact that there is no solvent dependence for the decay of ESA is consistent with the fast internal conversion being dominated by *intra*molecular dynamics.

One can attempt to track the vibrational cooling process using transient absorption with an IR probe pulse. Panel (c) of Figure 13.11 shows spectrally-resolved transient absorption measurements with an IR probe that follow population decay from highly excited vibrational states down to the ground vibrational state. Due to the anharmonicity of the ground-state potential, lower IR frequencies probe absorption of excited vibrational states ("hot bands"), while higher IR frequencies probe absorption from lower-lying vibrational states (~ 1625 cm^{-1} probes absorption of the ground vibrational state of the NH$_2$ scissors mode). Looking at the window from 0 to 5 ps, there is an increased probe absorption shown by the faint red band that shifts from approximately 1590 cm^{-1} down to 1620 cm^{-1} as time progresses. In addition, the (blue) bleach in the ground-vibrational-state absorption at 1625 cm^{-1} recovers on the same timescale. Together these indicate population moving from highly excited vibrational states down to the ground vibrational state on a timescale of several picoseconds. We note that the IR transient absorption data in panel (c) is dominated by a spectrally broad absorption around time zero. This reflects an optically driven, instantaneous nonlinearity and does not yield information about the molecular dynamics. This process is typical for transient absorption measurements in solution, and its effects can be characterized by making measurements with different solvents.

Section 13.2 presented two examples of experiments measuring dynamics that include coupling between electronic states. For reviews of this (large) research field, see [91, 92, 93, 94, 95].

13.3 TWO-DIMENSIONAL SPECTROSCOPY

13.3.1 Introduction

In the IR-transient absorption experiments described in Section 13.1.1, a pump pulse excites a single vibrational mode, after which a broadband probe measures changes in absorption at frequencies corresponding to a variety of different modes. A change in probe absorption at a particular frequency indicates a coupling between that mode and the pump-excited mode. The basic idea for measuring coupling, whether between specific vibrational modes or a general system-bath coupling, is to "pump" one mode and "probe" another.

Note that spectrally resolving the broadband probe provided time-dependent transient absorption signals at many different probe colors. In this way, one essentially does many experiments in parallel ("multiplexing"). It is possible to extend this idea to the pump excitation step as well, where one pumps multiple modes simultaneously and looks for couplings between them and other modes in the system. We will find that transient absorption experiments that spectrally resolve *both* the pump and probe processes can uncover detailed information on mode coupling and inhomogeneous broadening in multidimensional systems.

Techniques following this approach fall under the umbrella of "two-dimensional spectroscopy," so named since the data is a function of two different frequency axes (a 2D spectrum). Typically, one of the two axes covers the frequency spectrum of the broadband probe pulse (just like in 1D transient absorption), while the other axis is formed by generating the pump as a pulse pair with variable time delay. By stepping through the time delay and Fourier transforming the result, one creates an additional frequency axis for the 2D spectrum. While the experimental setup and data tend to be more complicated than in one dimension, two-dimensional spectroscopy experiments are essentially a spectrally-resolved (in the pump) version of transient absorption.

Although 2D experiments can be implemented in a variety of ways, all the different approaches can be described in the simple picture of a pump and probe interaction with the molecular system and quantified mathematically in terms of a third-order nonlinear polarization (e.g. see Equation 6.41). In this framework, the pump creates a time-dependent vibrational wave function that evolves until the probe arrives. A measurement of the coherently emitted light from the sample after the probe (either collinear with the probe or not) provides information on the dynamics occurring between the pump and probe pulses. As earlier, the macroscopic third-order polarization is calculated from the microscopic, or single-molecule, polarization. The microscopic polarization involves many terms that contain details about anharmonic couplings in the molecular Hamiltonian. While these provide for a vast array of potential information from 2D experiments, it can be challenging to interpret the resulting data. Because of this, experiments are typically designed with specific formulations of pump and probe fields that pick out particular terms from the nonlinear polarization. Focusing on a limited number of interactions provides data that, while rich with information, are still interpretable.

Similar to the discussion in Section 8.2.5, one might ask whether a 2D spectrum measured with broadband, short pulses is equivalent to a 2D spectrum built up from multiple measurements using narrowband, long pulses. Both approaches give rise to diagonal and cross peaks in the spectrum, indicating a coupling between modes or excitations at different frequencies. However, the short-pulse experiment measures changes in the 2D spectrum with pump–probe delay; this information is not available in a long-pulse experiment, making the two measurements fundamentally different. In this section we discuss the basic formulation of 2D spectroscopy, along with some specific implementations. Before digging into the details, we consider a brief example to help motivate the technique.

13.3.2 Motivation

Consider an ensemble of one-dimensional systems that have a distribution of a particular property. For example, it could be quantum dots (QDs) of varying size, gas-phase atoms with a spread of Doppler shifts, or molecules in a liquid with slightly different local environments. In all these cases, there will be an "inhomogeneous" broadening of the absorption spectrum. In the case of QDs, variation in the size of the dots leads to differences in the energy-level spacings. The location of absorption peaks will be shifted due to these different spacings, and measuring the absorption spectrum of the ensemble leads to a broad peak. This is illustrated in the left column of Figure 13.12, where the addition of many shifted peaks leads to a broad, inhomogeneously broadened peak. The inhomogeneously broadened peak shows up along the diagonal in the 2D spectrum, while the homogeneous linewidth lies along the antidiagonal.

However, an ensemble of identical QDs may also have a broad absorption spectrum if the dots have extremely short lifetimes. In this case the spectrum is homogeneously broadened, and the diagonal and antidiagonal widths of the peak in the 2D spectrum will have the same widths. Based on measurements of the one-dimensional spectrum alone, it is impossible to discriminate between these two different scenarios. In other words, the frequency domain measurement does not yield an unambiguous estimate of the dynamics (or lifetime) for an excited state: the broad spectrum could be due to either fast decay dynamics of identical dots, or inhomogeneous broadening of variable dots with slower decay dynamics. As we shall see, 2D spectroscopy measurements are well suited for discriminating between cases of homogeneous and inhomogeneous broadening.

In addition to separating homogeneous and inhomogeneous linewidths, a 2D spectrum also provides a direct visualization of coupling between modes. This is illustrated in the right column of Figure 13.12, where modes b and c can both be excited from the common ground state (a). In the standard 1D spectrum, one sees absorption peaks at the two resonance frequencies. However, in the 2D spectrum, modes that are coupled show up with cross peaks off of the diagonal, where the amplitude of the cross peak is proportional to the strength of the coupling.

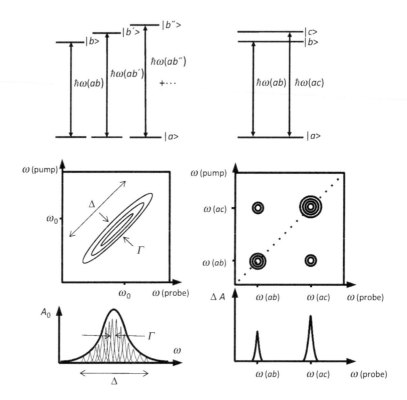

Figure 13.12 One- and two-dimensional spectra for inhomogeneously broadened (left column) and homogeneously broadened (right column) systems. The broad width along the diagonal in the case of an inhomogeneous sample is due to the addition of many shifted, narrow absorptions from each member of the ensemble. In the case of a homogeneously broadened sample, the 2D spectrum is composed of peaks associated with each transition in the system along the diagonal, as well as the appearance of cross (off-diagonal) peaks due to the inherent coupling of these transitions.

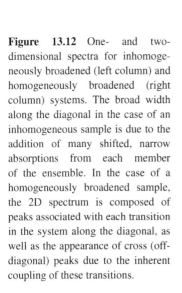

13.3.3 Experimental Setup

The concept of 2D spectroscopy in the IR and optical regions of the spectrum originated from nuclear magnetic resonance spectroscopy. In nuclear magnetic resonance, one has exquisite control over the temporal properties of the applied radio-frequency fields, but the k-vectors of the fields are not well defined due to their extremely long wavelengths. The situation is essentially reversed in the IR and visible portions of the spectrum: detailed temporal control of the applied fields is challenging, but it is quite simple to break up the pump and probe fields into subpulses with different k-vectors. This allows one to use phase matching (momentum conservation) of the pulses to pick out different energy-conserving terms from the nonlinear polarization (cf. Section A.8). In addition, developments in IR and optical pulse shaping have led to control over temporal aspects of the fields as well. For example, by using a pulse shaper, one can produce a pump-pulse pair with well-defined phases.

There are a number of different contributions to the third-order polarization that go by names such as photon echo, hole burning, and pump–probe absorption. These utilize different laser pulse geometries and sequences to measure some aspect of the sample's third-order response. Three of the primary beam geometries used in 2D spectroscopy are shown in the different panels of Figure 13.13: (a) collinear, (b) box-CARS, and (c) pump–probe. In the box-CARS geometry, temporal and spatial manipulation of the applied fields provide selection over the desired contributions to the measured signal (as discussed in Section 8.3). In the collinear and pump–probe geometries, it is primarily the temporal manipulation of the fields that accomplishes this purpose.

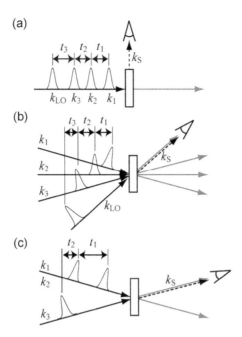

Figure 13.13 Beam geometries for three standard implementations of 2D spectroscopy. The wavevectors of the different pulses are represented by $\mathbf{k}_n$, where $n = 1, 2$ are the pump pulses, $n = 3$ is the probe pulse, $n = \text{LO}$ is the (optional) local oscillator, and $n = \text{S}$ is the signal. Time delays between the pulses are given by t_i. (a) Collinear geometry: All pulses arrive collinearly and fluorescence at $90°$ to the excitation wave vectors is collected. (b) box-CARS (four-wave mixing) geometry: beams arrive from different directions, and signals are emitted at the phase-matching angle (a local oscillator helps to retrieve its phase). (c) Pump–probe geometry (typically used with a pulse shaper): a pump-pulse pair arrives from one direction while the probe comes from another. The signal is emitted along the probe direction, similar to transient absorption. Figure from Ref. [96].

While the required pulses are often described in terms of a three-pulse sequence, we consider the two pump pulses as a single interaction arising from the combined pump field. The pump–probe geometry fits naturally into this framework, and for the purposes of this discussion, we focus on the situation where two, time-delayed subpulses comprise the total pump field (panel (c) of Figure 13.13). In the frequency domain, the pulse pair creates a modulated spectrum. Varying the delay between the subpulses and Fourier transforming the result with respect to this "pump–pump" delay is equivalent to spectrally resolving the excitation portion of the measurement. The pump-pulse pair is followed by the probe field, whose spectrally-resolved, differential transmission is measured as a function of pump–probe delay (essentially just transient absorption of the probe).

As with the other techniques, the pump and probe beams cross in the sample, allowing easy measurement of the differential probe transmission. By determining the spectrally-resolved absorption as a function of spectrally-resolved excitation, one can construct the final two-dimensional spectrum. Using a pulse shaper to generate the pump-pulse pair allows fine and stable control over the relative phases of the two pulses when collecting 2D spectra. After the shaper, the combined pump pulses travel along the same optical path enroute to the sample, ensuring that phase stability is well maintained despite any jitter in the optics. In addition, one can implement "phase cycling" with the pulse shaper to focus on particular contributions to the signal or to speed up data collection [97].

13.3.4 Data and Basic Interpretation

In this section we describe 2D spectroscopy experiments in four different physical systems. Although the measurements use different beam geometries and pulse sequences, the basic interpretation does not depend on the details of experimental implementation. We highlight the dynamics measured in each result before moving on to a general mathematical description.

13.3.4.1 Anharmonic Mode Coupling in RDC

Figure 13.14 shows a 2D-IR, photon-echo spectrum from the molecule dicarbonylacetylacetonato rhodium (RDC) in a hexane solvent [98]. The experiment focuses on the symmetric and asymmetric CO-stretch vibrations in the region of 2000–2100 cm^{-1}. A schematic energy-level diagram for the two, coupled modes is shown in Figure 13.15, where v_s represents the symmetric stretch, v_a the asymmetric stretch, and the ket $|n_a n_s\rangle$ represents a state with n_a quanta of vibrations in the asymmetric stretch and n_s quanta of vibrations in the symmetric stretch.

The two-dimensional spectrum of Figure 13.14 can be interpreted on an intuitive level. The two main peaks along the diagonal (around 2015 and 2084 cm^{-1}) correspond to exciting and probing vibrations along each of the quasi-normal modes. In particular, if the combined field of the pump-pulse pair contains spectral intensity at one of these two frequencies, this will modify absorption of the probe at the same frequency. These regions show up as peaks along the diagonal where the frequency values on the pump axis (ω_1) and probe axis (ω_3) are identical and equal to the frequency of the mode.

Upon closer inspection, one can see that these regions are each split into two separate peaks. This is due to the anharmonicity of the potential in each individual mode, resulting in the ground- to first-excited-state transition being at a different frequency than the first-excited to second-excited-state transition (see Figure 13.15). If there is excitation of more than one quanta of vibration, two peaks will appear. Thus, the two peaks near 2015 cm^{-1} correspond to $|00\rangle \rightarrow |10\rangle$ and $|10\rangle \rightarrow |20\rangle$ transitions in v_a,

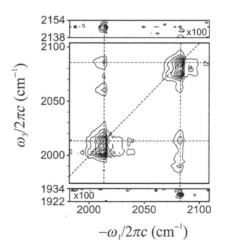

Figure 13.14 2D-IR spectrum in the region of symmetric and asymmetric CO stretch vibrations in the RDC molecule. Figure from Ref. [98].

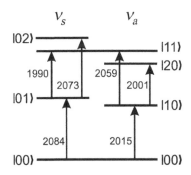

Figure 13.15 Energy-level diagram for symmetric (v_s) and asymmetric (v_a) CO-stretch vibrations in the RDC molecule. Figure from Ref. [98].

while the peaks near $2084\ \text{cm}^{-1}$ correspond to $|00\rangle \rightarrow |01\rangle$ and $|01\rangle \rightarrow |02\rangle$ transitions in v_s.

The off-diagonal peaks serve as a direct measurement of anharmonic coupling between the two modes. They show how excitation at one frequency results in a polarization (and emission) at another. In other words, resonantly exciting vibrations in one mode leads to motion along the other mode because they are coupled. The strengths of the off-diagonal peaks are directly related to the magnitude of the anharmonic coupling and can be used to determine these coefficients for the nuclear potential energy in the molecular Hamiltonian. Similar to Equation 10.1 from our treatment of coupled modes in Chapter 10, one can express the coupling potential to third-order in normal mode coordinates as

$$V(Q_s, Q_a) = \frac{1}{2}\omega_s^0 Q_s^2 + \frac{1}{2}\omega_a^0 Q_a^2$$
$$+ \frac{1}{6}\left(g_{sss}Q_s^3 + g_{aaa}Q_a^3 + 3g_{ssa}Q_s^2 Q_a + 3g_{saa}Q_s Q_a^2\right), \tag{13.14}$$

where Q_s and Q_a are the normal-mode coordinates for the symmetric and asymmetric CO-stretches and ω_i^0 is the (harmonic) frequency for mode i. The g's represent the anharmonic coefficients, with g_{sss}/g_{aaa} being the single-mode anharmonicities (not included in our Equation 10.1), and g_{ssa}/g_{saa} the anharmonic coefficients that describe the mode coupling. Based on the measured two-dimensional spectrum, the authors were able to determine values for these terms in the Hamiltonian: $g_{sss} = g_{aaa} = 32\ \text{cm}^{-1}$ and $g_{ssa} = g_{saa} = 22\ \text{cm}^{-1}$.

13.3.4.2 Intra- and Intermolecular Coupling in Water

Data in the previous section depicted measurements of intramolecular coupling between two vibrational modes in a molecule. In this section we consider measurements in water that demonstrate not only a strong intramolecular coupling between stretching and bending vibrations, but also intermolecular couplings with neighboring molecules. In fact, one sees a delocalization of vibrations over several molecules in the liquid that can be thought of in terms of a vibrational exciton.

Figure 13.16 shows both transient absorption and 2D-IR data from liquid water [99]. Panel (a) plots the steady-state absorption spectrum of water (black line), along with the spectra for the pump (blue shaded region) and probe (green shaded region) pulses. The energy levels of the participating intramolecular states are shown in panel (a) of Figure 13.17. The $1700\ \text{cm}^{-1}$ peak in the absorption spectrum is assigned to one quanta of the bending mode ($|0,1\rangle$), while the peak around $3400\ \text{cm}^{-1}$ is a mixture of the symmetric and asymmetric O–H stretches ($|1,0\rangle$) and the bending overtone ($|0,2\rangle$).

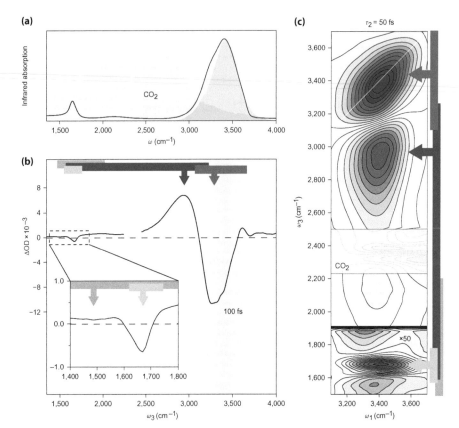

Figure 13.16 Broadband transient absorption and 2D-IR spectra of liquid water in the pump–probe geometry. (a) Steady-state absorption spectrum of water (black line), along with pump (blue) and probe (green) spectra. (b) Transient absorption spectrum of water at 100-fs delay after pump. (c) 2D-IR spectrum of water at 50-fs delay after pump. Figure from Ref. [99].

Panel (b) of Figure 13.16 shows the transient absorption spectrum when the probe pulse arrives 100 fs after the pump. The transient absorption spectrum is a projection of the 2D spectrum onto the probe frequency axis (ω_3) and highlights a number of features. Two of these are expected due to the population promoted to the excited state by the pump, including the ground-state bleach at just under 3400 cm^{-1} and a strong, excited-state absorption around 3000 cm^{-1} ($|1,0\rangle \to |2,0\rangle$, where the red-shift in frequency results from anharmonicity along the O–H stretching coordinate).

The data also include a weaker, broad absorption feature stretching from below 2000 cm^{-1} up to 3000 cm^{-1}. This broad ESA is attributed to the creation of "vibrational excitons," or vibrational excitations delocalized over several neighboring molecules in the liquid. The individual molecules are strongly coupled in the liquid, and it becomes useful to think about vibrations being delocalized over multiple molecules rather than in isolated ones. The potential energy as a function of vibrational coordinate for such delocalized vibrations is highly anharmonic, leading to a broad, lower-energy ESA feature in the liquid illustrated by the arrow and band labeled "2-Ex" in panel (b) of Figure 13.17

Finally, panel (c) of Figure 13.16 shows the 2D spectrum at a probe delay ("waiting time") of 50 fs. The fact that there are off-diagonal cross peaks in the 2D spectrum at such short delay times indicates very strong coupling between the O–H stretch and bend, such that the normal modes are not a good approximation to the true vibrational eigenstates. While these measurements nicely illustrate this point, the strong coupling

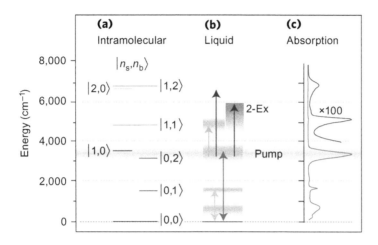

Figure 13.17 (See color insert.) Energy-level diagram with excitations for vapor and liquid water. Figure from Ref. [99].

had been observed in earlier transient absorption measurements [100]. However, the authors were able to follow the full, 2D spectrum as a function of probe delay time (panel (c) only shows the 2D spectrum at one delay time). The time-dependent evolution of the 2D spectrum allowed for the development of a detailed new picture of vibrational energy flow in water, which includes mixing of stretch, bend and intermolecular modes.

13.3.4.3 Complex Formation in Solution

The 2D measurements in the previous section uncovered intermolecular coupling between identical water molecules in a liquid sample. In this section we illustrate how 2D spectroscopy can highlight interactions between different molecules in a dense sample. In particular, we consider 2D-IR measurements in a room-temperature liquid sample of deuterated phenol–benzene [101]. These two molecules can form a complex, or a weak chemical bond between them, whose enthalpy of formation is so low that it is formed and broken repeatedly in solution at room temperature. 2D-IR measurements have the ability to probe the formation and dissociation of the complex in time.

Panel (a) of Figure 13.18 shows the 2D-IR "vibrational echo spectrum" for a pump–probe wait time of 200 fs. The term vibrational echo corresponds to a particular configuration of pump and probe k-vectors that select specific terms from the nonlinear, third-order polarization [102]. The red, diagonal peaks in panel (a) correspond to GSB of the phenol O–D (deuterated hydroxyl) stretch. The higher-frequency peak corresponds to O–D stretching in free phenol molecules, while the lower-frequency peak is O–D stretching in phenol molecules that have formed a complex with benzene. The blue, diagonally oriented peak corresponds to ESA of the same vibration. This peak is red-shifted down in the probe frequency due to the vibrational anharmonicity: absorption on the $1 \rightarrow 2$ transition occurs at a lower frequency than the pump-excited $0 \rightarrow 1$. Both peaks are broadened along the diagonal due to inhomogeneous broadening, as individual molecules experience different local environments that affect the transition frequency.

Extending the pump–probe delay time allows for the possibility of molecules transforming from the free to the complex state (or vice versa) in the time between the two pulses. When this occurs, molecules will have been "pumped" when they were free

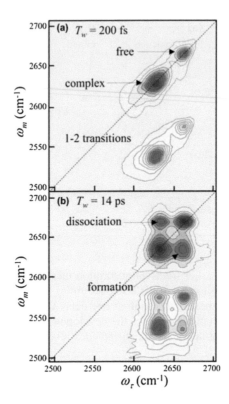

Figure 13.18 2D-IR spectra of deuterated phenol–benzene in solution at two different pump–probe delays. The spectra highlight both inhomogeneous broadening and the formation and dissociation of molecular complexes. Figure from Ref. [101].

and "probed" after they become part of a complex. Since the vibrational frequencies for the free and complex states are different, any molecules making a transformation from the free to the complex state will lead to the formation of cross peaks in the 2D spectrum (pumped at one frequency and probed at another). Panel (b) of Figure 13.18 show the 2D spectrum at a pump–probe delay of 14 ps. Cross peaks are indeed evident in both the GSB and ESA portions of the spectrum. This measurement nicely highlights how 2D-IR spectroscopy can be used to follow in time the formation and breaking of chemical bonds in solution.

13.3.4.4 Excitons in Solid State

As a final example of 2D spectroscopy, we consider dynamics of excitons in natural QDs coupled to excitons in a surrounding quantum well (QW).[7] An exciton is a bound state of an electron and a hole, where the electron is bound to the hole by static Coulomb attraction and has insufficient energy to escape as a free electron (similar to the way an electron is bound to a proton in a hydrogen atom). An exciton is generally free to move in a bulk semiconductor, but can be confined in one dimension in a QW, or all three dimensions in a QD. The confinement results in discrete states just below the bandgap of the bulk semiconductor. A QW is produced by growing a thin semiconductor layer (on the order of 10 nm) between two other materials. For example, a QW can be created by sandwiching GaAs between $Al_xGa_{1-x}As$, where x is a number between 0 and 1. A QD can be formed naturally in a QW if there are local monolayer

---

[7]In moving from coupled nuclear to electron dynamics, we note that the formalism required to describe the measurements and the insights derived from their interpretation are very similar.

fluctuations in the QW thickness. In this case, the QD is strongly coupled to the surrounding QW, and the interaction between the dot and well can be thought of in terms of a typical system-bath interaction.

Figure 13.19 shows 2D spectra of excitons in GaAs QDs/wells at different pump–probe delays using near-IR pulses tuned below the bandgap [103].[8] For the earliest wait time of 5 ps (upper-left panel), there are two peaks along the diagonal, corresponding to production of excitons in both the QDs and the QW. In the case of the dots, this is due to variations in the lateral dimensions among the various dots, while for the well the broadening is due to fluctuations in the QW thickness on length scales too short for localization of the exciton (i.e. too small to form a dot).

As the wait time is increased in the next three panels, two additional cross peaks become evident in the spectrum. These two cross peaks, labeled RP and EP, correspond to relaxation of a QW exciton and localization of the exciton at a QD (RP), and the converse transfer of a QD exciton to a QW (EP). Both of these correspond to absorption of radiation by either the dot or the well, followed by emission from the other, and thus involve a coupling between the dot and the well. The timescale for the cross peaks to form provides a measurement of the coupling dynamics. Note that the growth of the EP and RP peaks do not occur on the same timescale, indicating that the two coupling directions are not equivalent. This is not surprising given that the coupling is exothermic in one direction and endothermic in the other. In both cases, the coupling is mediated by phonons in the material.

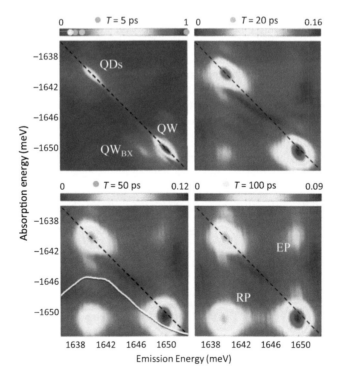

Figure 13.19 2D spectroscopy measurements of QDs and wells. QD labels the absorption peak associated with the dots, while QW labels the absorption peak associated with the wells. Figure from Ref. [103].

---

[8]Note in this figure the *y*-axis is reversed in comparison with earlier 2D data sets, and thus, the diagonal line corresponding to equal excitation and probing frequencies actually runs from the upper left to the lower right (black dashed line).

13.3.5 Mathematical Description

In this section we extend our mathematical treatment of the third-order polarization to include 2D spectroscopy measurements. We also connect our description to other approaches that use the density matrix formalism developed in Section 11.2. We focus on describing 2D measurements of nuclear dynamics, but note that the same mathematical formalism can be applied to dynamics of excitons in QWs (or other systems with coupled oscillators).

Before proceeding, it is worth reminding ourselves that while normal modes serve as a convenient basis for describing vibrations in polyatomic molecules, the actual vibrational eigenfunctions of any realistic molecule are nonseparable, multidimensional functions. As we have seen, the anharmonic terms in the molecular potential lead to coupling between normal modes, and one cannot accurately express the N-dimensional, vibrational wave function as a product of N, one-dimensional functions along normal mode coordinates. One could consider working with the actual eigenfunctions, but these are inevitably very cumbersome and offer little insight. Instead, we take a similar approach as we did with Born–Oppenheimer wave functions and use product states that are approximations to the true eigenstates (in the limit of zero coupling between vibrational modes, these product states are exact).

For example, consider the first few vibrational levels in a two-dimensional subspace of a large molecule. We have characterized the subspace in terms of approximate product states ($|00\rangle$, $|01\rangle$, $|10\rangle$, $|11\rangle$, etc.), where the different indices represent the quanta of vibration in each mode. However, none of these are actual eigenstates due to the coupling, and excitation into one of the modes always leads to energy flow into the other. In addition, the coupling implies that transition energies between our approximate states depend on excitation in the other mode. For instance, although both transitions involve one quantum of excitation in mode 1, the $|01\rangle \rightarrow |11\rangle$ transition frequency will generally be different than that for $|00\rangle \rightarrow |10\rangle$. In the completely decoupled limit, these would be the same. (In fact, in the absence of coupling, the state $|00\rangle$ could be factored as $|0\rangle|0\rangle$, and the two modes could be considered independently.)

In Section 13.1.1.3, we extended our mathematical treatment of the polarization to include multiple vibrational coordinates Q_1 and Q_2. In particular, we assumed that one quantum of excitation in mode one was nearly degenerate with two quanta in mode two, and that the pump pulse was tuned to initially excite the state $|10\rangle$ at time $t = 0$. This led to a first-order polarization *for the probe* given by Equation 13.6:

$$P_{\text{probe}}^{(1)} = \frac{-1}{i} \left\langle \chi_{10}^{(1)}(Q_1, Q_2, 0) \right| e^{+iHt} \mu_{nm,00} \right.$$
$$\left. \times \int_0^t dt_2 e^{-iH(t-t_2)} \left(\mu_{00,nm} E_2(t_2) \right) e^{-iHt_2} \left| \chi_{10}^{(1)}(Q_1, Q_2, 0) \right\rangle. \quad (13.15)$$

By initially exciting the state $|10\rangle$ with the pump and then probing both the $|00\rangle \leftrightarrow |10\rangle$ and the $|00\rangle \leftrightarrow |02\rangle$ transitions, one could study the anharmonic coupling between modes contained in the full Hamiltonian H. Of course, the pump pulse could have sufficient bandwidth to excite both states. With such a bandwidth-limited pulse, Equation 13.15 would generalize to

$$
\begin{aligned}
P_{\text{probe}}^{(1)} = \frac{-1}{i} &\left(\left\langle \chi_{10}^{(1)}(Q_1,Q_2,0) \right| + \left\langle \chi_{02}^{(1)}(Q_1,Q_2,0) \right| \right) e^{+iHt} \mu_{nm,00} \\
&\times \int_0^t dt_2 e^{-iH(t-t_2)} \left(\mu_{00,nm} E_2(t_2) \right) e^{-iHt_2} \left(\left| \chi_{10}^{(1)}(Q_1,Q_2,0) \right\rangle \right. \\
&\left. + \left| \chi_{02}^{(1)}(Q_1,Q_2,0) \right\rangle \right),
\end{aligned}
\tag{13.16}
$$

where the initial state prepared by the pump is a superposition of excited states $|10\rangle$ and $|02\rangle$.[9]

The basic idea is to simultaneously excite a superposition of excited states so that one can think of 2D spectroscopy as a multiplexed version of transient absorption. As noted, 2D experiments use different approaches to provide frequency resolution for the pump process, and it is important to consider the contributions of different pump frequencies leading to different initial states. In particular, in the pump–probe geometry of Figure 13.13, we consider excitation with two pump pulses separated by a time delay τ_{pump}. Each of the pump pulses creates a coherent superposition of states $|10\rangle$ and $|02\rangle$, and it is the time delay between the two pump pulses that controls the relative amounts of each state. For example, if τ_{pump} is chosen appropriately, the contribution to one of the states cancels while the other adds (as shown in Figure 13.20).

In 2D spectroscopy, one generally performs the experiment for a full range of τ_{pump} values, and the relative amplitudes in the initial superposition depend on the pump–pump delay. This produces a polarization that depends on both t and τ_{pump}:

$$
\begin{aligned}
P_{\text{probe}}^{(1)}(t, \tau_{\text{pump}}) = \frac{-1}{i} &\left(\left\langle \chi_{10}^{(1)}(Q_1,Q_2,0,\tau_{\text{pump}}) \right| \right. \\
&\left. + \left\langle \chi_{02}^{(1)}(Q_1,Q_2,0,\tau_{\text{pump}}) \right| \right) \\
&\times e^{+iHt} \mu_{nm,00} \int_0^t dt_2 e^{-iH(t-t_2)} \left(\mu_{00,nm} E_2(t_2) \right) e^{-iHt_2} \\
&\times \left(\left| \chi_{10}^{(1)}(Q_1,Q_2,0,\tau_{\text{pump}}) \right\rangle \right. \\
&\left. + \left| \chi_{02}^{(1)}(Q_1,Q_2,0,\tau_{\text{pump}}) \right\rangle \right),
\end{aligned}
\tag{13.17}
$$

where we keep the zero in the initial states to denote that the probe delay time is referenced to $t = 0$ (time of completion of the pump-pulse pair). As discussed in Section 8.2.3, it is this polarization that leads the measured signal in coherent optical techniques. In a 2D, pump–probe measurement, one spectrally resolves the transmitted probe pulse for a range of τ_{pump} values to yield $|E_{\text{probe}}(\omega_{\text{probe}}, \tau_{\text{pump}})|^2$. By Fourier transforming the measurement along the τ_{pump} axis, one produces the 2D spectrum $S_{\text{probe}}(\omega_{\text{probe}}, \omega_{\text{pump}})$, which can read as pumping at frequency ω_{pump} and probing at frequency ω_{probe}.

We note that with respect to the total applied field (pump + probe), Equation 13.17 corresponds to a third-order polarization. In this way, it is equivalent to expressions that one can derive for a third-order polarization based on the density matrix introduced in Chapter 11; this is an approach followed in many other texts and manuscripts (e.g. see

---

[9]We note that the $|00\rangle$ and $|02\rangle$ states are almost symmetric (for a harmonic potential, they would be perfectly symmetric). Since the dipole operator is antisymmetric, once expects a small transition dipole moment connecting the two states (i.e. the $|00\rangle \to |02\rangle$ transition is almost dark). However, the anharmonic mixing between $|10\rangle$ and $|02\rangle$ implies that the $|00\rangle \to |02\rangle$ transition is not actually dark and can be observed in the absorption spectrum.

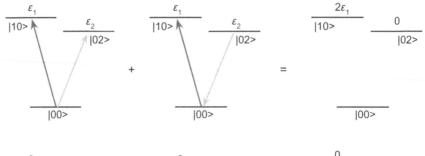

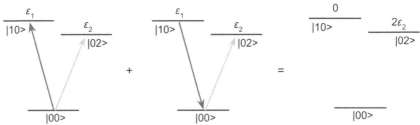

Figure 13.20 Illustration of pump–pulse pair excitation in 2D experiments with pump–pump delays chosen to produce either constructive addition in state $|10\rangle$ and destructive addition in $|02\rangle$ (upper panel), or vice versa (lower panel). ε_i represents the amplitude excited in mode i.

Refs. [44, 65]). Our approach has been to work with the wave function and the time-dependent Schrödinger equation to arrive at the third-order polarization responsible for the different experimentally-measured signals such as ESA and GSB in transient absorption, CARS, and the various 2D spectroscopies. In the absence of decoherence mechanisms, the Schrödinger approach using the wave function is equivalent to working with the Louiville-von-Neuman equation and the density matrix (as derived from the Schrödinger equation). These two different approaches amount to calculating the third-order polarization in these two ways:

$$P_{ab}^{(3)} = \left\langle \psi_a^{(1)} \middle| \mu \middle| \psi_b^{(2)} \right\rangle = \mathrm{Tr}(\mu \rho_{ab}^{(3)}) \tag{13.18}$$

where Equation 13.15 is essentially the first part of Equation 13.18. A primary reason for using the Louiville-von-Neuman equation is that it easily allows one to directly include relaxation processes in the expression for the third-order polarization (e.g. the Kubo model of Section 11.3.1). For reviews and more extended discussions of two-dimensional spectroscopy, see [44, 104, 65, 105, 106].

13.4 MEASURING MULTIELECTRON DYNAMICS

13.4.1 Charge Migration in Molecules

We introduced the idea of charge migration in Chapter 10, later discussing an experimental measurement in Section 12.5.3. That experiment used the incoherent technique of time-resolved ionization spectroscopy to measure multielectron dynamics in phenylalanine after an extreme ultraviolet (XUV) pump pulse created a superposition of cationic states. In this section, we discuss an experiment measuring charge migration in gas-phase iodoacetylene using the coherent technique of high-harmonic spectroscopy [107].

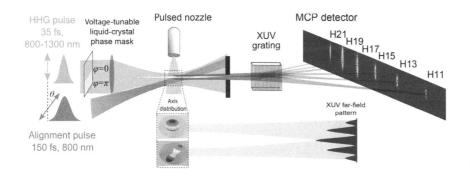

Figure 13.21 (See color insert.) Experimental approach for HHG spectroscopy of charge migration in an aligned, gas-phase sample. Figure from Ref. [107] supplemental material.

13.4.1.1 Experimental Setup

The experimental setup is more complicated than our standard arrangements and, for that reason, is shown in Figure 13.21. The experiment begins by inducing nonadiabatic alignment of the molecules using a strong, near-IR laser pulse. Once the molecules are aligned and oriented, a second pulse both ionizes the molecules and drives high-harmonic generation, or HHG (see Section B.4). The HHG pulse is actually split to form two, spatially separated foci, where only one of the foci overlaps the alignment pulse. Thus, there are two sources of harmonics, one from an aligned molecular sample and the other from an unaligned sample. These two sources can interfere in the far field, and the interference allows one to measure the phase between harmonics generated by the aligned and unaligned molecules. The spectrally resolved intensity of the combined harmonic light is detected in an XUV spectrometer, measuring the interference in each harmonic separately.

13.4.1.2 Data and Interpretation

As discussed in Section B.4, HHG is typically described in terms of a simplified picture known as the "three-step model:" (1) tunnel ionization in the strong field of the pulse on a timescale much shorter than the laser period, (2) acceleration of the electron away from the molecular ion and back again during one half cycle of the oscillating field, and (3) recombination of the colliding electron with the core, resulting in the emission of XUV radiation. For appropriate field strengths, the probability of removing an electron from the highest occupied molecular orbital (HOMO) and the HOMO−1 orbital is similar, projecting the molecule onto a superposition of the ground and excited states of the cation. This superposition constitutes an electronic wave packet in the ion, and the motion of this wave packet constitutes charge migration across the molecule on an attosecond timescale.[10]

The key measurement in HHG spectroscopy is characterizing the amplitude and phase of the states in the electronic wave packet prepared by tunnel ionization. This is accomplished by using the returning electron to interrogate the evolving wave packet in the molecular cation, as the XUV emission produced as the electron recollides with the core contains information about the constituent eigenstates. In particular, the phase and amplitude of the XUV emission are sensitive to the relative phases and amplitudes of the cationic states making up the electronic wave packet. This comes from the fact that the returning electron has an intrinsic chirp, or time dependence to its energy, known as

---

[10]It can be preferable to consider this wave packet a "hole wave packet," describing the motion of the hole left by the removed electron. This hole wave packet is actually an N−1 electron wave packet, involving the motion of many electrons.

the "attochirp." Therefore, the different harmonics contain information about different times during wave packet evolution in the cation.

Unfortunately, the relationship between the measured harmonic amplitudes/phases and eigenstates in the cationic wave packet is quite complicated and does not come from a simple retrieval procedure. This experiment involved an extensive fitting procedure that required some simplifying assumptions. The measurement does have an internal consistency check in that it is overdetermined: the phase and amplitude of many harmonics are measured, while the cationic wave packet is composed of only two eigenstates. Figure 13.22 shows the reconstructed charge migration dynamics in iodoacetylene as a function of time. The dynamics show attosecond rearrangement of the hole density from localization on the iodine atom, over the two carbons, and back to the iodine.

We note that HHG spectroscopy necessarily measures dynamics in the presence of a strong laser field, which itself can substantially affect the system. Furthermore, expanding the electronic wave function in terms of uncoupled electronic states with well-defined relative phases is only reasonable for very short times when nuclear dynamics play a minimal role. This procedure can fail dramatically for times greater than a few femtoseconds when nuclear dynamics lead to a loss of electronic coherence and the possibility of coupling between electronic states (e.g. internal conversion). Thus, it can be quite challenging to extract dynamics from HHG spectroscopy measurements.

13.4.2 Carrier Dynamics in Semiconductors

Multielectron dynamics play an important role in semiconductors, and many experiments follow relaxation dynamics after electron excitation by a pump pulse. In Section 8.4 we considered a measurement of 1D vibrational dynamics in few-layer graphene using transient reflectivity. In this section, we examine an optical transient absorption measurement of electron coupling in thin graphite films [108]. The experiment seeks to determine the prominent relaxation mechanism(s) following carrier excitation.

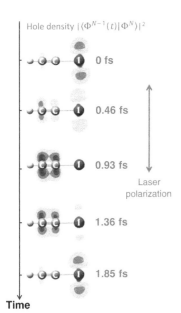

Figure 13.22 (See color insert.) Reconstructed charge migration dynamics in iodoacetylene using HHG spectroscopy. Figure adapted from [107].

13.4.2.1 Experimental Setup

The initial relaxation dynamics proceed very quickly, and the experiment uses sub-10 fs optical pulses for both the pump and probe to achieve the required time resolution. The left side of Figure 13.23 shows the excitation scheme. A pump pulse centered at 1.5 eV ($\sim$825 nm) promotes electrons from the valence band to the conduction band. A broadband, optical probe follows the pump, and the spectrally resolved, DT of the probe ($\Delta T / T_0$) is measured as a function of pump–probe delay.[11] The spectral resolution in the probe measurement should uncover any energy dependence in the charge dynamics. Lock-in detection with an optical chopper in the pump beam is employed to achieve the highest possible signal to noise.

13.4.2.2 Data and Interpretation

On the right side of Figure 13.23, panel (a) plots the DT as a function of both probe energy (frequency) and time delay. The initial, enhanced transmission across all probe

---

[11]Like many transient absorption experiments, this approach measures the DT of a probe. In fact, the technique is referred to as DT instead of transient absorption (the ideas are essentially the same). The name DT can serve as a reminder that many solid-state materials have nonnegligible reflection coefficients, and pump-induced changes can show up in reflectivity as well as transmission.

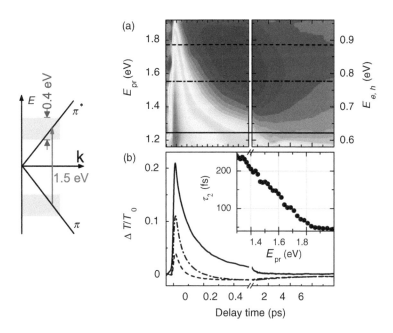

Figure 13.23 Left: Band structure and excitation scheme for ultrafast carrier generation in thin graphite film. Right panel (a): DT in a thin graphite film as a function of probe photon energy and pump–probe delay. Right panel (b): temporal lineouts at three different probe energies. Inset: decay time τ_2 as a function of probe energy (from fit to data). Figure adapted from [108].

energies is due to saturation of the optical transition by the pump; known as "state filling" in this context, it is the equivalent of the usual ground-state bleach in transient absorption. This bleach signal then decays as the carriers relax in the sample due to scattering into states in the conduction band that are outside the probe absorption bandwidth. Both carrier–carrier scattering and carrier–phonon scattering play a role in this relaxation; these processes are analogous to the electron–electron interactions and electron–nuclear coupling we discussed in the context of isolated molecules.

Panel (b) shows temporal lineouts at three different probe energies. After a minor, initial ultrafast decay ($\tau_1 \sim 13$ fs), the different probe energies all show a prominent decay with a timescale τ_2 that depends on the energy (see inset). This decay is associated with "cooling" of the carriers due to phonon scattering, where one can consider the electrons to be the "system" in contact with a phonon bath. The higher-energy states (probed with higher probe energies) cool faster, and thus show a faster decay than the lower-energy states (lower-probe energies). Detailed modeling of the experimental results allows the authors to extract separate, quasi-equilibrium distributions for electrons and holes as they cool down on femtosecond timescales. The decay of the differential probe transmission on longer (picosecond) timescales reflects recombination of electrons and holes, leading to a single equilibrium distribution.

CHAPTER 14

Coherent Diffractive Measurements in ND

14.1 INTRODUCTION

In Chapter 9 we discussed using diffractive imaging of a probe pulse to provide a more direct measure of time-dependent changes in molecular structure. In a typical time-resolved diffraction experiment, a laser pump pulse creates a time-dependent wave packet on an excited electronic potential energy surface. The probe is either an ultrafast electron bunch or an X-ray pulse, and to resolve structural changes, the electron or X-ray wavelength must be smaller than the length scale of interest. For typical angstrom separations between atoms in a polyatomic molecule, this requires photons with an energy greater than 10 keV or electrons with an energy greater than 100 eV. The elastically scattered electrons or X-rays are measured on a two-dimensional, position-sensitive detector, recording the far-field diffraction pattern that results from the probe interaction. This diffraction pattern is analyzed at different pump–probe delays to determine changes in molecule structure. As in Chapter 9, the mathematical description is the same for all measurements, and so we begin with it.

14.2 MATHEMATICAL DESCRIPTION

The equations describing diffractive measurements were derived in Section 9.2. Here we highlight only the primary results, no longer assuming the sample consists of diatomic (one-dimensional) molecules. The measured intensity on the detector as a function of the scattering vector s is given by the sum of atomic and molecular contributions:

$$I_{\text{averaged}}(s) \equiv I_A(s) + I_M(s), \tag{14.1}$$

where

$$I_A(s) = \frac{I_o}{R^2} \sum_{i=1}^{N} f_i(s) f_i^*(s) = \frac{I_o}{R^2} \sum_{i=1}^{N} |f_i(s)|^2, \tag{14.2}$$

and

$$I_M(s) = \frac{I_o}{R^2} \sum_{i=1}^{N} \sum_{j \neq i}^{N} |f_i(s)| \, |f_j(s)| \cos\left[\eta_i(s) - \eta_j(s)\right]$$
$$\times \int dr_{ij} P(r_{ij}) \frac{\sin(sr_{ij})}{sr_{ij}}. \tag{14.3}$$

The atomic scattering, I_A, can be calculated using form factors in the literature and does not depend on molecular structure. The measured intensity as a function of s is converted to the "modified molecular scattering intensity" $sM(s)$:

$$sM(s) \equiv s\frac{(I_{\text{averaged}} - I_A)}{I_A} = s\frac{I_M}{I_A} \tag{14.4a}$$

$$= \frac{1}{\sum_{i=1}^{N} |f_i(s)|^2} \sum_{i=1}^{N} \sum_{j \neq i}^{N} |f_i(s)||f_j(s)| \cos[\eta_i(s) - \eta_j(s)]$$

$$\times \int dr_{ij} P(r_{ij}) \frac{\sin(r_{ij}s)}{r_{ij}}. \tag{14.4b}$$

The modified molecular scattering intensity highlights the interference in scattering from different atomic centers in the molecule, and can be Fourier (sine) transformed to yield the radial distribution function $f(r)$:

$$f(r) = \int_0^{s_{\text{max}}} sM(s) \sin(rs) e^{-\kappa s^2} ds, \tag{14.5}$$

where s_{max} is the maximum momentum transfer measured on the detector, and κ is a damping constant chosen to avoid edge effects in the transform (for X-ray diffraction, the momentum transfer is usually represented as q rather than s). The radial distribution function gives the relative density of internuclear separations, and the measured $f(r)$ can be compared with calculated distribution functions for different molecular structures to follow changes in structure with time.

14.3 MOLECULAR EXAMPLES

14.3.1 Simulation in Thymine

In order to demonstrate the basic approach for diffractive measurements in polyatomic systems, we begin with simulated, static ultrafast electron diffraction (UED) results for the molecule thymine ($C_5H_6N_2O_2$). The top-left panel of Figure 14.1 shows the raw electron diffraction signal for a randomly aligned sample as a function of 2D position on the detector array. Note that the vast majority of electrons hit near the center of the detector, as the differential cross section favors low-angle scattering ($f(s)$ is strongly peaked in the forward direction). The top-right panel shows the modified molecular scattering intensity, $sM(s)$ (see Equation 14.4). The azimuthally averaged molecular scattering intensity, $sM(s)$, is plotted as a function of the scattering vector magnitude in the bottom-left panel. It is the modulations in this signal that yield the peaks in $f(r)$ (bottom-right panel) when performing a sine-transform of $sM(s)$.

For comparison, Figure 14.2 shows the experimentally determined, static radial distribution function in thymine (compare with the simulated data of Figure 14.1). Figure 14.2 also indicates the known atom-pair distances for the molecular ground state. Note that the contributions from pairs of atoms is independent of whether there is a molecular bond between them; the electrons scatter predominantly off the atomic centers and are not very sensitive to the bonding valence electrons. The multitude of overlapping atom-pair distances in Figure 14.2 demonstrates the challenge of trying to determine structural changes in polyatomic systems.

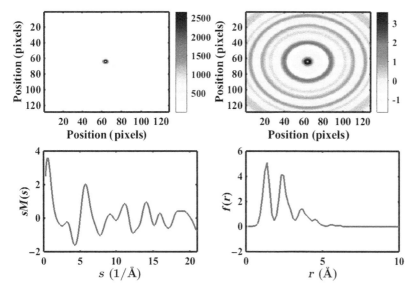

Figure 14.1 Simulations of electron diffraction from thymine molecules using 100-keV electrons. Top-left panel is the raw diffraction data. Top-right shows the modified molecular scattering intensity, $sM(s)$. Bottom-left is the azimuthally averaged $sM(s)$, while the bottom-right plots $f(r)$. Simulation courtesy of Alexander Johnson and Martin Centurion.

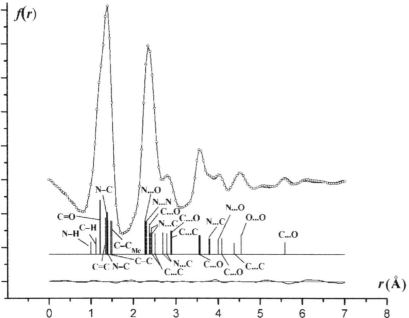

Figure 14.2 Measured static radial distribution function, $f(r)$, for thymine using gas-phase electron diffraction. Various atom-pair distances are identified beneath the curve. Figure from Ref. [109].

14.3.2 Aligned CF₃I

As noted in Section 9.2.3, the probe in a gas-phase, pump–probe experiment generally encounters a partially aligned molecular sample (either through active alignment or excitation-induced alignment). Unlike the case of one-dimensional systems, the multiple bond lengths present in polyatomic molecules make the issue of alignment quite important for multidimensional systems. In the absence of a perfectly aligned sample, the typical approach is to compare measured radial distribution functions with calculated ones in an attempt to interpret time-dependent changes in molecular structure.

This requires one to have reasonable guesses for the transient structures through which the molecules pass during their evolution following the pump pulse. If the measured and calculated radial distribution functions agree for a number of key geometries, experiments can confirm a proposed pathway. Experiments cannot, however, unambiguously establish a time-dependent structure without comparison to calculations.

Since molecular alignment plays such an important role in diffractive measurements of polyatomic molecules, we first consider a non-time-resolved experiment looking at diffraction from aligned molecules [110]. The measurements were performed in CF_3I, a symmetric-top molecule, with the sample undergoing supersonic expansion to rotationally cool the molecules. A 300-fs, 800-nm laser pulse was used to impulsively align the molecules, and a 500-fs electron pulse arrived shortly after the laser pulse, when the molecules came into field-free alignment. The results are shown in Figure 14.3. While it would be possible to holographically retrieve the molecular structure of a cylindrically symmetric molecule from a single diffraction measurement under perfect alignment conditions, the retrieval of the molecular structure with imperfect alignment required additional calculations. The authors used a genetic algorithm to construct a measured diffraction for perfectly aligned molecules, based on the measured one for imperfectly aligned molecules. They were then able to reconstruct the molecular structure using a holographic approach [111]. Time-resolved UED studies with atomic spatial resolution and a temporal resolution of 150 fs studied this same system to track wave packets dynamics through a conical intersection [112].

14.3.3 Experiments in a Pyridine Series

In this section we consider a series of UED measurements following excited-state dynamics in several aromatic molecules: pyridine (C_5H_5N), picoline (C_6H_7N – similar to pyridine but with an additional methyl group), and lutidine (C_7H_9N – with two additional methyl groups) [113]. These experiments follow radiationless decay from

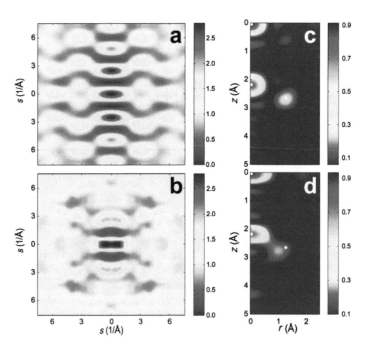

Figure 14.3 Diffraction patterns (left) and reconstructed molecular images (right) for UED measurements of CF_3I. Top row (panels a and c) are simulations, while the bottom row (panels b and d) shows experimental data. Figure from Ref. [110].

the excited state and are motivated by the fact that UED can provide information from dark states not normally accessible by spectroscopic means. For example, transient absorption experiments rely on being able to observe changes in optical absorption and relate them to dynamics on the excited state. However, if the energy separation between the states goes outside the laser bandwidth, or if the transition dipole moment between the two states becomes very small (i.e. a "dark state"), then a transient absorption measurement will lose information about the wave packet dynamics.

The experiments begin by measuring the static, ground-state diffraction signal. The top row of Figure 14.4 shows the diffraction pattern observed on the detector for pyridine (panel a), picoline (panel b), and lutidine (panel c). The bottom row shows the extracted radial distribution functions for each molecule, along with the ground-state structures. One can see clear similarities and small differences in the radial distribution functions, reflecting the variations in structure between molecules in this "homologous series" (the various bond lengths associated with these peaks are noted in Figure 14.6).

The experiment now applies a 266-nm pump pulse to excite the molecules to the first electronic bright state (near the middle of the first absorption band). Figure 14.5 shows difference diffraction data (both $\Delta sM(s)$ and $\Delta f(r)$) for the three molecules at large pump–probe delay. The idea is to look at long delay times to determine the final structure. Specifically, a $\Delta f(r)$ signal is simulated for a variety of different possible structures, and comparison with the measured data allows one to determine the most likely configuration. The best-fit calculation shows that pyridine and picoline evolve to ring-open structures, while lutidine does not. This result was not available from earlier ionization measurements, even though those provided more detailed information on the time evolution [114].

Finally, Figure 14.6 shows the evolution of the difference diffraction data as a function of pump–probe delay for both pyridine (left) and lutidine (right). While the time resolution is relatively coarse, there is a clear progression of features with increasing delay. The understanding of excited-state dynamics in this series of molecules was facilitated by comparing results from multiple experimental approaches. Previous time-resolved ionization measurements provided detailed quantitative information on the timescales involved [114], while the UED measurements uncovered specific transient structures. This highlights the fact that multiple measurements are frequently required to form a

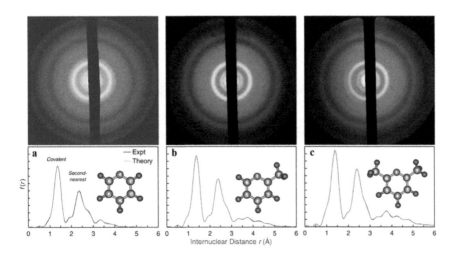

Figure 14.4 (See color insert.) Static, ground-state electron diffraction measurements and radial distribution curves for gas-phase pyridine (a), picoline (b), and lutidine (c). Figure from Ref. [113].

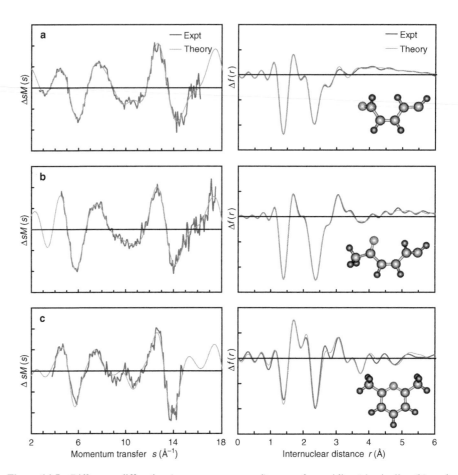

Figure 14.5 Difference-diffraction (pump vs. unpumped) curves for pyridine (a), picoline (b), and lutidine (c). Both experiment and theory are shown. Left panels plot the difference in the molecular scattering intensity, while right panels show the difference radial distribution functions and corresponding structures. Figure from Ref. [113].

complete picture of dynamics in a given system. In Chapter 15 we explore this idea in detail, studying the excited-state dynamics of one molecule using many different techniques.

14.3.4 Diffraction Measurements in Solution

Figure 14.7 illustrates how time-resolved X-ray diffraction can yield time-dependent structural information on molecules in solution [115]. The measurements were performed on gold nanoparticles ($[Au(CN)_2^-]_3$) in water. Van der Waals interactions between gold atoms lead to the formation of $[Au(CN)_2^-]_n$ complexes without any covalent bonds. Excitation by an ultraviolet laser pulse at 267 nm transfers an electron from an antibonding orbital to a bonding orbital, leading to the formation of new covalent bonds between gold atoms, as well as structural changes associated with the formation of these bonds. The left panel shows how the measurements were made, and the right panel highlights some of the experimental results. The X-ray diffraction signal at each

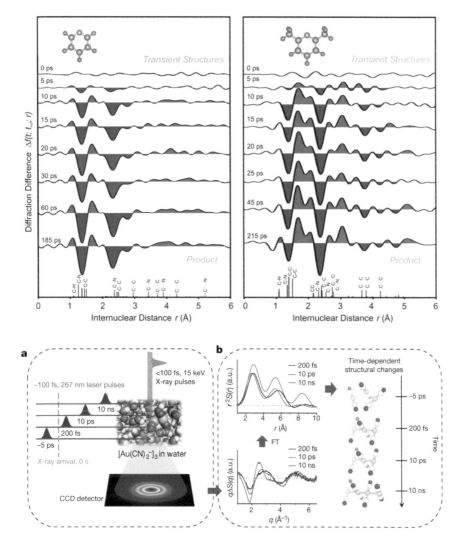

Figure 14.6 Difference-diffraction radial distribution curves as a function of pump–probe delay time for pyridine (left) and lutidine (right). Negative (blue) regions indicate less signal at particular atom–atom separations as compared to the unexcited molecule, while the positive (red) regions indicate enhanced signal at certain separations. Various atom-pair distances are indicated along the bottom. Figure from Ref. [113].

Figure 14.7 Time-resolved X-ray diffraction measurements of gold nanoparticles in solution. The left panel (a) illustrates the measurement, while the right panel (b) shows some experimental results, including the radial distribution function for selected delays, along with the inferred molecular geometries. Figure from Ref. [115].

delay was used to construct the radial distribution function, from which molecular geometries could be inferred (as shown on the far right of the figure).

14.4 CONCLUSIONS

In Section 9.4 we discussed the pros and cons of both electron and X-ray diffraction. While they can be described by the same mathematical formalism, the fact that X-rays scatter from only the *electron* charge density has consequences for the information content contained in the measured diffraction patterns. In particular, since X-rays do not scatter significantly from protons, they are less sensitive to lighter atoms such as Hydrogen for which the sole electron is delocalized and participates in molecular bonding. Having now considered diffractive measurements in multidimensional systems, we return to this idea using simulations in ethylene molecules [116].

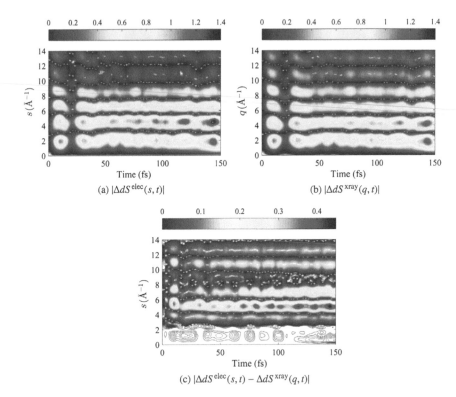

Figure 14.8 Simulated difference-diffraction (ΔdS) data as a function of pump–probe delay in ethylene using electrons (a), X-rays (b), and the difference between the two (c). Figure from Ref. [116].

The calculations make use of classical dynamics on potential energy surfaces that are calculated "on the fly." In this procedure, a series of classical trajectories are initiated at slightly different geometries based on the spread of the initial wave function in the ground state (mimicking the vertical transition of a pump pulse). Then, the time-independent Schrödinger equation (TISE) is used to calculate both the energy and gradient at a given point on the potential energy surface. The classical equations of motion are solved to propagate each trajectory to the next point on the surface, where the TISE is again used to calculate the energy and gradient. This procedure is iterated for each trajectory until the desired propagation time. Figure 14.8 shows the simulated difference-diffraction data as a function of pump–probe delay in ethylene for both electrons (panel a) and X-rays (panel b). Panel (c) shows the difference between the two measurements. Note that diffraction using electrons and X-rays yield quite similar results. However, there are differences seen in panel (c), and they vary with both momentum transfer (s/q) and time delay. These differences are largely attributed to the hydrogen atoms in ethylene, which do not scatter X-rays and whose contributions to the electron data are enhanced for larger momentum transfer and time delays. Therefore, it is generally preferable to use electrons when interested in structural dynamics of the lightest elements.

CHAPTER 15

One System, Multiple Approaches

15.1 INTRODUCTION

The previous three chapters examined a variety of experimental approaches for measuring different types of dynamics in multidimensional systems. Techniques included incoherent measurements such as time-resolved photoelectron spectroscopy, coherent approaches like transient absorption (TA), and diffractive imaging using ultrafast electrons (UED) or X-rays. Each technique has its particular advantages, depending on the type of system and what information one is most interested in obtaining. In this chapter we study a single molecular system using multiple experimental approaches, as well as supporting calculations. The goal is to show how different techniques provide different views, which can ultimately be combined to build a complete picture of the dynamics. The system we consider is cyclohexadiene (CHD, C_6H_8). Upon photoexcitation with ultraviolet (UV) radiation, the molecule undergoes internal conversion and possible ring-opening (isomerization) to form hexatriene (HT), or simply relaxation back to the ground state of CHD.[1] This system serves as a prototype for a variety of isomerization reactions in biologically-relevant molecules and has been extensively studied using a number of different techniques.

The basic picture of the reaction is shown in Figure 15.1 [117]. Excitation of CHD by a UV pump pulse at 266 nm initiates wave packet dynamics on the excited 1B-state. The wave packet passes through a conical intersection en route back to the ground potential energy surface (PES). Excited-state dynamics involve a reconfiguration of the molecular structure, such that upon return to the ground electronic state, there is some probability of bond cleavage and isomerization into the product HT. The excess electronic energy is deposited into nuclear motion, and the product is expected to go through a series of vibrationally hot conformers (tZc, tZt, etc.) before reaching thermal equilibrium. The series of experiments presented here seek to uncover details of the reaction process, including shapes of the PES, configurational changes in the molecule, and the timescales for both passage through the conical intersections and vibrational cooling in the product ground state in the case of measurements in solution.

---

[1] We note that the understanding of the exact path the molecule takes and the number of conical intersections involved in the internal conversion dynamics has evolved over the years. As with most systems studied using different approaches, our understanding of the dynamics is refined as new contributions provide a more comprehensive picture.

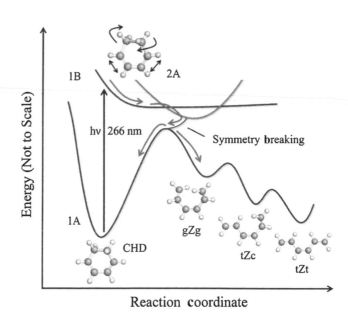

Figure 15.1 Illustration of the CHD-HT ring-opening dynamics. Figure from Ref. [117].

15.2 EXPERIMENTAL RESULTS

15.2.1 UED

We begin with gas-phase, electron diffraction experiments, where a UV pump pulse initiates the ring-opening reaction and is followed by a time-delayed probe consisting of a pulse of 30-kV electrons [118]. Figure 15.2 shows the measured, raw diffraction pattern from ground-state CHD molecules (no pump pulse present). A time-resolved measurement records similar diffraction patterns as a function of pump–probe delay and uses them to construct the radial distribution function, $f(r)$, at each time delay. The experimental radial distribution functions are compared with the calculated ones for molecular geometries suspected to be important in the dynamics (e.g. transition-state and end-point geometries in the isomerization reaction). For this particular experiment, geometries include the parent CHD, as well as the different, vibrationally-hot conformers of the product HT.

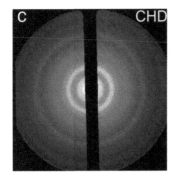

Figure 15.2 Raw electron diffraction pattern from ground-state CHD. Figure adapted from Ref. [118].

Panel (a) of Figure 15.3 shows the measured, *product-only* radial distribution function as a function of time delay (the $f(r,t)$ surface). Similar to the excited-state diffraction signal of Section 9.3.3, the product-only signal is obtained by adding back in the known reactant contribution. Panel (b) plots both measured and calculated radial distribution functions. The blue curves show lineouts of the experimental data at four different time delays. As expected, the product-only signal at negative time delay is essentially zero. As the time delay increases, peaks begin to appear at different atom-pair separations. The red curves show *calculated* radial distribution functions for the reactant CHD molecule (top), as well as for the vibrationally-excited reactant and different conformers of the product (bottom).

The experiments indicate that at long time delays, the measured radial distribution function corresponds to a mixture of different HT conformers (the green curves in panel (b) show fits to the data). While the measurements clearly evolve away from CHD toward HT, the data are not consistent with a single conformer, indicating that the molecule remains far from thermal equilibrium even at long delays after the pump

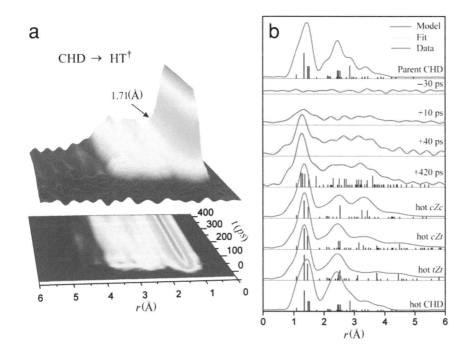

Figure 15.3 Time-resolved $f(r,t)$ surface (panel a) and $f(r)$ curves at different time delays for both experimental data and calculations (panel b). Figure from Ref. [118].

pulse. The time resolution in these measurements was limited by the duration of the electron pulse (approximately 4 ps), and thus information on a subpicosecond timescale was not available. While the fast, intermediate dynamics through the conical intersection could not be resolved, the longtime results are consistent with the picture of Figure 15.1, in which HT remains vibrationally hot for some time after formation.

15.2.2 X-ray Diffraction

The experimental procedure is essentially the same as UED: a UV pump pulse excites a gas-phase CHD sample and is followed by a time-delayed probe of 8.3 keV X-ray photons. The X-rays are generated by a free electron laser that provides bright, high-energy X-ray pulses with durations under 100 fs [119]. Raw X-ray diffraction patterns are collected at different pump–probe delays and plotted as a function of the momentum transfer q. We note that the momentum transfer range (up to 4 Å^{-1}) available from the X-ray diffraction measurements is not as large as for the UED measurements, leading to less information being available on short length scales. Figure 15.4 shows the percentage difference-diffraction signal, $\%\Delta I(t,q) = 100 \times (I_{\text{pump on}} - I_{\text{pump off}}/I_{\text{pump off}})$, as a function of both momentum transfer and time delay. The data show a turn on of two primary peaks at $q = 2$ and 4 Å^{-1} over approximately 200 fs. In addition, a dip is seen around $q = 1$ Å^{-1}.

The time-dependent diffraction differences are associated with changes in molecular structure as HT is formed from photoexcited CHD. The temporal behavior of the signal is seen most clearly in Figure 15.5, where panel (a) plots the time-dependent signal integrated over the regions $q = 2.1$–2.5 Å (rising curve) and $q = 2.9$–3.1 Å (falling curve). Panels (b) and (c) plot derivatives of the fits to the data in panel (a). These indicate a timescale for structural changes of approximately 80 fs.

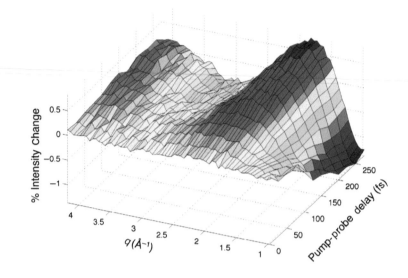

Figure 15.4 X-ray difference-diffraction measurements of isomerization of CHD. Figure from supplemental material of Ref. [119].

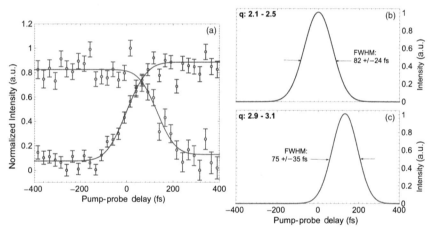

Figure 15.5 Panel (a): Temporal lineouts of X-ray diffraction measurements of isomerization of CHD. Panels (b) and (c): derivatives of the fits from panel (a). Figure from Ref. [119].

A more detailed interpretation of the experimental data is based on semiclassical dynamics calculations on the excited-state surface. The calculations begin with a distribution of initial molecular geometries on the excited state weighted by the probability density as a function of coordinate on the ground state (the pump initiates a "vertical transition," where the nuclei do not have time to move). As quantum dynamics in 36 dimensions (14 atoms with $3N-6$ degrees of vibrational freedom) are not feasible, the authors calculate semiclassical trajectories, where the molecular wave packet is propagated classically using a potential energy from quantum-mechanical electronic structure calculations and allowed to transition between PESs ("surface hopping"). They calculate over 100 trajectories, each of which yields positions of the constituent atoms as a function of time. From these atomic positions, the authors calculate the X-ray diffraction patterns for each point along the trajectories. They then fit the experimental diffraction data at each time delay to linear combinations of the calculated diffraction patterns for geometries along the trajectories.

They found that the data could be fit to a surprisingly small number of trajectories: only four trajectories captured 80% of the calculated change in diffraction intensity as

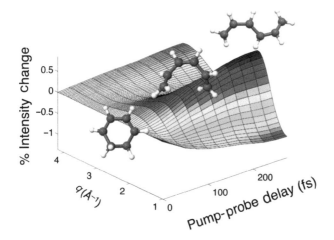

Figure 15.6 Simulations of X-ray diffraction data based on molecular dynamics calculations (with surface hopping) for isomerization of CHD. Figure from supplemental material of Ref. [119].

a function of time. For comparison with experiment, the calculated $\%\Delta I(t,q)$ for the top eight trajectories is shown in Figure 15.6 (compare with Figure 15.4). The relevant trajectories help provide a picture of the atomic rearrangement during the reaction. In particular, the results indicate a rapid expansion of the ring and breaking of the C1–C6 bond, with the majority of the population that undergoes the ring-opening process doing so on a 50-fs timescale (this is faster than some spectroscopic measurements that suggest 100 fs or more). The X-ray results also find that some of the remaining population is "trapped" in the excited state for longer times (145 fs) before returning to ground-state HT. These diffractive results complement the spectroscopic measurements discussed later by providing timescales for structural rearrangement of the molecule (which is not directly observable with spectroscopic techniques).

15.2.3 Transient Absorption

We next consider results showing broadband, TA in the UV. In diffractive measurements, one compares radial distribution functions at different time delays with measured or calculated distributions for relevant molecular geometries such as reactant or product. In TA, one compares the spectrally resolved absorption of the probe at different time delays with known or calculated spectra for relevant molecular geometries. In both cases, the time-resolved data need to be compared with calculations (or static measurements) for interpretation. While both rely on high-level electronic structure and dynamics calculations, TA further requires calculating the time-dependent absorption spectrum, which is more involved than calculations required for interpreting diffraction measurements. In diffraction measurements, calculation of the diffraction pattern follows straightforwardly from the time-dependent structure.

TA experiments in CHD are performed in solution with an ethanol solvent [120]. Figure 15.7 shows difference spectra (percentage change in optical density due to the pump pulse) as a function of pump–probe delay after excitation with 30-fs pump pulses at 250 nm. Above the 2D map are spectral lineouts at three different time delays, while the panel to the right shows temporal lineouts at three different colors. Immediately after time zero, the difference spectra show a single, broad enhanced absorption feature that shifts to bluer wavelengths as the time delay increases. By approximately 50 ps delay, the enhanced absorption has developed a multipeaked structure.

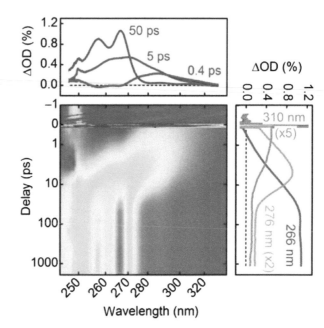

Figure 15.7 Broadband TA data from the CHD ring-opening reaction. Figure from Ref. [120].

Fifty picoseconds is a relatively long time delay in liquid phase due to collisions with the solvent, and one would expect the molecule to be approaching its steady state. It therefore makes sense to compare the spectrum at 50 ps to a static spectrum for the expected product HT (see Figure 15.8 [121]). The emergence of HT-like features in the TA spectra makes it clear the reaction has progressed significantly by this time. In the temporal lineouts at right, the 266 nm curve corresponds to the strong peak in the HT spectra; absorption at this color appears to grow steadily and be fully on by approximately 50 ps. On the other hand, the 310 nm and 276 nm curves indicate enhanced absorptions with different temporal behavior. The 310 nm curve (preferential CHD absorption) begins immediately before slowly turning off, while the 276 nm curve appears to track a transient, intermediate product.

One of the challenges of broadband TA experiments on large molecules is that while the vibrationally cold, ground-state absorptions of the reactants and products can easily be measured independently, the excited-state and vibrationally hot ground-state absorptions are often not well known. So while the data provide temporal dynamics across a wide range of probe colors, it is not immediately apparent how to interpret the measurements at intermediate times when the molecule is electronically excited

Figure 15.8 Static absorption spectrum from CHD (solid) and HT (dashed). The middle curve (dashed/dotted) shows a static difference spectrum assuming a 40% yield of HT. Figure from Ref. [121].

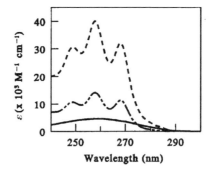

and/or vibrationally hot. In particular, the CHD-HT ring-opening reaction not only involves multiple excited PESs, but the molecule undergoes a significant structural rearrangement that leaves excess vibrational energy in the product. The path to equilibrium involves both changes in bond angles ("conformational relaxation") and vibrational redistribution and cooling ("thermalization"). For such processes, significant calculations and modeling are required to understand the intermediate dynamics. For example, one can attempt to calculate excited-state absorptions from relevant molecular geometries through which the molecule proceeds. However, it is often difficult to fully characterize the process, and additional information is required.

15.2.4 X-ray TA

A similar approach uses X-ray TA with gas-phase CHD molecules [122]. Like with UV TA, the near-edge X-ray absorption fine structure (NEXAFS) performed in this experiment is primarily sensitive to valence electronic structure.[2] In this way it is also complementary to techniques that are more sensitive to structural changes, such as diffractive measurements or X-ray absorption at higher photon energies. As before, a UV pump pulse (266 nm, 110 fs) photoexcites CHD to the 1B state. The time-delayed probe pulse consists of a broad distribution of high-harmonic peaks extending from 160–310 eV. The broadband, differential X-ray TA signal is monitored as a function of pump–probe delay. The data are shown in panel (b) of Figure 15.9; panel (a) shows the absorption spectrum for ground-state CHD in this region, as well as the spectrum from theory.

As with optical TA experiments, there is a wealth of potential information in the multichannel pump–probe data. The authors focus on three different time windows indicated by the shaded regions in panel (b): 0–40, 90–130, and 340–540 fs. X-ray absorption spectra averaged over each time window are compared to the static spectra for ground-state CHD and HT, as well as theoretical predictions for the dynamic absorption spectra. By comparing the observed changes in peak heights and spectral positions (as well as the emergence of new peaks) with theoretical calculations, the authors identify particular features indicative of progress along the reaction coordinate from CHD to HT. Temporal lineouts of features at particular photon energies then yield dynamical information about the ring-opening reaction.

For example, Figure 15.10 shows the differential absorption signal at 282.2 eV as a function of pump–probe delay. The emergent peak at this photon energy has been identified as a transition from the 1s carbon electron to a valence orbital of mixed character ($1s \rightarrow 2\pi/1\pi^*$) and is associated with motion around the conical intersection between the excited 2A state and the ground state. The center panel of Figure 15.11 shows the orbitals corresponding to this mixed state, where the 2π (HOMO) and $1\pi^*$ (LUMO) orbitals become nearly degenerate. The delayed rise of the 282.2 eV signal compared to time zero seen in Figure 15.10 is indicative of the time required for the wave packet to cross from the 1B to the 2A state (60 ± 20 fs). The decay of this signal measures the time it takes to pass through the conical intersection and return to the ground state (110 ± 60 fs).

---

[2]NEXAFS are atom-specific absorption measurements near the X-ray energy required to remove an inner shell electron. In this case, the measurements are probed below the ionization energy, promoting an electron from a carbon K shell (1s orbital) to valence orbitals involved in the internal conversion dynamics.

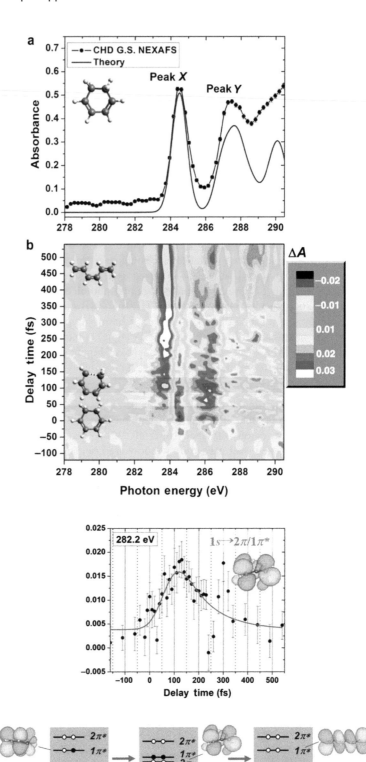

Figure 15.9 (a) Static X-ray absorption spectrum near the carbon K-edge for CHD (both experiment and theory). (b) Transient differential X-ray absorption spectrum as a function of pump–probe delay. Figure from Ref. [122].

Figure 15.10 Temporal dynamics of differential absorption at 282.2 eV. Figure adapted from Ref. [122].

Figure 15.11 Orbital diagrams for participating states in the CHD-HT ring-opening reaction. Figure adapted from Ref. [122].

15.2.5 Time-Resolved Ionization Spectroscopy

As a final experimental approach, we consider gas-phase, time-resolved ionization spectroscopy (TRIS) [123]. Ionization-based measurements generally provide the best time resolution, as there is essentially no dispersion propagating through gas-phase samples, there is no space charge effect as with electron pulses, and producing very short optical pulses with high flux is easier than very short X-ray pulses. Furthermore, the signal to noise is typically excellent, since the intensity of the probe pulse can be adjusted to achieve close to unit ionization probability (and ion detection can also be carried out with high efficiency). However, while ionization spectroscopy can provide excellent time resolution for a particular reaction pathway, there is rarely detailed information about the pathway taken (the signal is often a single number at each time delay). To interpret the measurements, calculations and/or other measurements must provide a preliminary understanding of the dynamics and a fairly well-established pathway.

As with the other approaches, the experiment uses a UV pump pulse (270 nm, 11 fs) to launch a vibrational wave packet on the excited 1B state. A strong-field probe pulse (810 nm, 12 fs) then ionizes the molecule at different time delays. The laser pulses interact with the molecules in a vacuum chamber equipped with a time-of-flight mass spectrometer and microchannel plate detector for measuring the ion yield. Panel (a) of Figure 15.12 plots the parent ion yield ($C_6H_8^+$) as a function of pump–probe delay. The data show a sharp increase in ion yield in going from negative to positive delays, followed by a decay with multiple features including a two time-step exponential decay and weak oscillations. The sharp increase in ion yield after time zero is due to the fact that the ionization potential in the excited state is much lower than from the ground state, leading to a higher yield after excitation by the pump pulse (the probe pulse intensity can be adjusted to minimize ionization from the ground state). Then, as the wave packet moves down the excited-state PES and undergoes internal conversion, the molecule progresses towards geometries where the ionization potential increases (leading to decay of the parent ion yield). Oscillations in the parent ion yield provide signatures of active vibrational modes along the reaction coordinate, and confirm the importance of these as predicted by theory.

The fit to the data in Figure 15.12 is modeled as a sum of exponentials convolved with the impulse response function of the apparatus (e.g. a Gaussian function representing the cross correlation of pump and probe pulses). In addition, the oscillations about the exponential decays are modeled using sine-wave modulations on top of the exponential decays. The initial decay of 21 fs (noted as L_1 in Figure 15.12) is assigned to the timescale for the wave packet to move from the Franck-Condon (FC) region to the roughly flat portion of the PES near the crossing between 1B and 2A states (see Figure 15.13). The subsequent decay of 35 fs is assigned to the transition between the 1B and 2A states (leading to a total time of 56 fs to reach the 2A state).

It is worth noting that these assignments are not obvious from the measurements alone, but rather are based on detailed calculations and comparison with prior experimental measurements. For example, oscillation frequencies observed during each stage of the ion yield decay can be compared to known vibrational modes at relevant locations on the excited-state PESs in an effort to confirm the interpretation. Panel (b) of Figure 15.12 shows the L_1 contribution to the decay, including a Fourier transform of the oscillations. The observed 950 cm^{-1} and 1430 cm^{-1} frequencies likely correspond to C–C and C=C stretches, which are expected to play a role in the initial motion away from the FC region. Similarly, the 310 cm^{-1} and 630 cm^{-1} oscillations

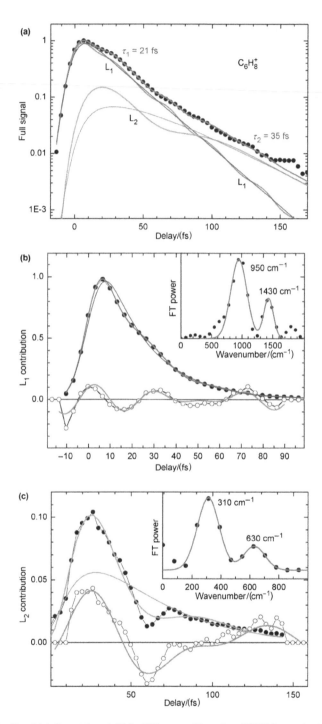

Figure 15.12   Panel (a): Parent ion yield for UV-pump, IR-probe of CHD isomerization. The signal is fit to a rate equation model which consists of a sum of exponential decays labeled L_1 and L_2 (shown both with and without the sinusoidal modulation). Panels (b) and (c) show the L_1 and L_2 contributions, along with the model. Insets plot the Fourier transforms of the modulated signals. Figure from Ref. [123].

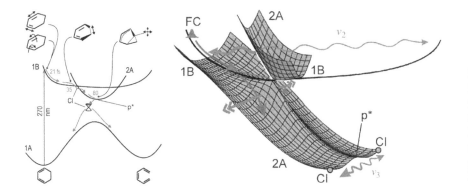

Figure 15.13 (See color insert.) Left: states and timescales involved in the isomerization of CHD as measured by ionization spectroscopy. Right: two-dimensional subspace of the PESs, along with the wave packet path and relevant vibrational modes. Figure adapted from Ref. [123].

in the subsequent L_2 decay (panel c) are assigned to the fundamental and second harmonic of ring puckering or twisting motion (C=C torsion). This is illustrated in the right panel of Figure 15.13, with v_2 corresponding to the C=C torsion mode. Analysis of the 79-amu fragment ion (not shown) uncovers the 80 fs timescale shown in Figure 15.13. This is assigned to decay through the conical intersection connecting the 2A state and the ground state. The total time to return to the ground electronic state is thus approximately 140 fs.

15.3 CONCLUSIONS

In concluding this chapter (and the book), we briefly compare the information available from different techniques and discuss how, in combination, they yield a more complete picture of the dynamics. Spectroscopic techniques such as TRIS and TA provide excellent time resolution using short optical pulses and can directly measure how state energies change in time. However, they both provide rather indirect information about changes in molecular structure. With TA, absorption spectra are not unique, and the connection between an absorption spectrum and molecular structure is not obvious and must come from calculations and auxiliary measurements. With TRIS, significant calculational effort is required to relate time-dependent ionization yields to changes in molecular structure. This is especially true when measuring only the total ion or electron yield (resolving the photoelectron spectrum provides information on energy differences between states, akin to spectroscopic measurements).

X-ray absorption spectroscopy can provide time-resolved information on valence orbital occupancy and geometry-dependent transition frequencies. While not a direct measurement of structure, it does provide a more "chemical" perspective on the dynamics, since orbital occupation encodes how electrons move and reorganize in the molecule, while energy shifts are related to the structure. Comparing absorption measurements with theory and diffractive probes of molecular structure is facilitated by the fact that electronic structure calculations can provide orbital occupations and energies for different geometries.

In contrast to spectroscopic measurements, UED and X-ray diffraction provide information that is more directly related to molecular structure. This usually comes with a cost of higher experimental overhead and/or poorer time resolution. With nonrelativistic UED, the problem is twofold: space charge leads to spreading of the electron beam as it propagates, and the group-velocity mismatch between electrons and photons in the sample leads to a smearing over time delays. While in principle, diffraction data can be directly transformed to provide detailed structural information, this requires a perfectly aligned sample. Thus, diffraction experiments also rely on comparison with calculations (e.g. radial distribution functions) to connect experimentally measured results to molecular structure.

Comparing the X-ray diffraction and spectroscopic (esp. TRIS) results provides interesting differences in the measured timescales. TRIS measured a timescale of approximately 50 fs for the wave packet to go from the FC region to the first PES intersection, with a total time of 140 fs required to relax back down to the HT (or CHD) ground state. These times are different from the 80 fs measured for structural changes using diffraction. At least in part, the different timescales for the two measurements reflect the fact that structural and chemical/spectroscopic changes are not necessarily synchronous. For instance, near a conical intersection, the electronic character of a state can change quite dramatically over a short timescale without much change in molecular structure. Conversely, as a molecule dissociates on a single electronic state, the character may not change very much while the structure changes significantly. Thus, it can be very useful to have multiple measurement approaches to characterize dynamics that involve both structural and chemical/spectroscopic changes. Together, the diffraction and spectroscopic measurements in CHD provide a more complete picture of the isomerization dynamics, with each highlighting different aspects of the reaction.

As seen in this chapter, a combination of multiple techniques can provide confirmation of a particular interpretation or produce a more comprehensive picture. For instance, the UED and X-ray diffraction measurements described earlier, in conjunction with electronic structure and trajectory surface-hopping calculations, provide detailed information about changes in molecular structure during isomerization of CHD into HT. On the other hand, the TA and TRIS experiments measured specific timescales for each step in the reaction process. Finally, the X-ray absorption measurements illustrate how the electronic structure of the molecule changes with time. Putting all of these measurements together provides a relatively complete picture of how both the electrons and nuclei move during the ring-opening reaction.

15.4 ADDITIONAL RESOURCES

In closing, we note that the field of time-resolved spectroscopy has been developing for some time, and we are indebted to the work of many others before us. A number of the theories we discuss were developed or described previously, and we encourage interested readers to pursue these (and other) sources for more detailed discussions of particular topics. David Tannor's book *Introduction to Quantum Mechanics: A Time-Dependent Perspective* inspired us to interpret measurements using both analytic and numerical methods, and it contains a much more extensive discussion of the time-domain approach to quantum mechanics. *Concepts and Methods of 2D Infrared Spectroscopy*, by Peter Hamm and Martin Zanni, is a detailed treatment of two-dimensional spectroscopy in the infrared region of the spectrum, while *Principles of Nonlinear Optical Spectroscopy* by Shaul Mukamel provides a comprehensive development of the coherent non-linear spectroscopies that form the foundation for many of the approaches discussed in the book. Andrei Tokmakoff's *Time-Dependent Quantum Mechanics and Spectroscopy*, along with his *Nonlinear and*

Two-Dimensional Spectroscopy Notes, describe time-domain approaches for studying dynamics in condensed-phase systems. On the technical side, there are a number of books covering the principles and applications of ultrafast laser pulses, including *Ultrafast Optics* by Andrew Weiner, *Ultrashort Laser Pulse Phenomena* by Jean-Claude Diels and Wolfgang Rudolph, and *Fundamentals of Attosecond Optics* by Zenghu Chang.

APPENDIX A

Quantum Mechanics Essentials

A.1 FOURIER TRANSFORMS

Fourier transforms are widely used in both analytic and numerical approaches in physics and chemistry. A Fourier transform relates two different representations of the same function. For example, two commonly used Fourier transform pairs include functions of time and frequency $[f(t)/f(\omega)]$, or functions of position and spatial frequency $[f(x)/f(k)]$, where t and ω are conjugate variables (or x and p). Mathematically, the Fourier transform of a time-dependent function $f(t)$ is given as

$$f(\omega) = \frac{1}{\sqrt{2\pi}} \int_{-\infty}^{+\infty} f(t)e^{-i\omega t}dt, \tag{A.1}$$

where $f(\omega)$ is the frequency-domain representation of the function (we note there are varying conventions for the leading factor). $f(t)$ and $f(\omega)$ represent the same function, just in different domains. The quantity $|f(\omega)|^2$ is often referred to as the "spectrum" of the signal. To switch back from $f(\omega)$ to $f(t)$, one applies the inverse Fourier transform:

$$f(t) = \frac{1}{\sqrt{2\pi}} \int_{-\infty}^{+\infty} f(\omega)e^{+i\omega t}d\omega. \tag{A.2}$$

In quantum mechanics, it is typically the position and momentum space representations of the wave function that are relevant. These Fourier transform pairs are usually defined to be (where we have used p instead of k since $p = k$ in atomic units)

$$\psi(x) = \frac{1}{\sqrt{2\pi}} \int_{-\infty}^{+\infty} \psi(p)e^{+ipx}dp \tag{A.3a}$$

$$\psi(p) = \frac{1}{\sqrt{2\pi}} \int_{-\infty}^{+\infty} \psi(x)e^{-ipx}dx. \tag{A.3b}$$

When one speaks of "the wave function," it tends to be the position-space version that is implied, as it is easier to visualize in terms of a physical probability density. However, the momentum-space version is equally valid, and considering both forms of the wave function often provides valuable physical insight (as well as numerical efficiency, see Section 5.2).

There are many important properties of Fourier transforms. One of these is known as the convolution theorem, and it relates the Fourier transform of a product to the convolution of the Fourier transforms. For two functions $f(t)$ and $g(t)$:

$$\mathcal{F}\{f(t)g(t)\} = \frac{1}{\sqrt{2\pi}} \int_{-\infty}^{+\infty} f(t)g(t)e^{-i\omega t} dt = f(\omega) * g(\omega), \qquad (A.4)$$

where $\mathcal{F}$ is the Fourier transform, and the $*$ operation represents a convolution, which is defined by

$$f(\omega) * g(\omega) = \int_{-\infty}^{+\infty} f(\omega')g(\omega - \omega')d\omega'. \qquad (A.5)$$

This result can be obtained by inserting the Fourier transform expressions for $f(t)$ and $g(t)$ given by Equation A.1.

A.1.1 Gaussian Wave Function Example

Consider the free-particle example discussed in Section 5.3. We saw that constraining the initial position of the particle in position led to a spreading of the wave function with time. This was interpreted in terms of the Heisenberg Uncertainly relation between x and p. We can formalize this behavior by considering the position and momentum representations of the wave function explicitly. Consider an initial, spatial wave function of a normalized Gaussian:

$$\Psi(x, t=0) = \left(\frac{2a}{\pi}\right)^{1/4} e^{-ax^2}. \qquad (A.6)$$

The momentum-space version of the wave function is calculated to be

$$\begin{aligned}\Psi(p,0) &= \frac{1}{\sqrt{2\pi}} \int_{-\infty}^{+\infty} \left(\frac{2a}{\pi}\right)^{1/4} e^{-ax^2} e^{-ipx} dx \\ &= \left(\frac{1}{2\pi a}\right)^{1/4} e^{-p^2/4a}.\end{aligned} \qquad (A.7)$$

As expected, the Fourier transform of a Gaussian is a Gaussian. The constant a, which determines the width of the Gaussian, appears in the numerator of the exponent in $\Psi(x,0)$ but the denominator in $\Psi(p,0)$. Therefore, changing a will change the widths of the two functions in an inverse manner, consistent with our discussion in Section 5.3. Specifically, narrowing the width in space (increasing a) to create a more localized position results in a larger width in the momentum wave function (and therefore more high-momentum components).

A.1.2 Discrete Fourier Transforms

When analyzing experimental data or performing numerical calculations, one often wishes to compute the Fourier transform of a discretely sampled data set. For example, suppose we have a spatial wave function $\psi(x)$ that has been discretely sampled at set of N points $x_j = j\Delta x$, where $j = 0, 1, 2, ..., N-1$ (referred to as $\psi(x_j)$). An example is given in Figure A.1. The representation of the momentum-space wave function is given by the discrete generalization of Equation A.3b:

$$\psi(p_k) = \sum_{j=0}^{N-1} \psi(x_j)e^{-i\, p_k x_j} \Delta x, \qquad (A.8)$$

where p_k represents the (discrete) momentum component and $\Delta x \equiv x_{j+1} - x_j$ is the point-to-point separation on the discretely sampled x-axis (assumed to be constant).

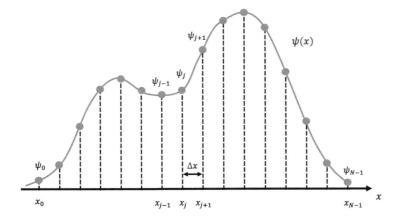

Figure A.1 Discretely sampled function for Fourier transform. The continuous function $\psi(x)$ is sampled at N total points.

The convention is for the inverse transform to have the $1/N$ normalization factor. While the discrete Fourier transform is readily implemented numerically in mathematical software packages, it is worth commenting on its use, as a discretely sampled initial function imposes constraints on the resulting transform. In particular, there is a minimum effective wavelength one can infer from the discretely sampled function $\psi(x_j)$, since one needs at least two samples per period of the wave for a meaningful reconstruction. This corresponds to a minimum wavelength $\lambda_{min} = 2\Delta x$, and therefore a maximum possible momentum component (in atomic units) of $|p_{max}| = 1/\lambda_{min} = 1/2\Delta x$. Note that this constraint is not present in the analytic case where one assumes a continuous sampling along the axis from $-\infty$ to $+\infty$.

Furthermore, the p_j in Equation A.8 are yet to be determined. In the continuous case, p in the exponent of Equation A.3b is simply the variable that appears in $\psi(p)$. In the discrete case, however, it is not immediately obvious at what values p_j one obtains the discrete function $\psi(p_j)$. In fact, numerical routines for computing discrete Fourier transforms typically operate only on the list of numbers representing the initial function, and do not concern themselves with the x_j axis when Fourier transforming the list of values representing $\psi(x_j)$. Recognizing there is a maximum value of momentum p_{max} that can come from the calculation, one typically defines the momentum axis with points at

$$p_j = \frac{j}{N\Delta x} \text{ for } j = -\frac{N}{2}, ..., +\frac{N}{2}, \tag{A.9}$$

where N is assumed even. This definition results in a uniformly spaced axis with extreme values of $\pm p_{max}$.

A.2 THE TIME-INDEPENDENT PICTURE

Throughout most of the book, we have used the time-dependent wave function $\Psi(x,t)$ as the fundamental description of the state. However, both the theoretical development of quantum mechanics and early spectroscopy experiments more naturally fit into a time-independent picture, a primary reason being that the Schrödinger equation is separable into spatial and temporal components when the potential is time independent. Although this treatment is included in most undergraduate quantum courses, for completeness we outline it here, starting with the time-dependent Schrödinger equation in atomic units

$$i\frac{\partial \Psi(x,t)}{\partial t} = -\frac{1}{2m}\frac{\partial^2 \Psi(x,t)}{\partial x^2} + V(x,t)\Psi(x,t), \qquad (A.10)$$

where m is the particle's mass, and, for simplicity, we write the expression in one spatial dimension. We look for solutions of the form

$$\Psi(x,t) = \psi(x)\phi(t). \qquad (A.11)$$

In the case when the potential is *independent* of time $(V \neq V(t))$, the time-dependent Schrödinger equation is transformed into two separate differential equations: one for $\phi(t)$ and one for $\psi(x)$. The equation for $\phi(t)$ can be immediately solved to yield

$$\phi(t) \propto e^{-iEt}, \qquad (A.12)$$

where E is a separation constant that will turn out to be the energy. The equation for $\psi(x)$ cannot be solved until a potential $V(x)$ is specified in the time-independent Schrödinger equation:

$$-\frac{1}{2m}\frac{\partial^2 \psi(x)}{\partial x^2} + V(x)\psi(x) = E\psi(x). \qquad (A.13)$$

This is often written more compactly by defining the Hamiltonian operator $\hat{H}$ such that:

$$\hat{H}\psi = E\psi. \qquad (A.14)$$

In the language of linear algebra, the eigenvalues of the Hamiltonian operator represent the energies of the system. For a given potential $V(x)$, one solves the time-independent equation and finds a set of solutions corresponding to different energies. These solutions are the energy eigenstates, $\psi_n(x)$, and are referred to as "stationary states," since both the probability density $|\Psi(x,t)|^2$ and the expectation value of any observable (represented by the operator $\hat{A}$) do not depend on time:

$$\langle A \rangle = \int_{-\infty}^{+\infty} \psi_n^*(x)e^{+iE_n t}\hat{A}\psi_n(x)e^{-iE_n t}dx$$
$$= \int_{-\infty}^{+\infty} \psi_n^*(x)\hat{A}\psi_n(x)dx. \qquad (A.15)$$

In addition, the linearity of the time-dependent Schrödinger equation ensures that all linear combinations of eigenstates are also solutions. This, along with the fact that the $\psi_n(x)$'s are assumed complete, allows one to write *any* solution to the time-dependent Schrödinger equation as a superposition of the energy eigenstates:

$$\Psi(x,t) = \sum_{n=1}^{\infty} c_n \psi_n(x) e^{-iE_n t}, \qquad (A.16)$$

where c_n's are constants that come from the initial condition. Such a superposition of states is often referred to as a "wave packet," implying a summation over the different constituent states. Figure A.2 illustrates how the superposition of eigenstates with different phases produces wave packets localized at different points in the potential. Each term in Equation A.16 has a temporal phase factor that evolves at a different rate (assuming nondegenerate energies). Therefore, in general, both the probability density and expectation values of operators change in time for such a superposition state.

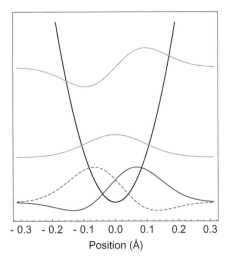

Figure A.2 Illustration of how eigenstates can be combined to form a wave packet. The wave functions for the first two states of the harmonic oscillator are shown in gray, along with the potential. At the bottom, the sum (solid) and difference (dashed) of the states are plotted, corresponding to a wave packet concentrated on the right and left side of the potential, respectively.

All these facts make the separation of variables approach very neat from a theoretical perspective, and this is how quantum mechanics is typically treated in most undergraduate textbooks. Early spectroscopy experiments using continuous-wave sources (first discharge lamps and later lasers) naturally measured the *energy differences* between the stationary states, which also fit nicely into the time-independent framework. For example, a single light-source, weak-field, absorption spectroscopy experiment shines light on a sample and measures the transmitted light intensity as a function of frequency. Dips in the spectral intensity indicate absorption of light due a transition in the quantum system from one state to another. By measuring the frequencies of the spectral dips, one measures the energy spectrum of the stationary states (the eigenvalues of the Hamiltonian). Knowing the eigenvalues allows easy comparison with quantum theory.

However, the time-independent approach minimizes the role of the time-dependent wave function, $\Psi(\mathbf{r},t)$, which represents the full quantum state of the system at any given time (now written in three spatial dimensions). Ideally, one could picture $\Psi(\mathbf{r},t)$ evolving in time separate from the underlying eigenstates, with a probe interaction measuring some aspects of $\Psi(\mathbf{r},t)$ at a specific, later time. It may be that the result of this measurement is conveniently interpreted in terms of the stationary states; when this is the case, we do so. However, often this measurement is more easily visualized outside the stationary state basis. Using computers, one can directly calculate $\Psi(\mathbf{r},t)$ numerically given a potential energy function $V(\mathbf{r})$, essentially bypassing the need to first solve the problem in terms of the stationary states (see Section 5.2 for a discussion of numerical implementation). Throughout the book, we attempt to keep the focus on the time-dependent wave function, $\Psi(\mathbf{r},t)$, relating it to the time-independent stationary states $\psi_n(\mathbf{r})$ when it's natural for the interpretation.

A.3 TIME-DEPENDENT PERTURBATION THEORY

It is desirable to derive analytic expressions to mathematically describe experimental results. This is most readily accomplished using perturbation theory, where the system

Hamiltonian H_0 is time-independent, and a time-dependent perturbation H_1 is due to an applied laser field (or some other interaction). As is typical, we use λ as a parameter that can range from zero to one (serving to tune the strength of the interaction); it also provides a convenient method for collecting terms of similar order. We write the total Hamiltonian for the system as

$$H = H_0 + \lambda H_1, \tag{A.17}$$

and we seek solutions to the time-dependent Schrödinger equation for the full Hamiltonian with a wave function of the form:

$$\Psi(x,t) = \Psi^{(0)}(x,t) + \lambda \Psi^{(1)}(x,t) + \lambda^2 \Psi^{(2)}(x,t) + \cdots = \sum_{n=0} \lambda^n \Psi^{(n)}(x,t), \tag{A.18}$$

where the superscript denotes the "order of correction" to the wave function from the unperturbed states (for simplicity, we assume only one spatial dimension). Inserting this into the Schrödinger equation, we have

$$i\frac{\partial \Psi(x,t)}{\partial t} = (H_0 + \lambda H_1)\Psi(x,t)$$

$$i\sum_{n=0} \lambda^n \frac{\partial \Psi^{(n)}(x,t)}{\partial t} = H_0 \sum_{n=0} \lambda^n \Psi^{(n)}(x,t) + H_1 \sum_{n=0} \lambda^{n+1} \Psi^{(n)}(x,t). \tag{A.19}$$

Since the expansion holds for arbitrary values of λ, we can equate terms with equal powers of λ:

$$i\frac{\partial \Psi^{(0)}(x,t)}{\partial t} = H_0 \Psi^{(0)}(x,t) \tag{A.20a}$$

$$i\frac{\partial \Psi^{(1)}(x,t)}{\partial t} = H_0 \Psi^{(1)}(x,t) + H_1 \Psi^{(0)}(x,t) \tag{A.20b}$$

$$i\frac{\partial \Psi^{(2)}(x,t)}{\partial t} = H_0 \Psi^{(2)}(x,t) + H_1 \Psi^{(1)}(x,t). \tag{A.20c}$$

The solution to Equation A.20a for $\Psi^{(0)}(x,t)$ is straightforward, as it is nothing more than the unperturbed solutions with the usual time propagator in front (H_0 is independent of time):

$$\Psi^{(0)}(x,t) = e^{-iH_0 t}\Psi^{(0)}(x,0). \tag{A.21}$$

The solution for $\Psi^{(1)}(x,t)$ is

$$\Psi^{(1)}(x,t) = \frac{1}{i}e^{-iH_0 t}\int_0^t dt' e^{iH_0 t'} H_1(t')\Psi^{(0)}(x,t') + e^{-iH_0 t}\Psi^{(1)}(x,0), \tag{A.22}$$

which can be differentiated to yield Equation A.20b.

In the typical case when the perturbation is off before some particular time, we assume that at $t = 0$ there is no higher-order correction to the wave function: $\Psi^{(1)}(x,t=0) = 0$. Making use of Equation A.21, we can rewrite our solution for $\Psi^{(1)}(x,t)$ as

$$\Psi^{(1)}(x,t) = \frac{1}{i}\int_0^t dt' e^{-iH_0(t-t')} H_1(t')e^{-iH_0 t'}\Psi^{(0)}(x,0). \tag{A.23}$$

Similarly, we can write out a formal solution for $\Psi^{(2)}(x,t)$ as

$$\Psi^{(2)}(x,t) = \frac{1}{i}e^{-iH_0 t}\int_0^t dt' e^{iH_0 t'} H_1(t')\Psi^{(1)}(x,t') + e^{-iH_0 t}\Psi^{(2)}(x,0). \tag{A.24}$$

If we substitute in our solution for $\Psi^{(1)}(x,t)$ and set $\Psi^{(2)}(x,0) = 0$, Equation A.24 can be written as

$$
\Psi^{(2)}(x,t) = \frac{1}{i^2} \int_0^t dt' \int_0^{t'} dt'' e^{-iH_0(t-t')} H_1(t') e^{-iH_0(t'-t'')}
$$
$$
\times H_1(t'') e^{-iH_0 t''} \Psi^{(0)}(x,0). \tag{A.25}
$$

While one can continue to higher orders in the expansion, it is rare that one needs to go beyond second order. We note that time-dependent perturbation theory does not preserve the normalization of the wave function, and this can lead to some counter-intuitive results (e.g. there is population in the excited state, even though the ground state is undepleted). It is especially important to remember this fact when considering energy conservation.

A.4 MATRIX NOTATION FOR MOLECULAR SYSTEMS

The matrix formalism of quantum mechanics, where the Hamiltonian is written as a square matrix and the system is represented as a state vector, provides a compact approach for keeping track of interactions in multistate systems such as molecules. In particular, we consider a diatomic molecule with two Born–Oppenheimer (BO) electronic states labeled a and b. The state vector $\Psi(x,t)$ is a column vector whose elements represent the *vibrational* wave functions in each of the electronic states:

$$
\Psi(x,t) \rightarrow \begin{pmatrix} \chi_a(R,t) \\ \chi_b(R,t) \end{pmatrix}, \tag{A.26}
$$

where, as usual, the spatial coordinate R represents the internuclear separation. As discussed in Section 6.2, this notation implies that the amplitude coefficients for the electronic potential energy surfaces (PESs) are contained in $\chi_a(R,t)$ and $\chi_b(R,t)$, the full, time-dependent vibrational wave functions on the two BO states.

In matrix notation, the zeroth-order state $\Psi^{(0)}(x,0)$ in Equation A.23 is simply a column vector whose elements are the vibrational wave functions in each of the electronic states at time $t = 0$. If the molecule is initially assumed to be in the ground electronic state a, the lower element is zero:

$$
\Psi^{(0)}(x,0) \rightarrow \begin{pmatrix} \chi_a(R,0) \\ 0 \end{pmatrix}. \tag{A.27}
$$

The unperturbed Hamiltonian operator H_0 is written in matrix notation as

$$
H_0 \rightarrow \begin{pmatrix} H_a & 0 \\ 0 & H_b \end{pmatrix}, \tag{A.28}
$$

where H_a is the *nuclear* Hamiltonian for evolution of the vibrational wave function on the PES corresponding to electronic state a (similarly for H_b). Note that the matrix operator H_0 must be exponentiated in Equation A.23. Defining the exponential of an operator through its Taylor series expansion, we see that we need to evaluate powers of the matrix H_0. For a diagonal matrix such as H_0, raising the matrix to any power n simply results in each element to the power n:

$$\begin{pmatrix} H_a & 0 \\ 0 & H_b \end{pmatrix}^n = \begin{pmatrix} H_a^n & 0 \\ 0 & H_b^n \end{pmatrix}. \tag{A.29}$$

Therefore, the exponential of a diagonal matrix can be written simply as a matrix of exponentials:

$$e^{-iH_0t} = \begin{pmatrix} e^{-iH_at} & 0 \\ 0 & e^{-iH_bt} \end{pmatrix}, \tag{A.30}$$

so that the e^{-iH_at} operator will be applied to the vibrational wave function in PES a in the state column vector (and e^{-iH_bt} to the wave function in PES b).

The interaction Hamiltonian H_1 in Equation A.23 comes from the applied light field and is typically assumed to resonantly couple the two BO states a and b with a single-photon transition. In this case, the matrix representing H_1 is off-diagonal and given in the dipole approximation as (see Section 3.2.2):

$$H_1(t) \rightarrow \begin{pmatrix} 0 & -\mu_{ab}E_1(t) \\ -\mu_{ba}E_1(t) & 0 \end{pmatrix}, \tag{A.31}$$

where $E_1(t)$ is the (time-dependent) electric field associated with the first pulse that interacts with the system (the pump). Putting these into the first-order perturbative result of Equation A.23 yields

$$\begin{pmatrix} \chi_a^{(1)}(R,t) \\ \chi_b^{(1)}(R,t) \end{pmatrix} = \frac{1}{i} \int_0^t dt' \begin{pmatrix} e^{-iH_a(t-t')} & 0 \\ 0 & e^{-iH_b(t-t')} \end{pmatrix}$$
$$\times \begin{pmatrix} 0 & -\mu_{ab}E_1(t') \\ -\mu_{ba}E_1(t') & 0 \end{pmatrix} \begin{pmatrix} e^{-iH_at'} & 0 \\ 0 & e^{-iH_bt'} \end{pmatrix} \begin{pmatrix} \chi_a(R,0) \\ 0 \end{pmatrix}. \tag{A.32}$$

Performing the matrix multiplication results in the following expression:

$$\begin{pmatrix} \chi_a^{(1)}(R,t) \\ \chi_b^{(1)}(R,t) \end{pmatrix} = \frac{1}{i} \int_0^t dt' \begin{pmatrix} 0 \\ e^{-iH_b(t-t')}(-\mu_{ba}E_1(t'))e^{-iH_at'}\chi_a(R,0) \end{pmatrix}. \tag{A.33}$$

As expected, to first-order in perturbation theory, there is no change in the wave function on the ground electronic state. The expression for the first-order change in the excited state wave function is given as

$$\chi_b^{(1)}(R,t) = \frac{1}{i} \int_0^t dt' e^{-iH_b(t-t')} \left(-\mu_{ba}E_1(t')\right) e^{-iH_at'} \chi_a(R,0). \tag{A.34}$$

A.5 MOLECULAR ORBITALS

Figures A.3 and A.4 illustrate the formation of molecular orbitals from atomic ones. Figure A.3 shows how the addition of two atomic s orbitals can form both σ (bonding) and σ^* (antibonding) orbitals, depending on whether they add in or out of phase (like in Figure 2.3). Similarly, Figure A.4 uses atomic p orbitals to form molecular π and π^* orbitals. The case of π orbital formation is complicated by the fact that atomic p orbitals can lie either along the interatomic axis (top two rows) or orthogonal to that direction (bottom four rows).

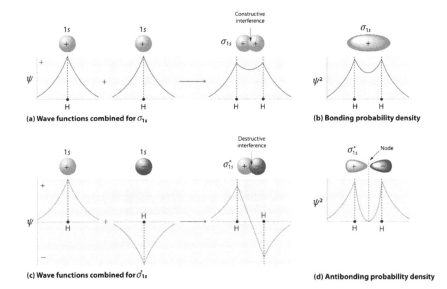

(a) Wave functions combined for σ_{1s}

(b) Bonding probability density

(c) Wave functions combined for σ_{1s}^*

(d) Antibonding probability density

Figure A.3 Illustration of how molecular σ orbitals are formed from two atomic s orbitals (using H_2 as an example). Depending on the relative phase of the atomic orbitals, the resulting molecular orbital can be either bonding (σ) or antibonding (σ^*). Figure from Ref. [124].

A.6 DRESSED-STATE TRANSFORMATIONS

In this section we consider two examples where the matrix formulation of Section A.4 is particularly useful. The first diagonalizes the Hamiltonian matrix for a model two-state system coupled by an interaction potential, while the second builds on the first to diagonalize the time-evolution operator to a form that is readily applied to numerical integration. We draw upon both of these results in main the text.

A.6.1 Two-Level System

Even when studying complex molecules, one often finds that a two state model provides substantial insight into the dynamics. Here we consider a two-level system with (uncoupled) energies E_1 and E_2, coupled by an interaction potential V. The Hamiltonian for the system, including the coupling, is given in general matrix form by

$$H = \begin{bmatrix} E_1 & V \\ V & E_2 \end{bmatrix}, \tag{A.35}$$

where H is assumed to operate on a state vector of the form $\Psi = (\Psi_1; \Psi_2)$ in the time-dependent Schrödinger equation. We begin by assuming an interaction potential V with a harmonic time dependence: $V = V_0 \cos(\omega_0 t)$. An example could be an oscillating applied field, where $V = -\mu E_0 \cos(\omega_0 t) \equiv -\chi_{\text{Rabi}} \cos(\omega_0 t)$. If the frequency ω_0 is close to the energy difference between levels 1 and 2, one can carry out a change of basis that transforms the energies along the diagonals to phases in the off-diagonals, make the rotating-wave approximation (RWA), and transform the phases back to the diagonal elements with a similar transformation.

The first transformation is given by

$$\Psi_1' = \Psi_1 e^{+iE_1 t} \tag{A.36a}$$

$$\Psi_2' = \Psi_2 e^{+iE_2 t}. \tag{A.36b}$$

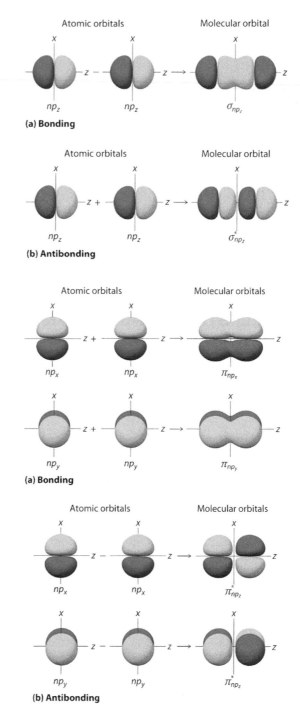

Figure A.4   Illustration of how molecular π orbitals are formed from two atomic p orbitals. The atomic p orbitals can be oriented along the internuclear axis (top two rows), or orthogonal to the internuclear axis (bottom four rows). As with σ orbitals, the relative phase between the atomic orbitals determines whether the molecular orbitals are bonding (π) or antibonding (π^*). Figure from Ref. [124].

After carrying out this transformation and discarding the counterrotating term in the RWA, we transform the phases in the off-diagonal terms to the diagonals using the transformation given by

$$\tilde{\Psi}_1 = \Psi'_1 e^{+i(E_2 - E_1 - \omega_0)t/2} \tag{A.37a}$$

$$\tilde{\Psi}_2 = \Psi'_2 e^{-i(E_2 - E_1 - \omega_0)t/2}. \tag{A.37b}$$

The Hamiltonian for the near-resonant harmonic case can then be written as

$$H = \frac{1}{2} \begin{bmatrix} -\Delta & V_0 \\ V_0 & \Delta \end{bmatrix} = \frac{1}{2} \begin{bmatrix} -\Delta & -\mu E_0 \\ -\mu E_0 & \Delta \end{bmatrix}, \tag{A.38}$$

where $\Delta \equiv E_2 - E_1 - \omega_0$. Note that in the case when V has no harmonic time dependence (i.e. $\omega_0 = 0$, with $V \equiv V_{DC}$), there is an additional factor of two in the off-diagonal terms of the Hamiltonian since one can no longer make the RWA:

$$H_{DC} = \begin{bmatrix} -\frac{\Delta}{2} & V_{DC} \\ V_{DC} & \frac{\Delta}{2} \end{bmatrix}. \tag{A.39}$$

The Hamiltonians of Equation A.38 and A.39 can be diagonalized by a rotation in the Hilbert space. To accomplish this, we rotate by an angle θ, and then solve for the value of θ that leads to zeros in the off-diagonal entries. The Hamiltonian matrix can be represented in the eigenstate basis using the similarity transformation $H' = UHU^{-1}$, where U is the two-dimensional rotation matrix:

$$\begin{aligned} H' &= \frac{1}{2} \begin{bmatrix} \cos\theta & -\sin\theta \\ \sin\theta & \cos\theta \end{bmatrix} \begin{bmatrix} -\Delta & V_0 \\ V_0 & \Delta \end{bmatrix} \begin{bmatrix} \cos\theta & \sin\theta \\ -\sin\theta & \cos\theta \end{bmatrix} \\ &= \frac{1}{2} \begin{bmatrix} -\Delta\cos^2\theta - 2V_0\sin\theta\cos\theta + \Delta\sin^2\theta & -2\Delta\cos\theta\sin\theta + V_0\cos^2\theta - V_0\sin^2\theta \\ -2\Delta\cos\theta\sin\theta + V_0\cos^2\theta - V_0\sin^2\theta & +2V_0\cos\theta\sin\theta + \Delta\cos^2\theta - \Delta\sin^2\theta \end{bmatrix}. \end{aligned} \tag{A.40}$$

This matrix is diagonal if

$$-2\Delta\sin\theta\cos\theta + V_0(\cos^2\theta - \sin^2\theta) = -\Delta\sin(2\theta) + V_0\cos(2\theta) = 0, \tag{A.41}$$

which is satisfied when $\tan(2\theta) = \frac{V_0}{\Delta}$. The diagonal eigenvalues, $\lambda_\pm$, of the matrix H' can be found by setting $\tan(2\theta) = \frac{V_0}{\Delta}$, and considering a right triangle with sides V_0, Δ, and $\sqrt{V_0^2 + \Delta^2}$. In this case, $\sin(2\theta) = \frac{V_0}{\sqrt{\Delta^2 + V_0^2}}$, $\cos(2\theta) = \frac{\Delta}{\sqrt{\Delta^2 + V_0^2}}$, and the eigenvalues are given by

$$\lambda = \pm\sqrt{\Delta^2 + V_0^2}. \tag{A.42}$$

The eigenvectors of H are given by rotating the basis vectors by the angle θ:

$$|1'\rangle = \cos\theta |1\rangle + \sin\theta |2\rangle \tag{A.43a}$$

$$|2'\rangle = -\sin\theta |1\rangle + \cos\theta |2\rangle. \tag{A.43b}$$

A.6.2 Time-Evolution Operator for Numerical Implementation

In Section 6.2.2 we showed how one could numerically propagate the molecular wave function in the presence of an applied field. In that treatment, we relied on rewriting the off-diagonal potential matrix in a form more easily incorporated into a split-operator integration. Here we show how to rewrite the potential energy operator (with off-diagonal couplings) such that the exponentiated matrices have only diagonal terms. The approach builds on that of the two-level system considered in Section A.6.1.

We start by considering a unitary transformation, U, which diagonalizes the potential energy matrix:

$$\begin{pmatrix} \lambda_1 & 0 \\ 0 & \lambda_2 \end{pmatrix} = U \begin{pmatrix} V_{11} & V_{12} \\ V_{12} & V_{22} \end{pmatrix} U^{-1}. \tag{A.44}$$

Here we have assumed that the off-diagonal elements of the potential matrix are real and therefore equal. Given that U represents a rotation in a two-dimensional space, we can write U as

$$U = \begin{pmatrix} \cos\theta & -\sin\theta \\ \sin\theta & \cos\theta \end{pmatrix}. \tag{A.45}$$

Taking out the average energy, $(V_{11} + V_{22})/2$, from the diagonals and defining $\tilde{V} = (V_{22} - V_{11})/2$, our potential matrix can be written as

$$\begin{pmatrix} V_{11} & V_{12} \\ V_{12} & V_{22} \end{pmatrix} = \begin{pmatrix} -\tilde{V} & V_{12} \\ V_{12} & \tilde{V} \end{pmatrix} + \frac{V_{11} + V_{22}}{2}\mathbf{I}, \tag{A.46}$$

where $\mathbf{I}$ represents the identity matrix. Now, making use of Equation A.42, we can write the eigenvalues of our potential matrix as

$$\lambda_1 = -\sqrt{V_{12}^2 + \tilde{V}^2} + \frac{V_{11} + V_{22}}{2} \tag{A.47a}$$

$$\lambda_2 = +\sqrt{V_{12}^2 + \tilde{V}^2} + \frac{V_{11} + V_{22}}{2}. \tag{A.47b}$$

The exponential of the potential matrix is therefore

$$\exp\left(-i\begin{pmatrix} V_{11} & V_{12} \\ V_{12} & V_{22} \end{pmatrix}\Delta t\right) = U\begin{pmatrix} e^{-i\lambda_1\Delta t} & 0 \\ 0 & e^{-i\lambda_2\Delta t} \end{pmatrix}U^{-1}$$

$$= U\begin{pmatrix} e^{-i(V_{11}+V_{22})\Delta t/2} & 0 \\ 0 & e^{-i(V_{11}+V_{22})\Delta t/2} \end{pmatrix}\begin{pmatrix} e^{i\sqrt{D}\Delta t} & 0 \\ 0 & e^{-i\sqrt{D}\Delta t} \end{pmatrix}U^{-1}, \tag{A.48}$$

where $\sqrt{D} \equiv \sqrt{\tilde{V}^2 + V_{12}^2}$.

Continuing, this can be rewritten as

$$
\exp\left(-i\begin{pmatrix} V_{11} & V_{12} \\ V_{12} & V_{22} \end{pmatrix}\Delta t\right) = e^{-i(V_{11}+V_{22})\Delta t/2}
$$
$$
\begin{pmatrix} \cos(\sqrt{D}\Delta t) - i\sin(\sqrt{D}\Delta t) & 0 \\ 0 & \cos(\sqrt{D}\Delta t) + i\sin(\sqrt{D}\Delta t) \end{pmatrix} U^{-1}
$$
$$
U\left(\cos(\sqrt{D}\Delta t)\begin{pmatrix} 1 & 0 \\ 0 & 1 \end{pmatrix} - i\sin(\sqrt{D}\Delta t)\begin{pmatrix} 1 & 0 \\ 0 & -1 \end{pmatrix}\right) U^{1}
$$
$$
\left(\cos(\sqrt{D}\Delta t)\begin{pmatrix} 1 & 0 \\ 0 & 1 \end{pmatrix} - i\sin(\sqrt{D}\Delta t)U\begin{pmatrix} 1 & 0 \\ 0 & -1 \end{pmatrix} U^{1}\right). \tag{A.49}
$$

Multiplying out the matrix product in the second term yields

$$
U\begin{pmatrix} 1 & 0 \\ 0 & -1 \end{pmatrix} U^{-1} = \begin{pmatrix} \cos(2\theta) & \sin(2\theta) \\ \sin(2\theta) & -\cos(2\theta) \end{pmatrix}. \tag{A.50}
$$

As before, we can rewrite $\sin(2\theta)$ and $\cos(2\theta)$ in terms of V_{12} and $\tilde{V}$:

$$
\sin(2\theta) = \frac{V_{12}}{\sqrt{V_{12}^2 + \tilde{V}^2}} \tag{A.51a}
$$

$$
\cos(2\theta) = \frac{\Delta}{\sqrt{V_{12}^2 + \tilde{V}^2}}. \tag{A.51b}
$$

Thus, we arrive at the result

$$
\exp\left(-i\begin{pmatrix} V_{11} & V_{12} \\ V_{12} & V_{22} \end{pmatrix}\Delta t\right) = e^{-i(V_{11}+V_{22})\Delta t/2}\left(\cos\left(\sqrt{D}\Delta t\right)\begin{pmatrix} 1 & 0 \\ 0 & 1 \end{pmatrix}\right.
$$
$$
\left. + \frac{i}{2\sqrt{D}}\sin\left(\sqrt{D}\Delta t\right)\begin{pmatrix} V_{22}-V_{11} & -V_{12} \\ -V_{12} & V_{11}-V_{22} \end{pmatrix}\right). \tag{A.52}
$$

A.7 SINGLE-ACTIVE-ELECTRON APPROXIMATION

In this section we justify the "single-active-electron" approximation introduced in Chapter 3, where the laser field is assumed to interact with only one electron. Formally, for an N-electron system, the dipole operator representing the laser-molecule interaction can be expressed in terms of a sum over single-electron operators for each of the N total electrons in the system:

$$
\mu = \sum_{i=1}^{N} -\mathbf{r}_i, \tag{A.53}
$$

where $\mathbf{r}_i$ represents the position of the ith electron. In general, this operator acts on an N-electron atomic or molecular wave function. To illustrate why a single-active-electron approach is sufficient, we consider the simple case of a three-electron lithium atom exposed to a light field whose central frequency ω_o is resonant with the $2s$–$2p$ transition energy.

As discussed in Section 2.3, the $1s^2 2s$ ground-state wave function (two electrons in the $1s$ orbital and one in the $2s$ orbital) is well described by a single Slater determinant.[1] The two electrons in the 1s orbital must have opposite spin (they are in the spin-singlet configuration), and for simplicity, we assume that the electron occupying the 2s orbital is spin up. In this case, the Slater determinant is written as

$$\Psi_{gs}(\mathbf{r}_i, s_i) = \frac{1}{\sqrt{3!}} \begin{vmatrix} \tilde{\psi}_{1s}(\mathbf{r}_1)\chi_+(s_1) & \tilde{\psi}_{1s}(\mathbf{r}_1)\chi_-(s_1) & \tilde{\psi}_{2s}(\mathbf{r}_1)\chi_+(s_1) \\ \tilde{\psi}_{1s}(\mathbf{r}_2)\chi_+(s_2) & \tilde{\psi}_{1s}(\mathbf{r}_2)\chi_-(s_2) & \tilde{\psi}_{2s}(\mathbf{r}_2)\chi_+(s_2) \\ \tilde{\psi}_{1s}(\mathbf{r}_3)\chi_+(s_3) & \tilde{\psi}_{1s}(\mathbf{r}_3)\chi_-(s_3) & \tilde{\psi}_{2s}(\mathbf{r}_3)\chi_+(s_3) \end{vmatrix}, \qquad (A.54)$$

where $\tilde{\psi}$ represents the (hydrogenic) spatial orbital, $\chi_{\pm}$ spin up/down, and the subscript on the spatial and spin coordinates refers to the electron number (1, 2, or 3). For the excited $1s^2 2p$ state of lithium, we also write the multielectron wave function in terms of a single Slater determinant:

$$\Psi_{es}(\mathbf{r}_i, s_i) = \frac{1}{\sqrt{3!}} \begin{vmatrix} \tilde{\psi}_{1s}(\mathbf{r}_1)\chi_+(s_1) & \tilde{\psi}_{1s}(\mathbf{r}_1)\chi_-(s_1) & \tilde{\psi}_{2p}(\mathbf{r}_1)\chi_+(s_1) \\ \tilde{\psi}_{1s}(\mathbf{r}_2)\chi_+(s_2) & \tilde{\psi}_{1s}(\mathbf{r}_2)\chi_-(s_2) & \tilde{\psi}_{2p}(\mathbf{r}_2)\chi_+(s_2) \\ \tilde{\psi}_{1s}(\mathbf{r}_3)\chi_+(s_3) & \tilde{\psi}_{1s}(\mathbf{r}_3)\chi_-(s_3) & \tilde{\psi}_{2p}(\mathbf{r}_3)\chi_+(s_3) \end{vmatrix}. \qquad (A.55)$$

We now consider the interaction Hamiltonian from Equation 3.14, taking $\psi_m = \Psi_{gs}$ and $\psi_n = \Psi_{es}$:

$$\left\langle \Psi_{es} \left| H_{\text{int}} \right| \Psi_{gs} \right\rangle = -\frac{1}{2} \left(e^{-i\omega_0 t} + e^{+i\omega_0 t} \right) E_0(t) \left\langle \Psi_{es} \left| \sum_{i=1}^{3} -\mathbf{r_i} \cdot \hat{\varepsilon} \right| \Psi_{gs} \right\rangle. \qquad (A.56)$$

Writing out the Slater determinants would produce six terms for each of the two configurations. For notational simplicity, we replace the electron labels $(\mathbf{r}_i)$ and (s_i) with a single label $(i = 1, 2, 3)$. The full expression takes the form

$$\begin{aligned} \left\langle \Psi_{es} \left| H_{\text{int}} \right| \Psi_{gs} \right\rangle &= \frac{1}{12} \left(e^{-i\omega_0 t} + e^{+i\omega_0 t} \right) E_0(t) \int d\mathbf{r}_1 d\mathbf{r}_2 d\mathbf{r}_3 \\ &\times \left(\tilde{\psi}_{1s}\chi_+(1)\tilde{\psi}_{1s}\chi_-(2)\tilde{\psi}_{2p}\chi_+(3) + \ldots \right)(\mathbf{r_1} + \mathbf{r_2} + \mathbf{r_3}) \cdot \hat{\varepsilon} \\ &\times \left(\tilde{\psi}_{1s}\chi_+(1)\tilde{\psi}_{1s}\chi_-(2)\tilde{\psi}_{2s}\chi_+(3) + \ldots \right), \end{aligned} \qquad (A.57)$$

where, for example, $\tilde{\psi}_{1s}\chi_+(1)$ refers to electron one as spin up in the 1s orbital.

While there are many terms in Equation A.57, almost all will be zero. To see why, we consider the first third of the Slater determinant. In particular, Equation A.57 contains the following:

$$\begin{aligned} \left\langle \tilde{\psi}_{1s}\chi_+(1)\left(\tilde{\psi}_{1s}\chi_-(2)\tilde{\psi}_{2p}\chi_+(3) - \tilde{\psi}_{2p}\chi_+(2)\tilde{\psi}_{1s}\chi_-(3) \right) \right| \mathbf{r}_i \\ \times \left| \tilde{\psi}_{1s}\chi_+(1)\left(\tilde{\psi}_{1s}\chi_-(2)\tilde{\psi}_{2s}\chi_+(3) - \tilde{\psi}_{2s}\chi_+(2)\tilde{\psi}_{1s}\chi_-(3) \right) \right\rangle, \end{aligned} \qquad (A.58)$$

where $\mathbf{r}_i$ can be $\mathbf{r}_1$, $\mathbf{r}_2$, or $\mathbf{r}_3$. Each $\mathbf{r}_i$ operator acts only on the spatial portion of the wave function of the corresponding numbered electron. In fact, the spin portions of

---

[1]We again note that for the more general case of molecules, while it may be possible to write the multielectron wave function for the ground state at a given nuclear geometry in terms of a single Slater determinant, it is generally not possible to do this with the same basis set for different geometries or excited states of the system.

the wave function, as well as the spatial portions corresponding to different electron numbers, all factor. From this, we can see immediately that all terms including $\mathbf{r}_1$ go to zero, since they all contain the factor $\langle \tilde{\psi}_{1s}(1) | \mathbf{r}_1 | \tilde{\psi}_{1s}(1) \rangle$ (which is zero due to the symmetry of the hydrogenic orbitals).

For $\mathbf{r}_2$, there are four terms to consider. Two of them are zero due to the symmetry of the spatial orbitals:

$$\langle \tilde{\psi}_{1s}(2) | \mathbf{r}_2 | \tilde{\psi}_{1s}(2) \rangle = \langle \tilde{\psi}_{1s}(2) | \mathbf{r}_2 | \tilde{\psi}_{2s}(2) \rangle = 0. \tag{A.59}$$

A third term can be ignored due to the fact that it will be off-resonance from the light field (essentially energy conservation): $\langle \tilde{\psi}_{2p}(2) | \mathbf{r}_2 | \tilde{\psi}_{1s}(2) \rangle = 0$. It is the fourth term from this portion of the Slater determinant that is on-resonance with the $2s \to 2p$ transition and has a nonzero contribution. Including all the factors, this term reads

$$\langle \tilde{\psi}_{1s}\chi_+(1) | \tilde{\psi}_{1s}\chi_+(1) \rangle \langle \tilde{\psi}_{2p}\chi_+(2) | \mathbf{r}_2 | \tilde{\psi}_{2s}\chi_+(2) \rangle$$
$$\times \langle \tilde{\psi}_{1s}\chi_-(3) | \tilde{\psi}_{1s}\chi_-(3) \rangle. \tag{A.60}$$

The fact that the spatial and spin wave functions are normalized means that all parts of this expression are one except for the spatial matrix element: $\langle \tilde{\psi}_{2p}(2) | \mathbf{r}_2 | \tilde{\psi}_{2s}(2) \rangle$.

This process follows similarly for $\mathbf{r}_3$, except that it is the $\langle \tilde{\psi}_{2p}(3) | \mathbf{r}_3 | \tilde{\psi}_{2s}(3) \rangle$ term that survives. Therefore, the full interaction Hamiltonian can be expressed in terms of sums of integrals over products of *single-electron wave functions* and single-electron $\mathbf{r}$ operators:

$$\langle \Psi_{es} | H_{\mathrm{int}} | \Psi_{gs} \rangle \propto \int d\mathbf{r}_2 \tilde{\psi}_{2p}(\mathbf{r}_2) \mathbf{r}_2 \tilde{\psi}_{2s}(\mathbf{r}_2) + \dots \tag{A.61}$$

This fact allows us to interpret the interaction of an applied optical field and a multi-electron atomic or molecular system in terms of a "single active electron."

A.8 THE NONLINEAR POLARIZATION

A.8.1 Introduction

In Section 6.4.2 we developed the primary equations governing coherent optical measurements, which relied on the induced polarization in the medium. In this section, we begin by considering the general nonlinear polarization before showing how it can be incorporated into Maxwell's wave equation to determine changes to the propagating field. Finally, we connect our notation for the polarization to that of the macroscopic susceptibility.

On the microscopic level, it is useful to think of the induced, time-dependent polarization of an individual molecule ($\mathbf{P}_{nm}$) being due to the interaction of the molecule and the optical radiation, as described by the transition dipole moment μ_{nm} between two electronic states, ψ_n and ψ_m. In general, for an individual molecule in a superposition of eigenstates, we refer to the time-dependent dipole moment of the molecule as its polarization:

$$\mathbf{P}(t) \equiv \langle \Psi(t) | \mu | \Psi(t) \rangle, \tag{A.62}$$

where $\Psi(t)$ is the total wave function of the molecule and μ the dipole moment operator.[2]

---

[2]Note that for a symmetric molecule entirely in the ground state (e.g. assuming no field is present), the symmetry of $|\Psi|^2$ implies that the polarization is zero. In other words, a symmetric molecule has no permanent dipole moment.

Since the induced polarization depends on the strength of the applied optical field, it is natural to consider a power-series expansion of the polarization (similar to the wave function in the time-dependent perturbation theory of Section A.3):

$$P(t) \equiv P^{(1)}(t) + P^{(2)}(t) + P^{(3)}(t) + \cdots . \tag{A.63}$$

Connecting to Equation A.62, we identify the different orders of the induced polarization by the total order of the wave functions involved. Explicitly writing out the perturbative wave function:

$$P(t) = \Big\langle \Psi^{(0)}(t) + \Psi^{(1)}(t) + \Psi^{(2)}(t) + \cdots \Big| \mu$$
$$\Big| \Psi^{(0)}(t) + \Psi^{(1)}(t) + \Psi^{(2)}(t) + \cdots \Big\rangle , \tag{A.64}$$

leads us to identify the orders as

$$P^{(0)}(t) \equiv \Big\langle \Psi^{(0)}(t) \Big| \mu \Big| \Psi^{(0)}(t) \Big\rangle \tag{A.65a}$$

$$P^{(1)}(t) \equiv \Big\langle \Psi^{(1)}(t) \Big| \mu \Big| \Psi^{(0)}(t) \Big\rangle + c.c. \tag{A.65b}$$

$$P^{(2)}(t) \equiv \Big\langle \Psi^{(1)}(t) \Big| \mu \Big| \Psi^{(1)}(t) \Big\rangle + \Big\langle \Psi^{(0)}(t) \Big| \mu \Big| \Psi^{(2)}(t) \Big\rangle + c.c. \tag{A.65c}$$

$$P^{(3)}(t) \equiv \Big\langle \Psi^{(1)}(t) \Big| \mu \Big| \Psi^{(2)}(t) \Big\rangle + c.c. + \Big\langle \Psi^{(0)}(t) \Big| \mu \Big| \Psi^{(3)}(t) \Big\rangle + c.c., \tag{A.65d}$$

where $c.c.$ indicates the complex conjugate of the preceding term. The induced polarization gives rise to a radiated field that is measurable in the lab. As discussed in Section 8.2, it is typically the third-order contribution, $P^{(3)}(t)$, that is relevant in most coherent optical measurements, as the macroscopic, second-order polarization typically vanishes due to symmetry in any centrosymmetric medium (e.g. unaligned molecules in a gas or liquid).

A.8.2 Coupling to the Field

We begin by noting that the macroscopic polarization is related to the microscopic polarization, $P_{nm}(t)$, by the molecular density in the sample:

$$P^{\mathrm{macro}}(t) = \rho \left(P_{nm}(t) + c.c \right), \tag{A.66}$$

where P_{nm} and its complex conjugate are added since the macroscopic polarization is a real quantity. It is the macroscopic polarization, $P^{\mathrm{macro}}(t)$, that acts as a source term in Maxwell's wave equation. We note that the phase of the microscopic polarization will vary spatially, since the external field that drives the microscopic polarization has a phase that varies spatially. This leads to spatial variation in the phase of the macroscopic polarization as well. Denoting this polarization as $P(z,t)$ for compactness and assuming 1D propagation with plane waves, Maxwell's wave equation in a dielectric medium is written in atomic units as

$$\left(\frac{\partial^2}{\partial z^2} - \frac{1}{c^2} \frac{\partial^2}{\partial t^2} \right) E(z,t) = \frac{4\pi}{c^2} \frac{\partial^2}{\partial t^2} P(z,t), \tag{A.67}$$

where $E(z,t)$ is the electric field. We assume plane-wave solutions propagating in the z-direction and ignore the vector aspect of the fields:

$$E(z,t) = E_0(z,t) e^{-\iota \omega_0 t + ikz} + E_0(z,t) e^{\iota \omega_0 t - ikz} \tag{A.68}$$

and

$$P(z,t) = P_0(z,t)e^{-i\omega_0 t + ikz} + P_0^*(z,t)e^{i\omega_0 t - ikz}, \tag{A.69}$$

where E_0 and P_0 are the space- and time-dependent field envelopes (we have taken the envelope for the electric field to be real). We first consider the second-order time derivative of the field E (where we have suppressed the complex conjugate terms for simplicity):

$$\frac{\partial^2}{\partial t^2}E(z,t) = e^{-i(\omega_0 t - kz)}\frac{\partial^2}{\partial t^2}E_0(z,t) - 2i\omega_0 e^{-i(\omega_0 t - kz)}\frac{\partial}{\partial t}E_0(z,t)$$
$$- \omega_0^2 e^{-i(\omega_0 t - kz)}E_0(z,t). \tag{A.70}$$

If the envelope varies slowly in time as compared to an optical period (the "slowly varying envelope approximation"), we see that

$$\omega_0^2 E_0(z,t) \gg \omega_0 \frac{\partial}{\partial t}E_0(z,t) \gg \frac{\partial^2}{\partial t^2}E_0(z,t), \tag{A.71}$$

and we thus keep only the last term in Equation A.70. Similarly expanding the spatial derivative yields

$$\frac{\partial^2}{\partial z^2}E(z,t) = e^{-i(\omega_0 t - kz)}\frac{\partial^2}{\partial z^2}E_0(z,t) + 2ike^{-i(\omega_0 t - kz)}\frac{\partial}{\partial z}E_0(z,t)$$
$$- k^2 e^{-i(\omega_0 t - kz)}E_0(z,t). \tag{A.72}$$

In order for the polarization to modify the field envelope as the pulse propagates, we keep the last two terms in Equation A.72 (the first term is much smaller assuming the envelope varies slowly in space as compared to the wavelength). The last term from Equation A.72 cancels with the term from Equation A.70, and the left side of Equation A.67 can be written as

$$\left(\frac{\partial^2}{\partial z^2} - \frac{1}{c^2}\frac{\partial^2}{\partial t^2}\right)E(z,t) \approx 2ike^{-i(\omega_0 t - kz)}\frac{\partial}{\partial z}E_0(z,t). \tag{A.73}$$

Next, we similarly expand the right-side Equation A.67:

$$\frac{\partial^2}{\partial t^2}P(z,t) = e^{-i(\omega_0 t - kz)}\frac{\partial^2}{\partial t^2}P_0(z,t) - \omega_0^2 e^{-i(\omega_0 t - kz)}P_0(z,t)$$
$$- 2i\omega_0 e^{-i(\omega_0 t - kz)}\frac{\partial}{\partial t}P_0(z,t). \tag{A.74}$$

Again making use of the slowly varying envelope approximation:

$$\omega_0^2 P_0(z,t) \gg \omega_0 \frac{\partial}{\partial t}P_0(z,t) \gg \frac{\partial^2}{\partial t^2}P_0(z,t), \tag{A.75}$$

we obtain for the right side of Equation A.67:

$$\frac{4\pi}{c^2}\frac{\partial^2}{\partial t^2}P(z,t) \approx \frac{-4\pi\omega_0^2}{c^2}P_0 e^{-i(\omega_0 t - kz)}. \tag{A.76}$$

Equating the two sides, we arrive at

$$\frac{\partial}{\partial z}E_0(z,t) = \frac{2\pi\omega_0 i}{c}P_0(z,t). \tag{A.77}$$

This differential equation provides a simple and direct connection between the polarization in a sample and the change in a propagating macroscopic field. In particular, the result illustrates how a macroscopic field can be absorbed as it propagates through a molecular sample.

It is instructive to compare this equation with an intuitive, microscopic picture of absorption. In the microscopic picture, an electron being driven by an oscillating field radiates energy, and how the radiated field from the electron adds to the driving field determines the effect on the macroscopic, propagating field. Consider a single atom or molecule in an oscillating field. On resonance, the electron has a $\pi/2$ phase lag relative to the driving field. The electron radiates energy as a result of its acceleration, and the acceleration is proportional to the second derivative of its displacement. This leads to a $3\pi/2$ overall phase shift for the emission of the electron relative to the driving field. Note that this does not appear to be consistent with the notion of absorption being due to destructive interference of the driving field and the emitted radiation. The resolution of this apparent paradox comes from considering the net effect of the ensemble of atomic or molecular absorbers. If one considers a "sheet" of molecules that lie in a plane orthogonal to the propagation direction of the driving field, the total radiated field from this ensemble of emitters has a phase of π relative to the drive field (as one expects for absorption). This makes it clear that absorption of a plane wave is best described as a macroscopic phenomenon using Maxwell's wave equation. For a detailed treatment of this topic, see Appendix A of Reference [125].

A.8.3 Macroscopic Susceptibility

In Section A.8.1, we considered a power-series expansion of the polarization. Here we relate the expansion of the polarization to the material susceptibility, which is commonly used in nonlinear optics. Specifically, a material is assumed to have a nonlinear response to an applied field $E(t)$:

$$P(t) = P^{(1)}(t) + P^{(2)}(t) + P^{(3)}(t) + \cdots$$
$$\equiv \frac{1}{4\pi}\left[\chi^{(1)}E(t) + \chi^{(2)}E^2(t) + \chi^{(3)}E^3(t) + \cdots\right], \qquad (A.78)$$

where $\chi^{(n)}$ is the nth-order susceptibility of the medium, and we have assumed the response to be instantaneous and the susceptibilities to be constants.[3] In this formalism, each order of susceptibility leads to different physical processes (e.g. $\chi^{(2)}$ can produce second-harmonic generation). In each case, the total number of fields involved in the process is one more than the order of the process, so a second-order $\left(\chi^{(2)}\right)$ process involves the interaction of three different electric fields, a third-order process $\left(\chi^{(3)}\right)$ involves the interaction of four different electric fields, etc. The term "mixing" is often used to describe the process: for example, four-wave mixing is an interaction of four fields, and therefore, a $\chi^{(3)}$ process. This is consistent with our results for transient absorption, where the $P^{(3)}(t)$ process involved four total fields: two of the pump and two of the probe.

Having a physical picture that corresponds to each of these processes can help with the understanding. Consider the $\chi^{(2)}$ process of sum-frequency generation, where two

---

[3] We note that the notation is potentially confusing, given our use of the symbol $\chi(R,t)$ to denote the vibrational portion of a molecular wave function. Unfortunately, $\chi^{(n)}$ is standard notation for the susceptibility, and we use it here.

input fields at frequencies ω_1 and ω_2 are incident on a medium possessing a nonzero $\chi^{(2)}$. The $\chi^{(2)}$ term in the polarization generates a field at frequency $\omega_3 = \omega_1 + \omega_2$; by energy conservation, one photon of the incident light at each frequency is converted into a single photon at the sum frequency. All three of these fields propagate in the medium, and energy can be exchanged between them (hence the term three-wave mixing). Note that second-harmonic generation is a special case of sum-frequency generation, where the two input fields have the same frequency ω (they are degenerate). A $\chi^{(3)}$ interaction can lead to a host of different four-wave-mixing processes depending on the input frequencies and k-vectors, and they include such names as transient absorption and coherent anti-Stokes Raman scattering (see Chapter 8).

Finally, we note that this physical picture can be extended to lower-order processes as well. In a $\chi^{(1)}$ process the polarization is described by a purely linear interaction with the electric field. Nevertheless, given the discussion earlier, there must be $1 + 1 = 2$ fields present. One of the fields is the incident light, while the second is the reradiated field, *at the same frequency*, by the linear polarization. Thus, there can be no new frequencies in the output light, but instead, only those effects such as dispersion that occur with a linear susceptibility assumed to be real. In the absence of any material susceptibility (i.e. the vacuum), we could consider a "$\chi^{(0)}$" process with a single field: that of the incident light propagating unchanged.

APPENDIX B

Experimental Considerations

B.1 ENSEMBLE MEASUREMENTS

While it is possible to isolate and study single molecules (e.g. see Section 12.4.2), time-resolved spectroscopy experiments are almost always performed on an *ensemble* of many, identical quantum systems. The ensemble could be a dilute gas of molecules or a small volume of liquid containing both target and solvent molecules. The fact that the experiment is essentially carried out many times in parallel on each laser shot enhances the signal, but it can also lead to a number of complications. We address some of the principle ones here.

B.1.1 Finite temperature and orientation averaging

In general, the molecules in an ensemble are at a finite temperature and thus will be in a dynamic state, having an energy on the order of $1/2\, kT$ in each degree of freedom. The energy gap between the ground and first-excited electronic state is typically quite large in the equilibrium geometry ($\sim$eV, see Section 2.5.1), and thus the Boltzmann probability for finding a molecule in anything other than the ground electronic state is essentially zero at room temperature and below. Thus all molecules in the ensemble can safely be assumed to be in the same initial electronic state.

However, the spacing between vibrational energy levels is typically more than an order of magnitude smaller, and molecules have a finite probability of being in an initial superposition of vibrational states. While the phase between different vibrational eigenstates is well defined in any given molecule, it is generally not the same across different molecules. Thus, dynamics associated with the superposition of vibrational eigenstates is not coherent across the ensemble. An experiment starting with such a "mixed state" incoherently averages over dynamics originating from the different thermally populated vibrational states of the system. For example, in an ensemble of room-temperature I_2 molecules with a vibrational level spacing in the electronic ground state of 214 cm^{-1}, the probabilities of finding the molecule in $v = 0, 1, 2, 3$ are roughly 64%, 23%, 8%, and 3%, respectively. Thus, if a pump–probe experiment does not intentionally select an initial vibrational eigenstate, a measurement will average over the dynamics for each of the thermally populated eigenstates (potentially washing out interesting time-dependent behavior).

However, the measurement can still reveal coherent dynamics if the motion generated by the pump pulse is larger than the thermal motion of the initial ensemble. For example, if the molecules are initially in an incoherent superposition of two eigenstates, but the laser excites a superposition of 5 or even 10 states, the coherent motion generated by the pump pulse will be significantly larger than the incoherent thermal motion. This will lead to a measurable, time-dependent signal. This situation is actually the typical case, with pump-induced motion on the excited potential energy surface generally much larger than any initial thermal motion, because the laser bandwidth is generally greater than kT.

Rotational energy-level spacings are even less than for vibrations, and there are typically many rotational levels thermally populated. Unless special care is taken, measurements average over contributions from molecules in all rotational quantum states. Classically, one can think of molecules as having some orientation in space, and the fact that molecules in a gas or liquid ensemble are randomly orientated with respect to each other presents two basic complications in a pump–probe measurement.[1]

The first is a varying interaction strength between the pump/probe pulses and different molecules in the ensemble due to the different projections of the molecular transition dipole moments onto the pump/probe polarization vector. Molecules in the ensemble whose transition dipole moments lie along the laser polarization axis are preferentially excited (the probability for excitation in a single-photon process is proportional to $\cos^2 \theta$, where θ is the angle between the molecular dipole moment and laser polarization vector). This has the effect of preparing a partially-aligned molecular sample in the excited state.

The second complication is that alignment generated by the pump pulse tends to dephase on timescales of a few picoseconds in gas-phase experiments because the excited molecules are still rotating (in random directions) with the thermal energy they had before the pump pulse arrived. Only probe measurements within that dephasing time experience an aligned sample. Any rotational dynamics occurring between pump and probe pulses contribute to a change in the signal as a function of time that may, or may not, be of experimental interest. Often the rotational dynamics are not particularly relevant to the process under investigation, in which case one would like to mitigate the effect of rotations on the signal. This can be accomplished by working with the pump and probe pulses polarized along axes aligned $57°$ with respect to each other, which cancels out the rotational effects (this is known as the "magic angle"). One way to think of this is that it is the angle for which there are an equal number of molecules rotating into alignment with the probe as rotating out.

Alternatively, for gas-phase experiments sensitive to molecular orientation, one can prepare the ensemble in an initial configuration where the rotational state is well known. This can be accomplished by actively aligning or orienting the molecular sample using the electric field of an applied laser pulse. Two different techniques, known as adiabatic and impulsive alignment, prepare the molecules in states with transient degrees of alignment [126]. In adiabatic alignment, a laser pulse is slowly turned on (typically on nanosecond timescales – slower than a rotational period). The molecules adiabatically come into alignment with the polarization axis of the laser, where the

---

[1]Here we use the term orientation to refer to alignment with direction or polarity. For example, in an asymmetric diatomic molecule such as hydrogen fluoride, orientation includes not only the direction of the internuclear axis relative to some other axis in space (perhaps defined by the laser polarization), but also the direction of permanent dipole moment of the molecule (it points "up" or "down").

alignment is along the molecular axis with the highest degree of polarizability. This approach is very effective at producing highly-aligned molecular ensembles, but comes at the price of not being field free: the time-resolved measurement is carried out in the presence of the alignment field, potentially altering dynamics relative to the field-free case.

The impulsive approach makes use of a laser pulse much shorter than the rotational period of the molecules, giving them a "kick" that produces an aligned sample after the pulse is off. This produces field-free alignment, but has the disadvantage that the alignment is transient and only lasts for a short time. Thus, any time-resolved measurement is confined to a relatively short duration during which the molecules are aligned. In addition, the degree of alignment is typically not as high as for adiabatic alignment, and the situation is complicated for asymmetric molecules that do not display regular rotational dynamics.

B.1.2 Volume averaging

A different complication arises in experiments where the molecular response to either the pump or probe pulse *individually* is nonlinear in the field strength. Different portions of the focal volume experience different laser intensities, and unless the signal is collected from a region significantly smaller than laser focus, a measurement averages over a range of pulse intensities. For a nonlinear process, this effect can average over any intensity-dependent structures one hopes to measure. For example, multiphoton ionization measurements can experience resonant enhancement due to Stark shifts that bring intermediate states into resonance. This will lead to different spatial portions of the beam producing different ionization signals. One approach to mitigating this problem is to collect signal from only a specific region of the focus.

B.2 DETECTION TECHNIQUES

B.2.1 Heterodyne measurements

Throughout the book, we have made a distinction between incoherent and coherent detection techniques. Another distinction worth discussing, especially for coherent optical techniques such as transient absorption and coherent anti-Stokes Raman spectroscopy (CARS), is between homodyne and heterodyne measurements. For example, in transient absorption, we are usually interested in measuring changes in some small optical field whose strength we denote by ε. Typical electronics do not have sufficient bandwidth to measure the extremely rapid oscillations of an optical field: 500 nm light has a field that oscillates at a frequency of 600 THz. Instead, photodetectors for optical frequencies are typically optimized to measure average power or pulse energy, which is the temporal- and spatially integrated *intensity* of the light.[2] Therefore, an intensity-sensitive detector measures a signal proportional to $|\varepsilon|^2$. One can infer the field from the intensity measurement, but if ε is small, $|\varepsilon|^2$ will be even smaller, and measuring small changes will be challenging. Measuring $|\varepsilon|^2$ alone with an intensity-sensitive detector is referred to as a "homodyne measurement."

---

[2]These detectors are typically referred to as intensity-sensitive detectors, even when they don't spatially or temporally resolve the measurement (i.e. even if they simply measure the energy per unit time on slow timescales relative to a short laser pulse).

An alternative to measuring the weak field alone is to add the weak field to a larger, well-characterized field at the same frequency. This is known as a "heterodyne measurement," and the additional field is sometimes known as the "local oscillator." If one adds a local oscillator with unit amplitude to ε, the total field becomes $1 + \varepsilon$, and the signal measured by an intensity-sensitive detector is $|1 + \varepsilon|^2 = 1 + \varepsilon + \varepsilon^* + |\varepsilon|^2$. Since ε is small, the ε terms in the expansion due to the local oscillator are much larger than the $|\varepsilon|^2$ term available in a homodyne measurement.

A heterodyne measurement also allows access to the phase of the field. Writing the field as $\varepsilon = |\varepsilon|e^{i\phi}$ shows that a term in the signal goes like $2|\varepsilon|\cos(\phi)$; from this, one obtains phase information relative to the local oscillator. Note, however, that the amplitude and phase stability of the local oscillator is essential, as amplitude fluctuations in the local oscillator on the order of ε can mask the signal one is trying to measure.

As an aside, we briefly note this heterodyne discussion is related to the optimal phase at which to measure small changes in an interference pattern. For example, at what phase of a fringe should one sit to measure small changes in an optical interferometer? (Alternatively, at what phase of a Rabi oscillation should one sit to optimally measure small changes in state population?) Consider the case of an interferometer where one sits at "a zero" (perfect destructive interference) and looks for small changes in the output signal due to variations in phase delay between the two arms. Since the signal is zero in the absence of any shift, it is a background-free measurement. While this seems ideal for measuring small changes in a signal, note that the actual signal you must measure is also small. Specifically, if the phase change one is trying to measure is $\delta\phi$, the total intensity at the exit of the Michelson interferometer is proportional to $\sin^2(\delta\phi) \sim (\delta\phi)^2$ for small $\delta\phi$. So the signal is roughly proportional to the square of what you hope to measure, just as in the case of the homodyne intensity measurement. In particular, if $\delta\phi$ is small, the signal will be very small.

However, consider looking for small changes in phase delay *away* from the zero point at some phase-angle ϕ_0. Here the total intensity at the exit is proportional to

$$\begin{aligned}
\sin^2(\phi_0 + \delta\phi) &\approx (\sin(\phi_0) + \cos(\phi_0)\delta\phi)^2 \\
&= (\sin(\phi_0))^2 + 2 \cdot \sin(\phi_0) \cdot \cos(\phi_0) \cdot \delta\phi + (\cos(\phi_0)\delta\phi)^2.
\end{aligned} \tag{B.1}$$

Note that this form is similar to the heterodyne measurement. The middle term is only first order in $\delta\phi$ (and therefore larger), but it is also sensitive to any fluctuations in the background. Ideally, one seeks to achieve a balance between the small, background-free signal and the larger, non-background-free situation. Given the relative sizes of the signal and noise on the background, one can determine the optimal phase value to maximize the signal-to-noise ratio in the experiment.

B.2.2 Lock-in detection

A technique frequently used to improve signal to noise is lock-in detection. The basic idea is to modulate the measurement with some electrical or mechanical mechanism, and then take only the Fourier component of the signal at the modulation frequency (the modulation frequency can be chosen to minimize the noise). This effectively removes most of the noise, while retaining most of the signal.

A lock-in amplifier takes as input both the modulated signal from a detector (together with whatever noise is present) and the modulation signal itself (e.g. a sine wave from a chopper that modulates the intensity of the pump pulse in an experiment). The lock-in amplifier multiplies the two inputs and integrates their product over a time long compared to the modulation period. This operation is equivalent to taking the Fourier component of the signal at the modulation frequency:

$$V_{\text{out}} = \langle V_s(t) \cos(\omega_m t + \phi_m) \rangle, \tag{B.2}$$

where the brackets indicate a time-average, $V_s(t)$ is the modulated signal (along with any noise present), ω_m is the modulation frequency, ϕ_m is the phase of the modulation, and V_{out} is the desired Fourier component of the signal at the modulation frequency. Note that only the noise component at the modulation frequency survives.

B.3 NONLINEAR OPTICAL FREQUENCY CONVERSION

In this section we briefly discuss the basics of nonlinear optical frequency conversion, which plays an important role in generating pump and probe pulses with specific frequencies and pulse durations. In addition, nonlinear optics can be used as a form of measurement with optical gating (see Section 7.2.7). A simple approach to understanding the nonlinear response of a material to an applied electric field is the Lorenz model, which considers the microscopic motion of a bound electron in an oscillating electric field. The electron motion generates a microscopic dipole moment for the atom, and this can be used to calculate a macroscopic polarization. As outlined in Section A.8, the macroscopic polarization in Maxwell's wave equation determines the optical response of the medium.

The Lorenz model assumes that an electron in an applied field behaves like a driven harmonic oscillator [9]. Near equilibrium, the potential energy as a function of displacement can be considered harmonic. This can be seen by expanding the potential in a Taylor series:

$$\begin{aligned}
V(r) &= V(0) + r \left(\frac{dV}{dr} \right)_{r=0} + \frac{1}{2!} r^2 \left(\frac{d^2V}{dr^2} \right)_{r=0} + \cdots \\
&\approx \frac{1}{2!} r^2 \left(\frac{d^2V}{dr^2} \right)_{r=0},
\end{aligned} \tag{B.3}$$

where r represents the electron displacement from equilibrium and $V(r)$ is the potential energy. In the second line we have used the fact that the linear term is zero at equilibrium and that $V(0)$ is an unimportant offset that can be set to zero.

If the electron is driven by a relatively strong field (e.g. from an intense laser pulse), it will explore regions of the potential that are not well described by the second-order term in Taylor series. In this case, higher-order terms need to be considered:

$$V(r) = \frac{1}{2!} r^2 \left(\frac{d^2V}{dr^2} \right)_{r=0} + \frac{1}{3!} r^3 \left(\frac{d^3V}{dr^3} \right)_{r=0} + \cdots. \tag{B.4}$$

These higher-order terms lead to a nonlinear restoring force on the electron. In particular, the equation of motion for the electron in an electric field $E(t)$ is given by (with $m = 1$ and $q = -1$ in atomic units):

$$\ddot{r} = F(r) = -E(t) - \frac{dV(r)}{dr}$$

$$\frac{d^2r}{dt^2} = -E(t) - r\left(\frac{d^2V}{dr^2}\right)_{r=0} - \frac{1}{2}r^2\left(\frac{d^3V}{dr^3}\right)_{r=0} + \cdots \tag{B.5}$$

$$\equiv -E(t) - \omega_0^2 r - \beta r^2 + \cdots,$$

where $\omega_0^2 \equiv \left(\frac{d^2V}{dr^2}\right)_{r=0}$ is the square of the harmonic spring frequency and $\beta \equiv \frac{1}{2}\left(\frac{d^3V}{dr^3}\right)_{r=0}$.

In general, it is not possible to solve this equation analytically if one includes nonlinear terms in the restoring force. If we assume the cubic term in the potential is small (a small nonlinearity), we can find an approximate solution. For a driving field given by $E(t) = E_0\cos(\omega t)$, the correction to the usual harmonic solution is given by

$$r(t) = -\frac{E_0}{\omega_0^2 - \omega^2}\cos(\omega t) - \frac{\beta E_0^2}{2\omega_0^2\left(\omega_0^2 - \omega^2\right)^2} - \frac{\beta E_0^2}{2\left(\omega_0^2 - 4\omega^2\right)\left(\omega_0^2 - \omega^2\right)^2}$$
$$\times \cos(2\omega t). \tag{B.6}$$

Note that in addition to the usual displacement following the oscillations of the driving field at frequency ω, the electron also responds with a static displacement and one that oscillates at twice the frequency of the applied field (2ω). If one considers even higher-order terms in the potential expansion, the solution of $r(t)$ will contain higher-order harmonics of the driving field. The microscopic response gives rise to a macroscopic polarization at the second harmonic (or higher-order harmonics) of the driving field, which can lead to coherent build up of light at multiples of the fundamental frequency.

The most important factor affecting the coherent buildup of harmonic emission from the ensemble is how contributions from individual emitters add to the total field at the harmonic frequency. In particular, for a macroscopic buildup of the harmonic field, the emission from each atom or molecule in the ensemble must add coherently. The phase of harmonic emission is tied to the phase of the driver, and constructive addition of the harmonic emission requires that the phases of the driving and harmonic fields stay matched as the beams propagate through the sample (i.e. the relative phase between the drive laser and harmonic evolve by an amount small compared to π). This condition is known as "phase matching."

Due to normal material dispersion, the phase velocities of the driver and harmonic fields are typically sufficiently different so that there is limited distance over which the phases stay matched. However, there are steps one can take to match the phases of the drive and harmonic fields. The most common technique uses birefringent crystals, where the fundamental (drive) and harmonic fields are polarized along different axes of the crystal. This allows one to match the phase velocities of the two fields by cutting the crystal in such a way that the projection of one polarization vector onto the crystal axes leads to a phase velocity that matches that of the other field (polarized along a different axis). When combining multiple fields in a gas or liquid, one can adjust the propagation directions such that addition of the k-vectors for the different beams leads to a coherent buildup of the desired field in a particular direction (e.g. CARS in Section 8.3). As discussed, phase matching essentially corresponds to momentum conservation, whereas the fact that one generates harmonics of a fundamental frequency by absorbing n photons of the fundamental to generate one photon of the nth harmonic represents energy conservation.

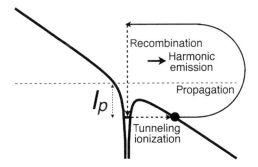

Figure B.1 Three-step model for high-harmonic generation. Figure from Ref. [128].

B.4 HIGH-HARMONIC GENERATION

For very strong driving fields, the electron response can be highly nonlinear. Much of attosecond science is driven by the application of high-harmonic generation, in which a strong applied field produces high-order harmonics of a drive laser. The harmonic spectrum can be quite broad, spanning from the near infrared to the extreme ultraviolet or soft X-ray regions of the spectrum. In most high-harmonic generation arrangements, a strong-field, near-infrared laser pulse is focused into a noble gas at high intensity (typically $\sim 1\%$ of the atomic unit of intensity, or a few times 10^{14} W/cm^2). At such high intensities, an electron is removed from the atom and tunnels into the continuum in less than a laser cycle. This is in some sense the ultimate exploration of the anharmonic portion of the potential, and thus a perturbative description of the electron motion is invalid.

A simple, nonperturbative description known as the three-step model provides a nice intuitive picture of high-harmonic generation [127]. The first step involves tunnel ionization of an electron in a fraction of a laser period. In step two, the electronic wave packet evolves in the continuum under the influence of the laser field, where the electron is accelerated away from and back toward the ionic core as the electric field of the laser changes sign in crossing through zero. Step three is recollision, in which the electron recombines with the ion and emits radiation (Figure B.1).

The classical picture of recollision is that the electron very rapidly decelerates, emitting a burst of radiation in doing so.[3] Because the process takes place with a range of accelerations, a broad spectrum of radiation is emitted. The sequence is repeated each half cycle of the laser field, and so the spectrum develops a structure corresponding to the odd harmonics of the laser driving frequency. Since all atoms in a plane perpendicular to the laser propagation direction experience the same acceleration simultaneously, emission from each atom is in phase with the others and produces a coherent burst of radiation emitted from the atomic ensemble. In fact, this can be thought of as the coherent version of an X-ray tube, in which electrons are decelerated rapidly in an unsynchronized way as they collide with a metallic target, thus emitting incoherent radiation.

---

[3]In the quantum picture, recollision involves the electron recombining with the ionic core and emitting an energetic photon as part of energy conservation.

APPENDIX C

Additional Problems

1. Write a program to simulate a pump–probe measurement of wave packet dynamics in the B state of I_2, probed by monitoring laser-induced fluorescence from a further-excited state (e.g. F state), as described in *Nature* **343**, 737 (1990). Make use of Equation 6.30 to numerically integrate the TDSE, including the fields of both the pump and probe pulses. Start with the molecule in the third vibrational state of the X-state potential. Use a pump pulse with a central frequency of 0.0735 a.u. and a pulse duration of 40 fs (full width at half maximum) to promote population to the B state. Assume a probe pulse with a central frequency twice that of the pump and the same pulse duration excites population to the F state. Use the Morse potential $V(R) = D_e(1 - e^{(-\beta(R-R_e))})^2$, with parameters for the X, B, and F states given as:

   X-state: $D_e = 0.0567$ a.u. $\beta = 0.9872$ a.u. $R_e = 5.04$ a.u.

   B-state: $D_e = 0.020$ a.u. $\beta = 0.97$ a.u. $R_e = 5.72$ a.u.

   F-state: $D_e = 0.0577$ a.u. $\beta = 0.4751$ a.u. $R_e = 6.75$ a.u.

   (a) Begin by running your calculation at a pump–probe delay of 500 fs. Make two-dimensional plots of the wave function amplitude squared on both the B and F states as a function of R and t. Hint: Don't save the wave function at every time and space point required for the calculation, as that will use a lot of memory. For the two-dimensional plot, sample your wave function coarsely using time steps of 10 fs and a total of 100 samples in R.

   (b) Now run your calculation for a series of pump–probe delays from 100 to 350 fs in steps of 25 fs. Plot the total (spatially integrated) F-state wave function as a function of pump–probe delay.

2. Calculate the diffraction pattern (what you would measure on a camera in the far-field), as well as $sM(s)$, for an *unaligned* ensemble of nitrogen molecules. You can find the atomic form factors for nitrogen atoms at http://lampx.tugraz.at/hadley/ss1/crystaldiffraction /atomicformfactors/electronatomicformfactors.php.

3. Do the same calculation as Problem 2 for an *aligned* sample of nitrogen molecules. Note that you need to start with an expression that does not average over molecular alignment.

4. Do the same calculation as Problem 2 for an *unaligned* ensemble of thymine molecules. You can find the atomic form factors for the atoms at http://lampx. tugraz.at/hadley/ss1/crystaldiffraction /atomicformfactors/electronatomicform factors.php.

5. Numerically calculate the photoelectron spectrum as a function of time delay between an attosecond XUV pulse and a strong-field, IR laser pulse for Auger decay of a model atom such as argon. Assume the XUV pulse has 100 as duration and a central energy of 40 eV, while the IR pulse has 5 fs duration and a peak intensity of 10^{12} W/cm^2. Assume an Auger lifetime of 10 fs.

6. Numerically solve the TDSE for an electron emerging from a one-dimensional, "soft-Coulomb" atomic potential ($V(x) = -1/\sqrt{x^2 + \beta^2}$ with $\beta = 1$) immediately after ionization. Compare your calculation with the propagation of a free electron with the same asymptotic energy. What is the difference in the group delays for the two wave packets? How is this related to the Wigner-Smith delay?

Bibliography

[1] R. Schoenlein, L. Peteanu, R. Mathies, and C. Shank, "The first step in vision: femtosecond isomerization of rhodopsin," *Science*, vol. 254, no. 5030, pp. 412–415, 1991.

[2] D. Polli, P. Altoè, O. Weingart, K. M. Spillane, C. Manzoni, D. Brida, G. Tomasello, G. Orlandi, P. Kukura, R. A. Mathies, *et al.*, "Conical intersection dynamics of the primary photoisomerization event in vision," *Nature*, vol. 467, no. 7314, pp. 440–443, 2010.

[3] T. F. Gallagher, *Rydberg Atoms*, vol. 3. Cambridge: Cambridge University Press, 2005.

[4] C. T. Middleton, K. de La Harpe, C. Su, Y. K. Law, C. E. Crespo-Hernández, and B. Kohler, "DNA excited-state dynamics: from single bases to the double helix," *Annual Review of Physical Chemistry*, vol. 60, pp. 217–239, 2009.

[5] S. Matsika and P. Krause, "Nonadiabatic events and conical intersections," *Annual Review of Physical Chemistry*, vol. 62, pp. 621–643, 2011.

[6] J. C. Tully, "Molecular dynamics with electronic transitions," *The Journal of Chemical Physics*, vol. 93, no. 2, pp. 1061–1071, 1990.

[7] K. Rzazewski and R. W. Boyd, "Equivalence of interaction Hamiltonians in the electric dipole approximation," *Journal of Modern Optics*, vol. 51, no. 8, pp. 1137–1147, 2004.

[8] P. Bucksbaum, M. Bashkansky, and T. McIlrath, "Scattering of electrons by intense coherent light," *Physical Review Letters*, vol. 58, no. 4, pp. 349–352, 1987.

[9] P. W. Miloni and J. H. Eberly, *Lasers*. New York: Wiley, 1988.

[10] P. H. Bucksbaum, "An atomic dimmer switch," *Nature*, vol. 396, pp. 217–219, 1998.

[11] D. Tannor, *Introduction to Quantum Mechanics: A Time-Dependent Perspective*. University Science Books, 2007.

[12] J. Fleck, J. Morris, and M. Feit, "Time-dependent propagation of high energy laser beams through the atmosphere," *Applied Physics*, vol. 10, no. 2, pp. 129–160, 1976.

[13] M. Feit, J. Fleck Jr, and A. Steiger, "Solution of the Schrödinger equation by a spectral method," *Journal of Computational Physics*, vol. 47, no. 3, pp. 412–433, 1982.

[14] J. Crank and P. Nicolson, "A practical method for numerical evaluation of solutions of partial differential equations of the heat-conduction type," *Mathematical Proceedings of the Cambridge Philosophical Society*, vol. 43, no. 1, pp. 50—-67, 1947.

[15] D. Griffiths, *Introduction to Quantum Mechanics*. 2nd ed. Cambridge: Cambridge University Press, 2016.

[16] M. Dantus, R. Bowman, and A. Zewail, "Femtosecond laser observations of molecular vibration and rotation," *Nature*, vol. 343, no. 6260, pp. 737–739, 1990.

[17] P. Cong, G. Roberts, J. Herek, A. Mohktari, and A. Zewail, "Femtosecond real-time probing of reactions. 18. Experimental and theoretical mapping of trajectories and potentials in the NaI dissociation reaction," *The Journal of Physical Chemistry*, vol. 100, no. 19, pp. 7832–7848, 1996.

[18] A. H. Zewail, *Femtochemistry: Ultrafast Dynamics of the Chemical Bond*. World Scientific Publishing Company, 1994.

[19] J. Gallagher, C. Brion, J. Samson, and P. Langhoff, "Absolute cross sections for molecular photoabsorption, partial photoionization, and ionic photofragmentation processes," *Journal of Physical and Chemical Reference Data*, vol. 17, no. 1, pp. 9–153, 1988.

[20] G. Scoles, D. Bassi, U. Buck, and D. C. Laine, *Atomic and Molecular Beam Methods*, vol. 1. Oxford: Oxford University Press, 1988.

[21] I. Fischer, M. J. Vrakking, D. Villeneuve, and A. Stolow, "Femtosecond time-resolved zero kinetic energy photoelectron and photoionization spectroscopy studies of I_2 wavepacket dynamics," *Chemical Physics*, vol. 207, no. 2-3, pp. 331–354, 1996.

[22] I. Fischer, D. Villeneuve, M. J. Vrakking, and A. Stolow, "Femtosecond wave-packet dynamics studied by time-resolved zero-kinetic energy photoelectron spectroscopy," *The Journal of Chemical Physics*, vol. 102, no. 13, pp. 5566–5569, 1995.

[23] A. Assion, M. Geisler, J. Helbing, V. Seyfried, and T. Baumert, "Femtosecond pump-probe photoelectron spectroscopy: Mapping of vibrational wave-packet motion," *Physical Review A*, vol. 54, no. 6, pp. R4605–R4608, 1996.

[24] T. Frohnmeyer and T. Baumert, "Femtosecond pump-probe photoelectron spectroscopy on Na_2: a tool to study basic coherent control schemes," *Applied Physics B: Lasers and Optics*, vol. 71, no. 3, pp. 259–266, 2000.

[25] D. M. Neumark, "Time-resolved photoelectron spectroscopy of molecules and clusters," *Annual Review of Physical Chemistry*, vol. 52, no. 1, pp. 255–277, 2001. PMID: 11326066.

[26] Gräfe, S. and Scheidel, D. and Engel, V. and Henriksen, N. E. and Møller, K. B., "Approaches to Wave Packet Imaging Using Femtosecond Ionization Spectroscopy," *The Journal of Physical Chemistry A*, vol. 108, no. 41, pp. 8954–8960, 2004.

[27] M. Wollenhaupt, V. Engel, and T. Baumert, "Femtosecond laser photoelectron spectroscopy on atoms and small molecules: Prototype studies in quantum control," *Annual Review of Physical Chemistry*, vol. 56, no. 1, pp. 25–56, 2005. PMID: 15796695.

[28] G. Wu, P. Hockett, and A. Stolow, "Time-resolved photoelectron spectroscopy: from wavepackets to observables," *Physical Chemistry Chemical Physics*, vol. 13, pp. 18447–18467, 2011.

[29] J. A. Yeazell and C. Stroud Jr, "Observation of fractional revivals in the evolution of a Rydberg atomic wave packet," *Physical Review A*, vol. 43, no. 9, pp. 5153–5156, 1991.

[30] T. Okino, Y. Furukawa, Y. Nabekawa, S. Miyabe, A. A. Eilanlou, E. J. Takahashi, K. Yamanouchi, and K. Midorikawa, "Direct observation of an attosecond electron wave packet in a nitrogen molecule," *Science Advances*, vol. 1, no. 8, p. e1500356, 2015.

[31] J. W. Cooper, "Photoionization from outer atomic subshells. a model study," *Physical Review*, vol. 128, pp. 681–693, Oct 1962.

[32] P. B. Corkum, "Plasma perspective on strong field multiphoton ionization," *Physical Review Letters*, vol. 71, pp. 1994–1997, Sep 1993.

[33] M. Hentschel, R. Kienberger, C. Spielmann, G. A. Reider, N. Milosevic, T. Brabec, P. Corkum, U. Heinzmann, M. Drescher, and F. Krausz, "Attosecond metrology," *Nature*, vol. 414, no. 6863, p. 509, 2001.

[34] P. B. Corkum and F. Krausz, "Attosecond science," *Nature Physics*, vol. 3, no. 6, p. 381, 2007.

[35] F. Krausz and M. Ivanov, "Attosecond physics," *Reviews of Modern Physics*, vol. 81, pp. 163–234, Feb 2009.

[36] Z. Chang, *Fundamentals of Attosecond Optics*. Taylor & Francis, 2011.

[37] H.-Y. Chen, I.-R. Lee, and P.-Y. Cheng, "Gas-phase femtosecond transient absorption spectroscopy," *Review of Scientific Instruments*, vol. 77, no. 7, p. 076105, 2006.

[38] M. A. Reber, Y. Chen, and T. K. Allison, "Cavity-enhanced ultrafast spectroscopy: ultrafast meets ultrasensitive," *Optica*, vol. 3, no. 3, pp. 311–317, 2016.

[39] Z. Wei, J. Li, L. Wang, S. T. See, M. H. Jhon, Y. Zhang, F. Shi, M. Yang, and Z.-H. Loh, "Elucidating the origins of multimode vibrational coherences of polyatomic molecules induced by intense laser fields," *Nature Communications*, vol. 8, p. 735, 2017.

[40] W. T. Pollard, S. Lee, and R. A. Mathies, "Wave packet theory of dynamic absorption spectra in femtosecond pump–probe experiments," *The Journal of Chemical Physics*, vol. 92, no. 7, pp. 4012–4029, 1990.

[41] C. Bressler and M. Chergui, "Molecular structural dynamics probed by ultrafast X-ray absorption spectroscopy," *Annual Review of Physical Chemistry*, vol. 61, no. 1, pp. 263–282, 2010. PMID: 20055677.

[42] K. Ramasesha, S. R. Leone, and D. M. Neumark, "Real-time probing of electron dynamics using attosecond time-resolved spectroscopy," *Annual Review of Physical Chemistry*, vol. 67, no. 1, pp. 41–63, 2016. PMID: 26980312.

[43] T. Siebert, M. Schmitt, S. Gräfe, and V. Engel, "Ground state vibrational wave-packet and recovery dynamics studied by time-resolved CARS and pump-CARS spectroscopy," *Journal of Raman Spectroscopy*, vol. 37, no. 1-3, pp. 397–403, 2006.

[44] S. Mukamel, *Principles of Nonlinear Optical Spectroscopy*. Oxford series in optical and imaging sciences, Oxford: Oxford University Press, 1999.

[45] A. M. Zheltikov, "Coherent anti-Stokes Raman scattering: from proof-of-the-principle experiments to femtosecond CARS and higher order wave-mixing generalizations," *Journal of Raman Spectroscopy*, vol. 31, no. 8-9, pp. 653–667.

[46] von Vacano, Bernhard and Motzkus, Marcus, "Time-resolving molecular vibration for microanalytics: single laser beam nonlinear Raman spectroscopy in simulation and experiment," *Physical Chemistry Chemical Physics*, vol. 10, no. 5, pp. 681–691, 2008.

[47] W. Shockley, *Electrons and Holes in Semiconductors*. Van Nostrand, 1950.

[48] D. Boschetto, L. Malard, C. H. Lui, K. F. Mak, Z. Li, H. Yan, and T. F. Heinz, "Real-time observation of interlayer vibrations in bilayer and few-layer graphene," *Nano Letters*, vol. 13, no. 10, pp. 4620–4623, 2013.

[49] M. Berger, J. Hubbell, S. Seltzer, J. Chang, J. Coursey, R. Sukumar, D. Zucker, and K. Olsen, "XCOM: Photon cross section database (version 1.5)," in *NIST Photon Cross Section Database*, National Institute of Standards and Technology, Gaithersburg MD: National Institute of Standards and Technology, 2010. http://physics.nist.gov/xcom.

[50] M. Centurion, "Ultrafast imaging of isolated molecules with electron diffraction," *Journal of Physics B: Atomic, Molecular and Optical Physics*, vol. 49, no. 6, p. 062002, 2016.

[51] J. Yang, J. Beck, C. J. Uiterwaal, and M. Centurion, "Imaging of alignment and structural changes of carbon disulfide molecules using ultrafast electron diffraction," *Nature Communications*, vol. 6, p. 8172, 2015.

[52] J. Yang, M. Guehr, X. Shen, R. Li, T. Vecchione, R. Coffee, J. Corbett, A. Fry, N. Hartmann, C. Hast, *et al.*, "Diffractive imaging of coherent nuclear motion in isolated molecules," *Physical Review Letters*, vol. 117, no. 15, p. 153002, 2016.

[53] J. Glownia, A. Natan, J. Cryan, R. Hartsock, M. Kozina, M. Minitti, S. Nelson, J. Robinson, T. Sato, T. van Driel, *et al.*, "Self-referenced coherent diffraction X-ray movie of Angstrom-and femtosecond-scale atomic motion," *Physical Review Letters*, vol. 117, no. 15, p. 153003, 2016.

[54] J. M. Glownia, A. Natan, J. P. Cryan, R. Hartsock, M. Kozina, M. P. Minitti, S. Nelson, J. Robinson, T. Sato, T. van Driel, G. Welch, C. Weninger, D. Zhu, and P. H. Bucksbaum, "Glownia et al. reply:," *Physical Review Letters*, vol. 119, p. 069302, 2017.

[55] A. Barty, J. Küpper, and H. N. Chapman, "Molecular imaging using X-ray free-electron lasers," *Annual Review of Physical Chemistry*, vol. 64, no. 1, pp. 415–435, 2013. PMID: 23331310.

[56] R. D. Miller, "Mapping atomic motions with ultrabright electrons: The chemists' gedanken experiment enters the lab frame," *Annual Review of Physical Chemistry*, vol. 65, no. 1, pp. 583–604, 2014. PMID: 24423377.

[57] S. P. Weathersby, G. Brown, M. Centurion, T. F. Chase, R. Coffee, J. Corbett, J. P. Eichner, J. C. Frisch, A. R. Fry, M. Gühr, N. Hartmann, C. Hast, R. Hettel, R. K. Jobe, E. N. Jongewaard, J. R. Lewandowski, R. K. Li, A. M. Lindenberg, I. Makasyuk, J. E. May, D. McCormick, M. N. Nguyen, A. H. Reid, X. Shen, K. Sokolowski-Tinten, T. Vecchione, S. L. Vetter, J. Wu, J. Yang, H. A. Dürr, and X. J. Wang, "Mega-electron-volt ultrafast electron diffraction at SLAC national accelerator laboratory," *Review of Scientific Instruments*, vol. 86, no. 7, p. 073702, 2015.

[58] V. V. Nosenko, G. Y. Rudko, A. M. Yaremko, V. O. Yukhymchuk, and O. M. Hreshchuk, "Anharmonicity and Fermi resonance in the vibrational spectra of a CO_2 molecule and CO_2 molecular crystal: Similarity and distinctions," *Journal of Raman Spectroscopy*, vol. 49, no. 3, pp. 559–568.

[59] C. G. Elles, M. J. Cox, and F. F. Crim, "Vibrational relaxation of CH_3I in the gas phase and in solution," *The Journal of Chemical Physics*, vol. 120, no. 15, pp. 6973–6979, 2004.

[60] D. R. Yarkony, "Conical intersections: The new conventional wisdom," *The Journal of Physical Chemistry A*, vol. 105, pp. 6277–6293, 2001.

[61] P. Krause, S. Matsika, M. Kotur, and T. Weinacht, "The influence of excited state topology on wavepacket delocalization in the relaxation of photoexcited polyatomic molecules," *The Journal of Chemical Physics*, vol. 137, no. 22, p. 22A537, 2012.

[62] T. Sekikawa, T. Okamoto, E. Haraguchi, M. Yamashita, and T. Nakajima, "Two-photon resonant excitation of a doubly excited state in He atoms by high-harmonic pulses," *Optics Express*, vol. 16, no. 26, pp. 21922–21929, 2008.

[63] X. Tong and C. Lin, "Double photoexcitation of He atoms by attosecond XUV pulses in the presence of intense few-cycle infrared lasers," *Physical Review A*, vol. 71, no. 3, p. 033406, 2005.

[64] J. Breidbach and L. Cederbaum, "Universal attosecond response to the removal of an electron," *Physical Review Letters*, vol. 94, no. 3, p. 033901, 2005.

[65] P. Hamm and M. Zanni, *Concepts and methods of 2D infrared spectroscopy*. Cambridge: Cambridge University Press, 2011.

[66] R. Kubo, *Advances in Chemical Physics: Stochastic Processes in Chemical Physics*, ch. A Stochastic Theory of Line Shape, pp. 101–127. New York: John Wiley & Sons, Inc., 1969.

[67] M. Gruebele, "Intramolecular vibrational dephasing obeys a power law at intermediate times," *Proceedings of the National Academy of Sciences*, vol. 95, no. 11, pp. 5965–5970, 1998.

[68] Z.-H. Loh and S. R. Leone, "Capturing ultrafast quantum dynamics with femtosecond and attosecond X-ray core-level absorption spectroscopy," *The Journal of Physical Chemistry Letters*, vol. 4, no. 2, pp. 292–302, 2013.

[69] M. Levantino, H. Lemke, G. Schirò, M. Glownia, A. Cupane, and M. Cammarata, "Observing heme doming in myoglobin with femtosecond X-ray absorption spectroscopy," *Structural Dynamics*, vol. 2, no. 4, p. 041713, 2015.

[70] A. Stolow and J. G. Underwood, *Advances in Chemical Physics*, ch. Time-Resolved Photoelectron Spectroscopy of Nonadiabatic Dynamics in Polyatomic Molecules, pp. 497–584. Hoboken: Wiley, 2008.

[71] C. Melania Oana and A. I. Krylov, "Dyson orbitals for ionization from the ground and electronically excited states within equation-of-motion coupled-cluster formalism: Theory, implementation, and examples," *The Journal of Chemical Physics*, vol. 127, no. 23, p. 234106, 2007.

[72] T. Gustavsson, Á. Bányász, E. Lazzarotto, D. Markovitsi, G. Scalmani, M. J. Frisch, V. Barone, and R. Improta, "Singlet excited-state behavior of uracil and thymine in aqueous solution: a combined experimental and computational study of 11 uracil derivatives," *Journal of the American Chemical Society*, vol. 128, no. 2, pp. 607–619, 2006.

[73] R. Leon, S. Chaparro, S. Johnson, C. Navarro, X. Jin, Y. Zhang, J. Siegert, S. Marcinkevičius, X. Liao, and J. Zou, "Dislocation-induced spatial ordering of InAs quantum dots: Effects on optical properties," *Journal of Applied Physics*, vol. 91, no. 9, pp. 5826–5830, 2002.

[74] E. Van Dijk, J. Hernando, M. García-Parajó, and N. Van Hulst, "Single-molecule pump-probe experiments reveal variations in ultrafast energy redistribution," *The Journal of Chemical Physics*, vol. 123, no. 6, p. 064703, 2005.

[75] S. Pisharody and R. Jones, "Probing two-electron dynamics of an atom," *Science*, vol. 303, no. 5659, pp. 813–815, 2004.

[76] R. Pazourek, S. Nagele, and J. Burgdörfer, "Attosecond chronoscopy of photoemission," *Reviews of Modern Physics*, vol. 87, no. 3, pp. 765–802, 2015.

[77] P. Hockett, E. Frumker, D. M. Villeneuve, and P. B. Corkum, "Time delay in molecular photoionization," *Journal of Physics B: Atomic, Molecular and Optical Physics*, vol. 49, no. 9, p. 095602, 2016.

[78] M. Drescher, M. Hentschel, R. Kienberger, M. Uiberacker, V. Yakovlev, A. Scrinzi, T. Westerwalbesloh, U. Kleineberg, U. Heinzmann, and F. Krausz, "Time-resolved atomic inner-shell spectroscopy," *Nature*, vol. 419, no. 6909, pp. 803–807, 2002.

[79] P. M. Paul, E. S. Toma, P. Breger, G. Mullot, F. Augé, P. Balcou, H. G. Muller, and P. Agostini, "Observation of a train of attosecond pulses from high harmonic generation," *Science*, vol. 292, no. 5522, pp. 1689–1692, 2001.

[80] H. G. Muller, "Reconstruction of attosecond harmonic beating by interference of two-photon transitions," *Applied Physics B*, vol. 74, no. 1, pp. s17–s21, 2002.

[81] F. Calegari, D. Ayuso, A. Trabattoni, L. Belshaw, S. De Camillis, S. Anumula, F. Frassetto, L. Poletto, A. Palacios, P. Decleva, *et al.*, "Ultrafast electron dynamics in phenylalanine initiated by attosecond pulses," *Science*, vol. 346, no. 6207, pp. 336–339, 2014.

[82] K. Ramasesha, S. R. Leone, and D. M. Neumark, "Real-time probing of electron dynamics using attosecond time-resolved spectroscopy," *Annual Review of Physical Chemistry*, vol. 67, pp. 41–63, 2016.

[83] R. Locher, L. Castiglioni, M. Lucchini, M. Greif, L. Gallmann, J. Osterwalder, M. Hengsberger, and U. Keller, "Energy-dependent photoemission delays from noble metal surfaces by attosecond interferometry," *Optica*, vol. 2, no. 5, pp. 405–410, 2015.

[84] L. De Marco, K. Ramasesha, and A. Tokmakoff, "Experimental evidence of Fermi resonances in isotopically dilute water from ultrafast broadband IR spectroscopy," *The Journal of Physical Chemistry B*, vol. 117, no. 49, pp. 15319–15327, 2013.

[85] D. Bingemann, A. M. King, and F. F. Crim, "Transient electronic absorption of vibrationally excited CH_2I_2: Watching energy flow in solution," *The Journal of Chemical Physics*, vol. 113, no. 12, pp. 5018–5025, 2000.

[86] C. G. Elles and F. F. Crim, "Connecting chemical dynamics in gases and liquids," *Annual Review of Physical Chemistry*, vol. 57, pp. 273–302, 2006.

[87] P. Kukura, D. W. McCamant, and R. A. Mathies, "Femtosecond stimulated Raman spectroscopy," *Annual Review of Physical Chemistry*, vol. 58, pp. 461–488, 2007.

[88] P. Kukura, R. Frontiera, and R. A. Mathies, "Direct observation of anharmonic coupling in the time domain with femtosecond stimulated Raman scattering," *Physical Review Letters*, vol. 96, no. 23, p. 238303, 2006.

[89] D. J. Nesbitt and R. W. Field, "Vibrational energy flow in highly excited molecules: role of intramolecular vibrational redistribution," *The Journal of Physical Chemistry*, vol. 100, no. 31, pp. 12735–12756, 1996.

[90] C. T. Middleton, B. Cohen, and B. Kohler, "Solvent and solvent isotope effects on the vibrational cooling dynamics of a DNA base derivative," *The Journal of Physical Chemistry A*, vol. 111, no. 42, pp. 10460–10467, 2007.

[91] G. A. Worth and L. S. Cederbaum, "Beyond Born-Oppenheimer: Molecular dynamics through a conical intersection," *Annual Review of Physical Chemistry*, vol. 55, no. 1, pp. 127–158, 2004. PMID: 15117250.

[92] B. G. Levine and T. J. Martínez, "Isomerization through conical intersections," *Annual Review of Physical Chemistry*, vol. 58, no. 1, pp. 613–634, 2007. PMID: 17291184.

[93] S. Matsika and P. Krause, "Nonadiabatic events and conical intersections," *Annual Review of Physical Chemistry*, vol. 62, no. 1, pp. 621–643, 2011. PMID: 21219147.

[94] W. Domcke and D. R. Yarkony, "Role of conical intersections in molecular spectroscopy and photoinduced chemical dynamics," *Annual Review of Physical Chemistry*, vol. 63, no. 1, pp. 325–352, 2012. PMID: 22475338.

[95] M. S. Schuurman and A. Stolow, "Dynamics at conical intersections," *Annual Review of Physical Chemistry*, vol. 69, no. 1, pp. 427–450, 2018. PMID: 29490199.

[96] E. M. Grumstrup, S.-H. Shim, M. A. Montgomery, N. H. Damrauer, and M. T. Zanni, "Facile collection of two-dimensional electronic spectra using femtosecond pulse-shaping technology," *Optics Express*, vol. 15, pp. 16681–16689, Dec 2007.

[97] S.-H. Shim and M. T. Zanni, "How to turn your pump-probe instrument into a multidimensional spectrometer: 2D IR and Vis spectroscopies via pulse shaping," *Physical Chemistry Chemical Physics*, vol. 11, pp. 748–761, 2009.

[98] O. Golonzka, M. Khalil, N. Demirdöven, and A. Tokmakoff, "Vibrational anharmonicities revealed by coherent two-dimensional infrared spectroscopy," *Physical Review Letters*, vol. 86, no. 10, pp. 2154–2157, 2001.

[99] K. Ramasesha, L. De Marco, A. Mandal, and A. Tokmakoff, "Water vibrations have strongly mixed intra-and intermolecular character," *Nature Chemistry*, vol. 5, no. 11, pp. 935–940, 2013.

[100] J. Lindner, P. Vöhringer, M. S. Pshenichnikov, D. Cringus, D. A. Wiersma, and M. Mostovoy, "Vibrational relaxation of pure liquid water," *Chemical Physics Letters*, vol. 421, no. 4, pp. 329–333, 2006.

[101] J. Zheng, K. Kwak, J. Asbury, X. Chen, I. R. Piletic, and M. D. Fayer, "Ultrafast dynamics of solute-solvent complexation observed at thermal equilibrium in real time," *Science*, vol. 309, no. 5739, pp. 1338–1343, 2005.

[102] K. Kwak, J. Zheng, H. Cang, and M. Fayer, "Ultrafast two-dimensional infrared vibrational echo chemical exchange experiments and theory," *The Journal of Physical Chemistry B*, vol. 110, no. 40, pp. 19998–20013, 2006.

[103] G. Moody, M. Siemens, A. Bristow, X. Dai, A. Bracker, D. Gammon, and S. Cundiff, "Exciton relaxation and coupling dynamics in a GaAs/Al$_x$Ga$_{1-x}$As quantum well and quantum dot ensemble," *Physical Review B*, vol. 83, no. 24, p. 245316, 2011.

[104] M. Khalil, N. Demirdöven, and A. Tokmakoff, "Coherent 2d IR spectroscopy: molecular structure and dynamics in solution," *The Journal of Physical Chemistry A*, vol. 107, no. 27, pp. 5258–5279, 2003.

[105] A. M. Brańczyk, D. B. Turner, and G. D. Scholes, "Crossing disciplines - a view on two-dimensional optical spectroscopy," *Annalen der Physik*, vol. 526, no. 1-2, pp. 31–49.

[106] F. D. Fuller and J. P. Ogilvie, "Experimental implementations of two-dimensional Fourier transform electronic spectroscopy," *Annual Review of Physical Chemistry*, vol. 66, no. 1, pp. 667–690, 2015. PMID: 25664841.

[107] P. M. Kraus, B. Mignolet, D. Baykusheva, A. Rupenyan, L. Hornỳ, E. F. Penka, G. Grassi, O. I. Tolstikhin, J. Schneider, F. Jensen, *et al.*, "Measurement and laser control of attosecond charge migration in ionized iodoacetylene," *Science*, vol. 350, no. 6262, pp. 790–795, 2015.

[108] M. Breusing, C. Ropers, and T. Elsaesser, "Ultrafast carrier dynamics in graphite," *Physical Review Letters*, vol. 102, no. 8, p. 086809, 2009.

[109] N. Vogt, L. S. Khaikin, O. E. Grikina, A. N. Rykov, and J. Vogt, "Study of the thymine molecule: Equilibrium structure from joint analysis of gas-phase electron diffraction and microwave data and assignment of vibrational spectra using results of ab initio calculations," *The Journal of Physical Chemistry A*, vol. 112, no. 33, pp. 7662–7670, 2008.

[110] C. J. Hensley, J. Yang, and M. Centurion, "Imaging of isolated molecules with ultrafast electron pulses," *Physical Review Letters*, vol. 109, no. 13, p. 133202, 2012.

[111] P. J. Ho, D. Starodub, D. K. Saldin, V. L. Shneerson, A. Ourmazd, and R. Santra, "Molecular structure determination from X-ray scattering patterns of laser-aligned symmetric-top molecules," *The Journal of Chemical Physics*, vol. 131, no. 13, p. 131101, 2009.

[112] J. Yang, X. Zhu, T. J. A. Wolf, Z. Li, J. P. F. Nunes, R. Coffee, J. P. Cryan, M. Gühr, K. Hegazy, T. F. Heinz, K. Jobe, R. Li, X. Shen, T. Veccione, S. Weathersby, K. J. Wilkin, C. Yoneda, Q. Zheng, T. J. Martinez, M. Centurion, and X. Wang, "Imaging CF3I conical intersection and photodissociation dynamics with ultrafast electron diffraction," *Science*, vol. 361, no. 6397, pp. 64–67, 2018.

[113] R. Srinivasan, J. S. Feenstra, S. T. Park, S. Xu, and A. H. Zewail, "Dark structures in molecular radiationless transitions determined by ultrafast diffraction," *Science*, vol. 307, no. 5709, pp. 558–563, 2005.

[114] D. Zhong, E. W.-G. Diau, T. M. Bernhardt, S. De Feyter, J. D. Roberts, and A. H. Zewail, "Femtosecond dynamics of valence-bond isomers of azines: transition states and conical intersections," *Chemical Physics Letters*, vol. 298, no. 1, pp. 129–140, 1998.

[115] K. H. Kim, J. G. Kim, S. Nozawa, T. Sato, K. Y. Oang, T. W. Kim, H. Ki, J. Jo, S. Park, C. Song, *et al.*, "Direct observation of bond formation in solution with femtosecond X-ray scattering," *Nature*, vol. 518, no. 7539, pp. 385–389, 2015.

[116] M. Stefanou, K. Saita, D. V. Shalashilin, and A. Kirrander, "Comparison of ultrafast electron and X-ray diffraction–A computational study," *Chemical Physics Letters*, vol. 683, pp. 300–305, 2017.

[117] B. C. Arruda and R. J. Sension, "Ultrafast polyene dynamics: the ring opening of 1, 3-cyclohexadiene derivatives," *Physical Chemistry Chemical Physics*, vol. 16, no. 10, pp. 4439–4455, 2014.

[118] C.-Y. Ruan, V. A. Lobastov, R. Srinivasan, B. M. Goodson, H. Ihee, and A. H. Zewail, "Ultrafast diffraction and structural dynamics: The nature of complex molecules far from equilibrium," *Proceedings of the National Academy of Sciences*, vol. 98, no. 13, pp. 7117–7122, 2001.

[119] M. Minitti, J. Budarz, A. Kirrander, J. Robinson, D. Ratner, T. Lane, D. Zhu, J. Glownia, M. Kozina, H. Lemke, *et al.*, "Imaging molecular motion: Femtosecond X-ray scattering of an electrocyclic chemical reaction," *Physical Review Letters*, vol. 114, no. 25, p. 255501, 2015.

[120] "General discussion," *Faraday Discuss.*, vol. 163, pp. 243–275, 2013.

[121] S. H. Pullen, N. A. Anderson, L. A. Walker, and R. J. Sension, "The ultrafast photochemical ring-opening reaction of 1,3-cyclohexadiene in cyclohexane," *The Journal of Chemical Physics*, vol. 108, no. 2, pp. 556–563, 1998.

[122] A. R. Attar, A. Bhattacherjee, C. Pemmaraju, K. Schnorr, K. D. Closser, D. Prendergast, and S. R. Leone, "Femtosecond X-ray spectroscopy of an electrocyclic ring-opening reaction," *Science*, vol. 356, no. 6333, pp. 54–59, 2017.

[123] K. Kosma, S. A. Trushin, W. Fuß, and W. E. Schmid, "Cyclohexadiene ring opening observed with 13 fs resolution: coherent oscillations confirm the reaction path," *Physical Chemistry Chemical Physics*, vol. 11, no. 1, pp. 172–181, 2009.

[124] Brown et al., "LibreTexts: Chemistry," *Map: Chemistry - The Central Science*, 2017, August 7. Available from: https://chem.libretexts.org/Textbook_Maps/General_Chemistry/Map%3A_Chemistry_-_The_Central_Science

[125] M. Sargent III, M. O. Scully, and W. E. Lamb Jr, *Laser Physics*. Reading, MA: Addison-Wesley, 1974.

[126] H. Stapelfeldt and T. Seideman, "Colloquium: Aligning molecules with strong laser pulses," *Reviews of Modern Physics*, vol. 75, no. 2, pp. 543–557, 2003.

[127] M. Lewenstein, P. Balcou, M. Y. Ivanov, A. L'huillier, and P. B. Corkum, "Theory of high-harmonic generation by low-frequency laser fields," *Physical Review A*, vol. 49, no. 3, pp. 2117–2132, 1994.

[128] K. L. Ishikawa, *Advances in Solid State Lasers Development and Applications*, ch. High-Harmonic Generation. InTech, 2010.

Index

Note: Page numbers followed by "n" denote endnotes.

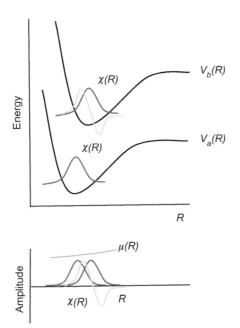

Figure 3.4 Illustration of the vibrational wave functions, $\chi(R)$, and transition dipole moment, $\mu_{ba}(R)$, for a diatomic molecule. The upper portion of the diagram shows the ground vibrational eigenstate in the lower PES, along with both the ground and first-excited eigenstates in the upper PES. The bottom portion of the figure shows all three wave functions projected down onto a single coordinate axis to visualize the Franck–Condon factors $\langle \chi_b(R) | \chi_a(R) \rangle$. The raised line shows how the dipole moment, $\mu_{ba}(R)$, might vary with R.

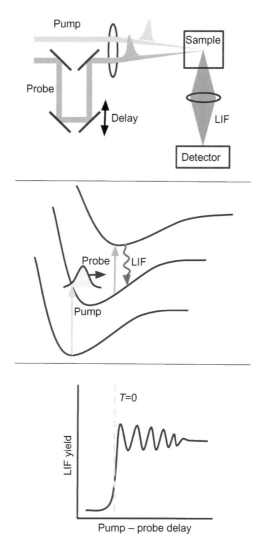

Figure 4.1 Cartoon diagram illustrating LIF. The top panel illustrates the experimental apparatus; the middle panel shows the interaction with the pump and probe pulses, along with evolution of the time-dependent molecular wave function, on molecular potential energy surfaces; and the bottom panel shows the LIF yield as a function of pump–probe delay.

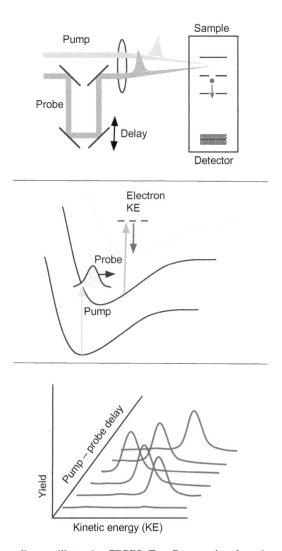

Figure 4.2 Cartoon diagram illustrating TRPES. Top: Pump and probe pulses are focused into a vacuum chamber where the atomic or molecular sample is ionized. The photoelectron energy can be measured via time-of-flight or velocity-map imaging, where the time/position of the electrons encodes their energy/momentum, respectively. Middle: After the pump pulse excites a vibrational wave packet on an excited state, the time-delayed probe pulse ionizes the molecule, resulting in the ejection of an electron. Bottom: The electron yield is plotted as a function of both pump–probe delay and the KE of the emitted electron.

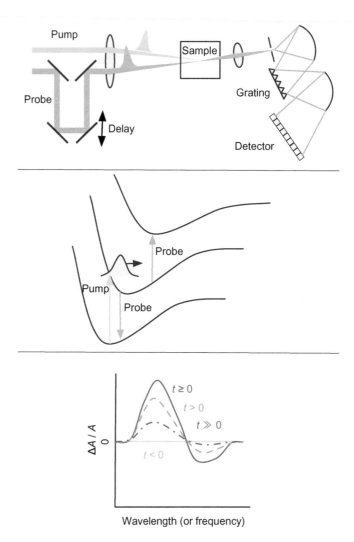

Figure 4.3 Cartoon diagram illustrating TA. Top: As with both LIF and TRPES, the pump and probe pulses are focused into the molecular sample with a variable time delay. In TA, the coherently scattered probe light is collected and spectrally resolved in a spectrometer. Middle: Once again the pump pulse promotes a portion of the ground-state wave function to an excited state, where it subsequently evolves in time. In TA, the time-delayed probe pulse interacts with the induced polarization of the molecule, resulting in either enhanced or suppressed absorption of the probe. Bottom: The change in absorption of the probe pulse is plotted as a function of wavelength for different time delays.

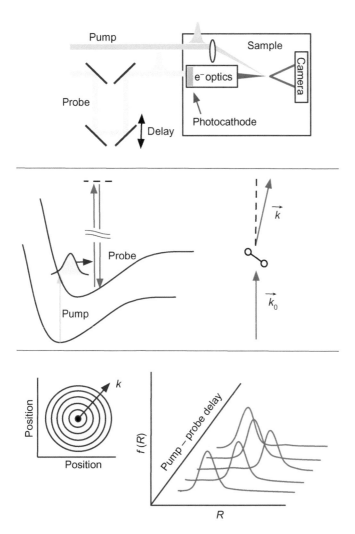

Figure 4.4 Cartoon diagram illustrating the technique of UED. Top: The time-delayed probe generates an ultrafast bunch of electrons that diffracts off the molecules. The spatial diffraction pattern as a function of pump–probe delay is recorded on a camera. Middle: Both energy (left) and momentum (right) pictures of the interaction are shown. In the case of diffractive imaging, the momentum viewpoint is typically more useful since the scattering is elastic (illustrated by the arrows of the same length on the left). Bottom: Typical data for a UED experiment, including the raw image on the 2D detector (left), and the radial distribution function, $f(R)$, as a function of pump–probe delay.

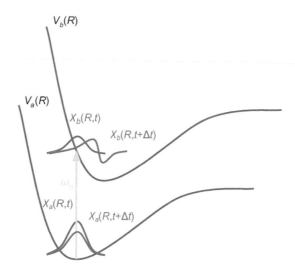

Figure 6.1 Illustration of the various components of Equation 6.25. $\chi_i(R,t)$ is the initial vibrational wave function on state i, while $\chi_i(R,t + \Delta t)$ is the new wave function after one time step. The ω_0 arrow represents the the field-induced coupling.

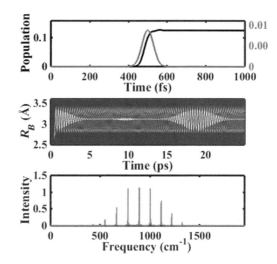

Figure 6.2 Solution of the TDSE for I_2 interacting with an ultrafast laser pulse resonant between the X and B states of the molecule. The top panel shows the electric field envelope (red, right axis), along with the population of the B state (black, left axis), versus time. The middle panel shows the time- and space-dependent probability density (blue shading) and expectation value (white curve) on the B state. Note that the time axes for the top and middle panels are different. Finally, the bottom panel shows the spectrum of the B-state wave packet.

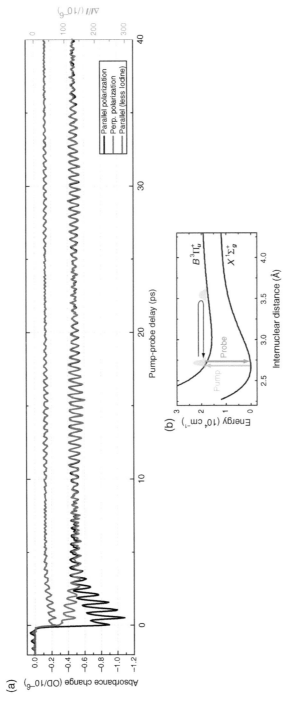

Figure 8.3 GSB data in I_2 using cavity-enhanced spectroscopy. Panel (a): long-time GSB signal for different polarization configurations. Panel (b): illustration of wave packet dynamics and pump/probe photons for the relevant PESs. Figure adapted from [26].

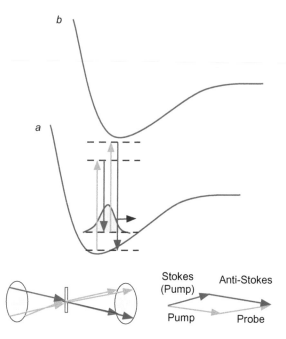

Figure 8.6 Illustration of energy and momentum conservation in a box-CARS experiment. The top panel shows the PESs involved, along with the pump and probe interactions (including the Stokes and anti-Stokes fields). The bottom-left depicts the experimental beam geometry: three beams are focused into the sample, and coherently scattered probe light at the anti-Stokes frequency serves as the signal. The bottom-right illustration shows a two-dimensional representation of the phase-matching condition, in which the wave vectors for the beams combine so that $\Delta\mathbf{k} = 0$.

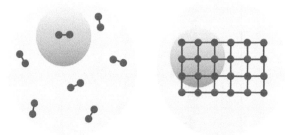

Figure 9.3 Illustration of diffraction from gas-phase molecules (left panel) and a crystal lattice (right panel). The small and large shaded areas indicate the typical transverse coherence widths of electron and X-ray beams, respectively. Note that in the case of the lattice, the atoms or molecules are arranged in a regular array such that the diffracted X-rays or electrons from each molecule or unit cell add coherently. This is in contrast with the gas-phase sample, for which the molecules are randomly arranged and oriented, leading to an incoherent addition of the signals from each molecule, independent of the coherence width of the beam.

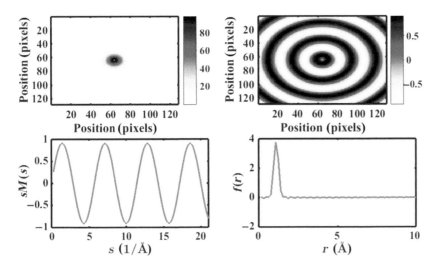

Figure 9.6 Simulations of electron diffraction from diatomic nitrogen molecules with an electron beam energy of 100 keV. The top left panel is the raw diffraction data, while the top right shows the molecular contribution $I_M(s)$. The bottom left panel shows $sM(s)$, and the bottom right panel plots the radial distribution function, $f(R)$. Simulation courtesy of Alexander Johnson and Martin Centurion.

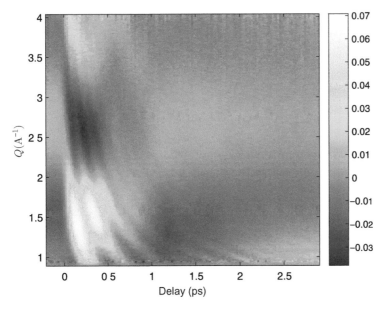

Figure 9.9 Time-resolved X-ray scattering data from excited I_2. Plot shows the β_2 parameter as a function of both the momentum transfer magnitude, Q, and time, t. Figure from Ref. [35].

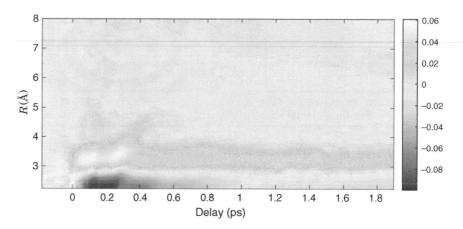

Figure 9.10 (Inverse) Fourier transform of the data in Figure 9.9 along the momentum-transfer coordinate. Figure from Ref. [36].

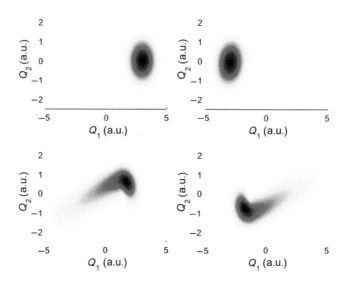

Figure 10.1 Snapshots of a wave function evolving on the 2D anharmonic potential of Equation 10.1. Panel (a) shows the initial wave function displaced along only one coordinate. Panel (b) shows the wave function one-half period later. Panel (c) shows the wave function many oscillations later when motion along one coordinate has coupled into motion along the other. Panel (d) shows the wave function one-half period after panel (c).

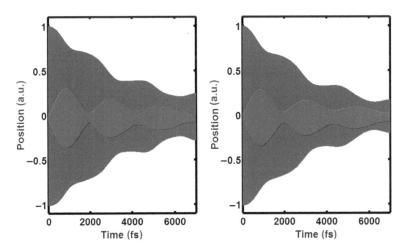

Figure 10.2 Quantum (left) and classical (right) solutions for two modes anharmonically coupled as in Equation 10.1. Graph shows the first moments $\langle Q_1(t)\rangle$ (blue) and $\langle Q_2(t)\rangle$ (red). Note that the fast harmonic oscillations (period of ~ 20 fs) under the envelope are not resolved due to the finite line thickness in the plot.

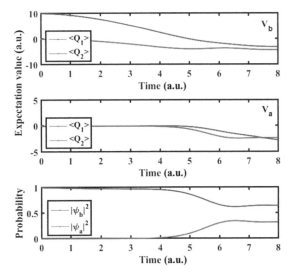

Figure 10.10 Expectation values of Q_1 and Q_2 as functions of time on upper PES V_b (top panel) and lower PES V_a (middle panel). Bottom panel plots adiabatic-state probabilities as a function of time.

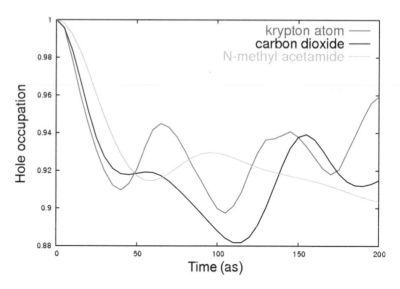

Figure 10.13 Attosecond charge migration following core electron removal in three different systems. All three show initial dynamics on an approximately 50 attosecond timescale. Figure from Ref. 42.

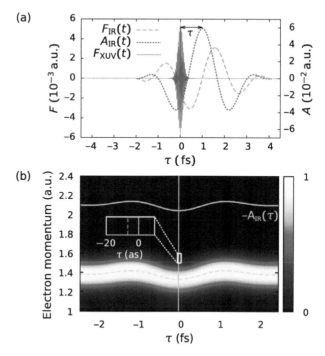

Figure 12.11 Illustration of a streaking measurement using a combination of femtosecond IR and attosecond XUV pulses. Panel (a) shows the IR field $F_{IR}(t)$, its associated vector potential $A_{IR}(t)$, and the XUV attosecond field $F_{XUV}(t)$ as a function of time delay. Panel (b) shows a simulation of the streaking diagram, where both $A_{IR}(t)$ and the electron momentum are plotted as a function of time delay. The ionization delay is seen by the phase shift between the $A_{IR}(t)$ and the electron momentum $p(t)$ as a function of time delay. Figure from Ref. [53].

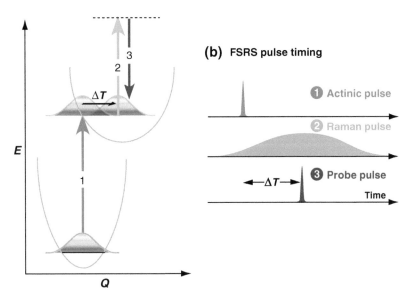

Figure 13.6 Timing and energy diagram illustrating the technique of FSRS. An "actinic" pump pulse initiates vibrational dynamics in an excited PES. Together, the Raman and probe pulses measure mode coupling. Figure adapted from [64].

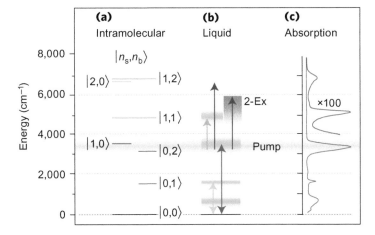

Figure 13.17 Energy-level diagram with excitations for vapor and liquid water. Figure from Ref. [70].

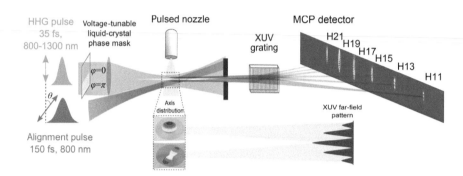

Figure 13.21 Experimental approach for HHG spectroscopy of charge migration in an aligned, gas-phase sample. Figure from Ref. [76] supplemental material.

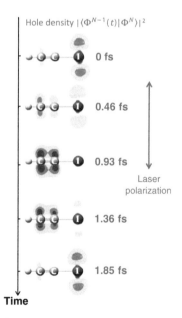

Figure 13.22 Reconstructed charge migration dynamics in iodoacetylene using HHG spectroscopy. Figure adapted from [76].

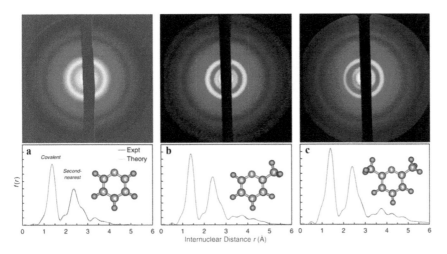

Figure 14.4 Static, ground-state electron diffraction measurements and radial distribution curves for gas-phase pyridine (a), picoline (b), and lutidine (c). Figure from Ref. [82].

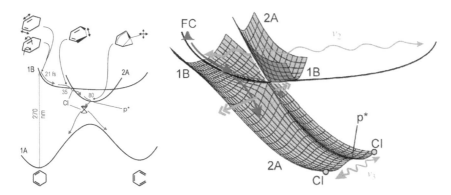

Figure 15.13 Left: states and timescales involved in the isomerization of CHD as measured by ionization spectroscopy. Right: two-dimensional subspace of the PESs, along with the wave packet path and relevant vibrational modes. Figure adapted from Ref. [92].

T - #0495 - 071024 - C372 - 254/203/16 - PB - 9780367780401 - Gloss Lamination